Gary B. C___

W9-APJ-578

TAYLOR, MICH
292-7965

16.50

FOURIER
SERIES
AND
BOUNDARY
VALUE
PROBLEMS

FOURIER
SERIES
AND
BOUNDARY
VALUE
PROBLEMS

THIRD EDITION

RUEL V. CHURCHILL
PROFESSOR EMERITUS OF MATHEMATICS
THE UNIVERSITY OF MICHIGAN

JAMES WARD BROWN
PROFESSOR OF MATHEMATICS
THE UNIVERSITY OF MICHIGAN–DEARBORN

McGRAW-HILL BOOK COMPANY

New York   St. Louis   San Francisco   Auckland   Bogotá
Düsseldorf   Johannesburg   London   Madrid   Mexico
Montreal   New Delhi   Panama   Paris   São Paulo
Singapore   Sydney   Tokyo   Toronto

Library of Congress Cataloging in Publication Data

Churchill, Ruel Vance, date
    Fourier series and boundary value problems.

    Bibliography: p.
    Includes index.
    1. Fourier series. 2. Functions, Orthogonal.
3. Boundary value problems. I. Brown, James Ward,
joint author. II. Title.
QA404.C6 1978   515'.2432   77-12779
ISBN 0-07-010843-9

FOURIER SERIES AND BOUNDARY VALUE PROBLEMS

Copyright © 1978, 1963 by McGraw-Hill, Inc. All rights reserved.
Copyright 1941 by McGraw-Hill, Inc. All rights reserved.
Copyright renewed 1959 by Ruel V. Churchill.
Printed in the United States of America. No part of this publication
may be reproduced, stored in a retrieval system, or transmitted, in any
form or by any means, electronic, mechanical, photocopying, recording, or
otherwise, without the prior written permission of the publisher.

1 2 3 4 5 6 7 8 9 0 D O D O 7 8 3 2 1 0 9 8 7

This book was set in Times Roman. The editor was A. Anthony Arthur;
the production supervisor was David Damstra; the cover was
designed by John Hite.
R. R. Donnelley & Sons Company was printer and binder.

CONTENTS

PREFACE

This is an introductory treatment of Fourier series and their applications to boundary value problems in partial differential equations of engineering and physics. It is designed for students who have completed the equivalent of one term of advanced calculus. The physical applications, explained in some detail, are kept on a fairly elementary level.

The *first objective* is to introduce the concept of orthogonal sets of functions and representations of arbitrary functions in series of functions from such sets. The most prominent special cases, representations of functions by trigonometric Fourier series, are given special attention. Fourier integral representations and expansions in series of Bessel functions and Legendre polynomials are also treated.

The *second objective* is a clear presentation of the classical method of separation of variables used in solving boundary value problems with the aid of those representations. Some attention is given to the verification of solutions and to uniqueness of solutions; for the method cannot be presented properly without such considerations. Other methods are treated in the first author's books "Operational Mathematics" and "Complex Variables and Applications."

This edition is a revision of the 1963 edition of the book, the original one having been published in 1941. Considerable attention has been given here to improving the exposition and including new problems. To note a few of the specific changes in this third edition, the motivation of the concept of inner products of functions has been amplified, the proofs of the convergence theorems for Fourier series and integrals have been completely rewritten, and there is more detail in the discussion of Bessel and Legendre functions of the second kind. There are more footnotes referring to other texts which give proofs of the more delicate results in advanced calculus that are occasionally needed, and the Bibliography has been updated. Some rearrangement of topics was found desirable. For example, the chapter on the applications of Fourier series to the solution of boundary value problems now precedes the chapter on Fourier integrals, thus allowing the student to reach those applications earlier.

The chapters on Bessel functions and Legendre polynomials, Chapters 8 and 9, are essentially independent of each other and can be taken up in either order. Much of Chapter 5, on further properties of Fourier series, and Chapter 10, on uniqueness of solutions, can be omitted to shorten the course; this also applies to some sections of other chapters.

In preparing this edition, the authors have benefited from the comments of a variety of people, including students. A number of specific suggestions were made by Craig Comstock and George Springer. Thanks are also due R. P. Boas, Jr. for providing the reference to Kronecker's extension of the method of integration by parts, which is often useful in evaluating integrals needed in finding Fourier coefficients; and the derivation of the laplacian in spherical coordinates that is given was suggested by a note of R. P. Agnew's in the *American Mathematical Monthly*, vol. 60 (1953). Finally, the authors are indebted to Catherine A. Rader for her careful typing of the manuscript.

RUEL V. CHURCHILL

JAMES WARD BROWN

FOURIER
SERIES
AND
BOUNDARY
VALUE
PROBLEMS

ONE

PARTIAL DIFFERENTIAL EQUATIONS OF PHYSICS

1. Two Related Problems

We shall be concerned here with two general types of problems. One type deals with the representation of arbitrarily given functions by infinite series of functions from a prescribed set. The other consists of boundary value problems in partial differential equations, with emphasis on equations that are prominent in physics and engineering.

Representations by series are encountered in solving boundary value problems. The theories of those representations can be presented independently. They have such attractive features as the extension of concepts of geometry, vector analysis, and algebra into the field of mathematical analysis. Their mathematical precision is also pleasing. But they gain in unity and interest when presented in connection with boundary value problems.

The set of functions that make up the terms in the series representation is determined by the boundary value problem. Representations or expansions in Fourier series, which are certain types of series of sine and cosine functions, are associated with the more common boundary value problems. We shall give special attention to the theory and application of Fourier series. But we shall also consider extensions and generalizations of such series, including Fourier integrals and series of Bessel functions and Legendre polynomials.

A boundary value problem is correctly set if it has one and only one solution. Physical problems associated with partial differential equations often suggest boundary conditions under which a problem may be correctly set. In fact, it is sometimes helpful to interpret a problem physically in order to judge whether the boundary conditions may be adequate. This is a prominent reason for associating such problems with their physical applications, aside from the opportunity to display interesting and important contacts between mathematical analysis and the physical sciences.

1

The theory of partial differential equations gives results on the existence of solutions of boundary value problems. But such results are necessarily limited and complicated by the great variety of features: types of equations and conditions and types of domains. Instead of appealing to general theory in treating a specific problem, our approach will be to actually find a solution, which can often be shown to be the only possible solution.

2. Linear Boundary Value Problems

Theory and applications of ordinary or partial differential equations in a dependent variable, or function, u usually require that u satisfy not only the differential equation throughout some domain of its independent variable or variables but also some conditions on boundaries of that domain. The equations that represent those boundary conditions may involve values of derivatives of u, as well as u itself, at points on the boundaries. In addition, some conditions on the continuity of u and its derivatives within the domain and at the boundaries are required.

Such a set of requirements constitutes a *boundary value problem* in the function u. We apply that term whenever the differential equation is accompanied by some boundary conditions, even though the conditions may not be adequate to ensure a unique solution of the problem.

The three equations

(1)
$$u''(x) - u(x) = -1 \qquad (0 < x < 1),$$
$$u'(0) = 0, \qquad u(1) = 0,$$

for example, constitute a boundary value problem in ordinary differential equations. The domain of the independent variable x is the interval $0 < x < 1$ whose boundaries consist of the two points $x = 0$ and $x = 1$. The solution of this problem which, together with each of its derivatives, is continuous everywhere is found to be

(2)
$$u(x) = 1 - \frac{\cosh x}{\cosh 1}.$$

Frequently it is convenient to indicate partial differentiation by writing independent variables as subscripts. If, for instance, u is a function of x and y, we may write

$$u_x \text{ or } u_x(x,y) \text{ for } \frac{\partial u}{\partial x}, \qquad u_{xx} \text{ for } \frac{\partial^2 u}{\partial x^2}, \qquad u_{xy} \text{ for } \frac{\partial^2 u}{\partial y \, \partial x},$$

and so on. We shall always assume that the partial derivatives of u satisfy conditions allowing us to write $u_{xy} = u_{yx}$.

Also, we shall be free to use the symbols $u_x(x_0,y)$ and $u_{xx}(x_0,y)$ to denote the values of the functions $\partial u/\partial x$ and $\partial^2 u/\partial x^2$, respectively, on the line $x = x_0$; and corresponding symbols will be used for boundary values of other derivatives.

The problem consisting of the partial differential equation

$$(3) \qquad\qquad u_{xx}(x,y) + u_{yy}(x,y) = 0 \qquad\qquad (x > 0,\ y > 0)$$

and the two boundary conditions

$$(4) \qquad\begin{aligned} u(0,y) &= u_x(0,y) & (y > 0), \\ u(x,0) &= \sin x + \cos x & (x > 0) \end{aligned}$$

is an example of a boundary value problem in partial differential equations. The domain is the first quadrant of the xy plane. The reader can verify that the function

$$(5) \qquad\qquad u(x,y) = e^{-y}(\sin x + \cos x)$$

is a solution of that problem. This function and its partial derivatives are everywhere continuous in the two variables x,y together and bounded in the domain $x > 0$, $y > 0$.

A differential equation in a function u, or a boundary condition on u, is *linear* if it is an equation *of the first degree in u and derivatives of u.* Thus the terms of the equation are either prescribed functions of the independent variables alone, including constants, or such functions multiplied by either u or one of the derivatives of u.

The differential equations and boundary conditions (1), (3), and (4) above are all linear. The differential equation

$$(6) \qquad\qquad zu_{xx} + xy^2 u_{yy} - e^x u_z = f(y,z)$$

in $u(x,y,z)$ is linear. But the equation

$$u_{xx} + uu_y = x$$

is nonlinear in $u(x,y)$ because the term uu_y is not of the first degree as an algebraic expression in the two variables u and u_y.

Let the letters A through G denote either constants or functions of the independent variables x and y only. Then the general *linear partial differential equation of the second order in $u(x,y)$* has the form

$$(7) \qquad\qquad Au_{xx} + Bu_{xy} + Cu_{yy} + Du_x + Eu_y + Fu = G.$$

A boundary value problem is *linear* if its differential equation and all its boundary conditions are linear. Problem (1) and the problem consisting of equations (3) and (4) are examples of linear boundary value problems.

Methods of solution presented in this book do not apply to nonlinear boundary value problems.

A linear differential equation or boundary condition is *homogeneous* if *each* of its terms, other than zero itself, is of the first degree in the function u and its derivatives.

Equation (7) is homogeneous in a domain if and only if $G = 0$ throughout that domain. Equation (6) is nonhomogeneous, unless $f(y,z) = 0$ for all values of y and z. Equation (3) and the first of conditions (4) are homogeneous, but the second of those conditions is not. Homogeneous equations will play a central role in our treatment of linear boundary value problems.

3. The Vibrating String

A tightly stretched string, whose position of equilibrium is some interval on the x axis, is vibrating in the xy plane. Each point of the string, with coordinates $(x,0)$ in the equilibrium position, has a transverse displacement $y(x,t)$ at time t. We assume that the displacements y are small relative to the length of the string, that slopes are small, and that other conditions are such that the movement of each point is essentially in the direction of the y axis. Then at time t the point has coordinates (x,y).

Let the tension T of the string be great enough that the string behaves as if it were perfectly flexible. That is, at each point (x,y) on the string the part of the string on the left of that point exerts a force of magnitude T in the tangential direction upon the part on the right, and the effect of bending moments at the point can be neglected. The magnitude of the x component of the tensile force is denoted by H. See Fig. 1, where that x component has the same positive sense as the x axis. Our final assumption is that H is constant; that is, the variation of H with x and t can be neglected.

Those idealizing assumptions are severe, but they are justified in many applications. They are adequately satisfied, for instance, by strings of musical instruments under ordinary conditions of operation. Mathematically, the

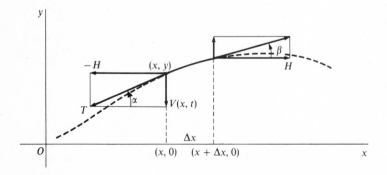

Figure 1

assumptions will lead us to a partial differential equation in $y(x,t)$ which is linear.

Now let $V(x,t)$ denote the y component of the tensile force exerted by the left-hand portion of the string on the right-hand portion at the point (x,y). We take the positive sense of V as that of the y axis. If α is the slope angle of the string at the point (x,y) at time t, then

$$\frac{-V(x,t)}{H} = \tan \alpha = y_x(x,t),$$

as indicated in Fig. 1. Thus *the y component $V(x,t)$ of the force exerted at time t by the part of the string on the left of a point (x,y) upon the part on the right* is given by the equation

(1) $$V(x,t) = -Hy_x(x,t) \qquad (H > 0),$$

which is basic for deriving the equation of motion of the string. Equation (1) is also used in setting up certain types of boundary conditions.

Suppose that all external forces such as the weight of the string and resistance forces, other than forces at the end points, can be neglected. Consider a segment of the string not containing an end point and whose projection on the x axis has length Δx. Since x components of displacements are negligible, the mass of the segment is $\delta \Delta x$ where the constant δ is the mass of the string per unit length. At time t the y component of the force exerted by the string on the segment at the left-hand end (x,y) is $V(x,t)$, given by equation (1). The tangential force exerted on the other end of the segment by the part of the string on the right of that end is also indicated in Fig. 1. Its y component is evidently

(2) $$H \tan \beta = Hy_x(x + \Delta x, t)$$

where β is the slope angle of the string at the right-hand end of the segment. Note that it is, in fact, $-V(x + \Delta x, t)$; the negative sign signifies that the force is exerted upon the part of the string on the left by the part on the right. The acceleration of the end (x,y) in the y direction is $y_{tt}(x,t)$. From Newton's second law of motion (mass times acceleration equals force), it follows that

(3) $$\delta \Delta x \, y_{tt}(x,t) = -Hy_x(x,t) + Hy_x(x + \Delta x, t),$$

approximately, when Δx is small. Hence

$$y_{tt}(x,t) = \frac{H}{\delta} \lim_{\Delta x \to 0} \frac{y_x(x + \Delta x, t) - y_x(x,t)}{\Delta x} = \frac{H}{\delta} y_{xx}(x,t)$$

at each point where the partial derivatives exist.

Thus the function $y(x,t)$, representing the transverse displacements in a stretched string under the conditions stated above, satisfies the one-dimensional *wave equation*

$$(4) \qquad\qquad y_{tt}(x,t) = a^2 y_{xx}(x,t) \qquad\qquad (a^2 = H/\delta)$$

at points where no external forces act on the string. The constant a has the physical dimensions of velocity.

4. Modifications and End Conditions

When external forces parallel to the y axis act along the string, let F denote the force per unit length of string, the positive sense of F being that of the y axis. Then a term $F \, \Delta x$ must be added on the right-hand side of equation (3), Sec. 3, and the equation of motion is

$$(1) \qquad\qquad y_{tt}(x,t) = a^2 y_{xx}(x,t) + \frac{F}{\delta}.$$

In particular, with the y axis vertical and its positive sense upward, suppose that the external force consists of the weight of the string. Then $F \, \Delta x = -\delta \, \Delta x \, g$, where the positive constant g is the acceleration of gravity; and equation (1) becomes the linear nonhomogeneous equation

$$(2) \qquad\qquad y_{tt}(x,t) = a^2 y_{xx}(x,t) - g.$$

In equation (1), F may be a function of x, t, y, or derivatives of y. If the external force per unit length is a damping force proportional to the velocity in the y direction, for example, F is replaced by $-By_t$, where the positive constant B is a damping coefficient. Then the equation of motion is linear and homogeneous:

$$(3) \qquad\qquad y_{tt}(x,t) = a^2 y_{xx}(x,t) - by_t(x,t) \qquad\qquad (b = B/\delta).$$

If one end $x = 0$ of the string is kept fixed at the origin at all times $t \ge 0$, the boundary condition there is clearly

$$(4) \qquad\qquad y(0,t) = 0 \qquad\qquad (t \ge 0).$$

But if that end is permitted to slide along the y axis and if the end is moved along that axis with a displacement $f(t)$, the boundary condition is the linear nonhomogeneous condition

$$(5) \qquad\qquad y(0,t) = f(t) \qquad\qquad (t \ge 0).$$

When the left-hand end is looped around the y axis and a force $g(t)$ $(t > 0)$ in the y direction is applied to that end, $g(t)$ is the limit of the force

$V(x,t)$ described in Sec. 3 as x tends to zero through positive values. The boundary condition is then

$$(6) \qquad\qquad\qquad -Hy_x(0,t) = g(t) \qquad\qquad\qquad (t > 0).$$

The negative sign disappears, however, if $x = 0$ is the *right-hand* end because $g(t)$ is then the force exerted on the part of the string to the left of that end.

5. Other Examples of Wave Equations

We can present further functions in physics and engineering which satisfy wave equations and still limit our attention to fairly simple physical phenomena.

(*a*) *Longitudinal Vibrations of Bars.* Let the coordinate x denote distances from one end of an elastic bar in the shape of a cylinder or prism to other cross sections when the bar is unstrained. Displacements of the ends or initial displacements or velocities of the bar, all directed lengthwise along the bar and uniform over each cross section involved, cause the sections of the bar to move in the direction of the x axis. At time t the longitudinal displacement of the section labeled x is denoted by $y(x,t)$. Thus the origin of the displacement y of that section is fixed outside the bar, in the plane of the original reference position of that section (Fig. 2).

At the same time a neighboring section labeled $x + \Delta x$, to the right of section x, has a displacement $y(x + \Delta x, t)$; thus the element of the bar with natural length Δx is stretched by the amount $y(x + \Delta x, t) - y(x,t)$. We assume that such an extension or compression of the element satisfies Hooke's law, so that the force exerted upon the element over its left-hand end is, except for the effect of the inertia of the moving element,

$$-AE\frac{y(x + \Delta x,\, t) - y(x,t)}{\Delta x}$$

where A is the area of a cross section and E is the *modulus of elasticity* of the material in tension and compression. When Δx tends to zero, it then follows that the total longitudinal force $g(x,t)$ exerted on the section x by the part of the bar on the left of that section is given by the basic equation

$$(1) \qquad\qquad\qquad g(x,t) = -AEy_x(x,t).$$

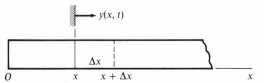

$y(x, t)$

Δx

$0 \qquad\qquad x \qquad x + \Delta x \qquad\qquad\qquad x$ **Figure 2**

Let δ denote the mass of the material per unit volume. Then, applying Newton's second law to the motion of an element of the bar of length Δx, write

$$(2) \qquad \delta A \, \Delta x \, y_{tt}(x,t) = -AEy_x(x,t) + AEy_x(x + \Delta x, \, t),$$

where the last term represents the force on the element at the end $x + \Delta x$. We find, after dividing by $\delta A \, \Delta x$ and letting Δx tend to zero, that

$$(3) \qquad y_{tt}(x,t) = a^2 y_{xx}(x,t) \qquad\qquad (a^2 = E/\delta).$$

Thus the longitudinal displacements $y(x,t)$ in an elastic bar satisfy the wave equation (3) when no external longitudinal forces act on the bar other than at the ends. We have assumed only that displacements are small enough that Hooke's law applies and that sections remain planar after being displaced. The elastic bar here may be replaced by a column of air; then equation (3) has applications in the theory of sound.

The boundary condition $y(0,t) = 0$ signifies that the end $x = 0$ of the bar is held fixed. If, instead, the end $x = 0$ is free when $t > 0$, then no force acts across that end; that is, $g(0,t) = 0$ and, in view of equation (1),

$$(4) \qquad y_x(0,t) = 0 \qquad\qquad (t > 0).$$

(b) *Transverse Vibrations of Membranes.* Let z denote small displacements in the z direction, at time t, of points of a flexible membrane stretched tightly over a frame in the xy plane. The tensile stress T, the tension per unit length across any line on the membrane, is large; and the magnitude H of its component parallel to the xy plane is assumed to be constant. Then the internal force in the z direction at a section $x = x_0$, per unit length of that line, is $-Hz_x(x_0,y,t)$, corresponding to the force V (Sec. 3) in the vibrating string. The force in the z direction at a section $y = y_0$, per unit length, is $-Hz_y(x,y_0,t)$.

Consider an element of the membrane whose projection onto the xy plane is a rectangle with opposite vertices (x,y) and $(x + \Delta x, \, y + \Delta y)$. When Newton's second law is applied to the motion of that element in the z direction, we find that $z(x,y,t)$ satisfies the two-dimensional wave equation

$$(5) \qquad z_{tt} = a^2(z_{xx} + z_{yy}) \qquad\qquad (a^2 = H/\delta).$$

Here δ is the mass of the membrane per unit area. Details of the derivation are left to the problems.

If an external transverse force $F(x,y,t)$ per unit area acts over the membrane, the equation of motion takes the form

$$(6) \qquad z_{tt} = a^2(z_{xx} + z_{yy}) + \frac{F}{\delta}.$$

PROBLEMS

1. Give details in the derivation of equation (1), Sec. 4, for the forced vibrations of a stretched string.

2. A tightly stretched string with its ends fixed at the points $(0,0)$ and $(2c,0)$ hangs at rest under its own weight. The y axis points vertically upward. State why the static displacements $y(x)$ of points on the string satisfy the boundary value problem

$$a^2 y''(x) - g = 0 \qquad\qquad (0 < x < 2c),$$

$$y(0) = y(2c) = 0.$$

By solving this problem, show that the string hangs in the parabolic arc

$$(x - c)^2 = \frac{2a^2}{g}\left(y + \frac{gc^2}{2a^2}\right) \qquad\qquad (0 \le x \le 2c)$$

and that the depth of the vertex of the arc varies directly with δ, the mass of the string per unit length, and with c^2 and inversely with H.

3. Use expression (1), Sec. 3, for the vertical force V and the equation of the arc in which the string in Problem 2 lies to show that the vertical force exerted on that string by either support is $g\delta c$, half the weight of the string.

4. A strand of wire 1 ft long, stretched between the origin and the point $(1,0)$, weighs 0.032 lb $(g\delta = 0.032, g = 32 \text{ ft/sec}^2)$ and $H = 10$ lb. At the instant $t = 0$ the strand lies along the x axis but has a velocity of 1 ft/sec in the direction of the y axis, perhaps because the supports were in motion and were brought to rest at that instant. Assuming that no external forces act along the wire, show why the displacements $y(x,t)$ should satisfy this boundary value problem:

$$y_{tt}(x,t) = 10^4 y_{xx}(x,t) \qquad\qquad (0 < x < 1, t > 0),$$

$$y(0,t) = y(1,t) = 0, \qquad y(x,0) = 0, \qquad y_t(x,0) = 1.$$

5. The physical dimensions of H, the magnitude of the x component of the tensile force in a string, are those of mass times acceleration: MLT^{-2} where M denotes mass, L length, and T time. Show that since $a^2 = H/\delta$, the quantity a has the dimensions of velocity LT^{-1}.

6. The end $x = 0$ of a cylindrical elastic bar is kept fixed, and a constant compressive force of magnitude F_0 units per unit area is exerted at all times $t > 0$ over the end $x = c$. The bar is initially unstrained and at rest with no external forces acting along it. Verify that the function $y(x,t)$ representing the longitudinal displacements of cross sections should satisfy this boundary value problem:

$$y_{tt}(x,t) = a^2 y_{xx}(x,t) \qquad\qquad (0 < x < c, t > 0; a^2 = E/\delta),$$

$$y(0,t) = 0, \qquad Ey_x(c,t) = -F_0, \qquad y(x,0) = y_t(x,0) = 0.$$

7. The left-hand end $x = 0$ of an elastic bar is elastically supported in such a way that the longitudinal force per unit area exerted on the bar at that end is proportional to the displacement of the end, but opposite in sign. Show that the end condition there has the form

$$Ey_x(0,t) = Ky(0,t) \qquad\qquad (K > 0).$$

8. Derive equation (6), Sec. 5. Also show that the *static* transverse displacements $z(x,y)$ of a membrane, over which a transverse force $F(x,y)$ per unit area acts, satisfy *Poisson's equation*

$$z_{xx} + z_{yy} + f = 0 \qquad\qquad (f = F/H).$$

6. Conduction of Heat

Thermal energy is transferred from warmer to cooler regions interior to a solid body by conduction. It is convenient to refer to that transfer as the *flow* of heat, as if heat were a fluid or gas which diffuses through the body from regions of high concentration into regions of low concentration.

Let P_0 denote a point (x_0, y_0, z_0) interior to the body and S a plane or smooth curved surface through P_0. At a time t_0 the *flux of heat* $\Phi(x_0, y_0, z_0, t_0)$ across S at P_0 is the quantity of heat per unit area per unit time that is being conducted across S at that point. Flux is measured in such units as calories per square centimeter per second.

If $u(x, y, z, t)$ denotes the temperatures at points of the solid at time t and if n is a coordinate that represents distance normal to the surface S at the interior point P_0 (Fig. 3), the flux across S will be in the positive direction of n if du/dn is negative and in the negative direction if du/dn is positive. The flux is accordingly positive or negative, or zero; and *a fundamental postulate of the mathematical theory of heat conduction* states that there is a thermal coefficient K of the material such that

$$(1) \qquad \Phi = -K\frac{du}{dn} \qquad (K > 0).$$

That is, the flux $\Phi(x_0, y_0, z_0, t_0)$ across S is proportional to the value at P_0, t_0 of the directional derivative of the temperature function u in a direction normal to S at P_0. The coefficient K is called the *thermal conductivity* of the material.

In particular, the flux across each of the planes $x = x_0$, $y = y_0$, and $z = z_0$ through the point P_0 has the values

$$(2) \qquad \Phi_1 = -Ku_x, \qquad \Phi_2 = -Ku_y, \qquad \Phi_3 = -Ku_z,$$

respectively, where the partial derivatives are evaluated at P_0, t_0. For a fixed value of t, the derivatives u_x, u_y, and u_z are the components of the gradient of u. Since the direction of the vector grad u is that in which u increases most rapidly, the vector

$$(3) \qquad \mathbf{J} = -K \text{ grad } u$$

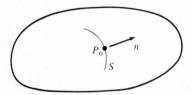

Figure 3

is called the *current of heat flow*, or total flux, at the point P_0.† It measures the flux across the isothermal surface

$$u(x,y,z,t_0) = u(x_0,y_0,z_0,t_0)$$

through P_0. The projection $J_{(n)}$ of **J** onto the normal to the arbitrary surface S is the flux Φ given by equation (1):

$$J_{(n)} = \Phi = -K \, du/dn.$$

Another thermal coefficient of the material is the *heat capacity* c_0 *per unit volume*. This is the quantity of heat required to raise the temperature of a unit volume of the material by a unit on the temperature scale.

Unless otherwise stated, *we shall always assume that the coefficients c_0 and K are constants.* With that assumption, *a second postulate* of the mathematical theory is that conduction leads to a temperature function u which, together with its derivative u_t as well as those of the first and second order with respect to x, y, and z, is a continuous function throughout each domain interior to the solid in which no heat is generated.

To derive the differential equation satisfied by u, first consider a spherical surface S_0 with center at P_0, small enough that all points within and on S_0 are interior to the solid body. The integral of $c_0 u(x,y,z,t)$ over the region R_0 bounded by S_0 is a measure of the instantaneous heat content $Q_0(t)$ of R_0. Hence

$$(4) \qquad Q_0'(t) = \frac{d}{dt} \iiint_{R_0} c_0 u \, dV = c_0 \iiint_{R_0} \frac{\partial u}{\partial t} \, dV,$$

where dV is written for the volume element in the integral.

Heat enters R_0 only by conduction through its boundary surface S_0. If n is distance normal to that surface, positive in the outward direction, and if dA denotes the element of area of S_0, then the time rate $Q_0'(t)$ of conduction of heat *into* the sphere has the alternative expression

$$(5) \qquad Q_0'(t) = \iint_{S_0} K \frac{du}{dn} \, dA = - \iint_{S_0} J_{(n)} \, dA.$$

According to Gauss's divergence theorem for transforming surface integrals into volume integrals,

$$(6) \qquad \iint_{S_0} J_{(n)} \, dA = \iiint_{R_0} \text{div } \mathbf{J} \, dV.$$

---

† Results from three-dimensional vector calculus that are used in this section are developed in, for example, Kaplan (1973, pp. 204 ff and 337 ff), listed in the Bibliography.

Thus expression (5) can be written

$$(7) \qquad Q_0'(t) = -\iiint_{R_0} \text{div } \mathbf{J} \, dV = K \iiint_{R_0} \text{div (grad } u) \, dV.$$

Now div (grad u) is the *laplacian*

$$\nabla^2 u = u_{xx} + u_{yy} + u_{zz};$$

so it follows from expressions (4) and (7) that

$$(8) \qquad \iiint_{R_0} (c_0 u_t - K \nabla^2 u) \, dV = 0.$$

Equation (8) holds for each spherical region R_0 about P_0, provided only that all points of R_0 are interior to the solid and that R_0 is free from sources. The integrand is a continuous function of x, y, and z in R_0 at time t. If the integrand has a positive value at P_0, its continuity requires that there be a spherical region R_1 about P_0, where R_1 is interior to R_0, such that $c_0 u_t - K \nabla^2 u > 0$ throughout R_1. The integral in equation (8), with R_0 replaced by R_1, would then have a positive value rather than the value zero. The corresponding contradiction arises if we assume that the integrand has a negative value at P_0. Consequently, at P_0

$$c_0 u_t - K \nabla^2 u = 0.$$

Since P_0 represents any interior point of the body in a domain free from sources, the temperature function $u(x,y,z,t)$ in every such domain satisfies the *heat equation*

$$(9) \qquad u_t = k(u_{xx} + u_{yy} + u_{zz}),$$

where

$$(10) \qquad k = \frac{K}{c_0}.$$

The coefficient k is called the *thermal diffusivity* of the material.

7. Discussion of the Heat Equation

Suppose that the temperatures within a solid are independent of z; that is, there is no flow of heat in the z direction. The heat equation then reduces to the equation for two-dimensional flow parallel to the xy plane:

$$(1) \qquad u_t = k(u_{xx} + u_{yy}).$$

For one-dimensional flow parallel to the x axis the equation becomes

$$(2) \qquad u_t(x,t) = k u_{xx}(x,t).$$

If temperatures are in a *steady state*, where u does not vary with time, the heat equation becomes *Laplace's equation* $\nabla^2 u = 0$:

$$(3) \qquad\qquad u_{xx} + u_{yy} + u_{zz} = 0.$$

We have assumed so far that heat is neither generated nor lost within the solid but only transferred by conduction. If there is a uniform source throughout the solid that generates heat at a constant rate q, where q denotes the quantity of heat generated per unit volume per unit time, the heat equation takes the form

$$(4) \qquad\qquad c_0 u_t = K \nabla^2 u + q.$$

The derivation of the heat equation in Sec. 6 can be modified to obtain equation (4) if the postulate on the continuity of u and its derivatives is extended to include cases in which a constant uniform source is present.

Boundary conditions which describe the thermal conditions on the surfaces of the solid and the initial temperature distribution throughout the solid must accompany the heat equation if we are to determine the temperature function u. The conditions on the surfaces may be other than just prescribed temperatures. Suppose, for example, that a surface is perfectly insulated. From Sec. 6 we know that if du/dn is the directional derivative of u in the direction of the outward normal at any point on the surface, the flux out of the solid through that point is the value of $-K\, du/dn$ there. Since the flux is zero through any point on an insulated surface, it follows that $du/dn = 0$ at such a point.

On the other hand, there may be surface heat transfer from a boundary surface into a medium whose temperature is a constant T. We assume that the flux of heat through that surface at each point is proportional to the difference between the temperature at the point and the temperature T. According to this approximate linear law of cooling, sometimes called *Newton's law*, there is a positive constant H, known as the *surface conductance* of the material, such that

$$(5) \qquad\qquad -K \frac{du}{dn} = H(u - T)$$

at points on the surface. Condition (5) is often written

$$(6) \qquad\qquad \frac{du}{dn} = h(T - u) \qquad\qquad (h = H/K > 0).$$

For an illustration, consider a slab occupying the region $0 \le x \le b$. If the surface $x = 0$ is insulated and if there is surface heat transfer across the surface $x = b$ into a medium at temperature zero, the desired boundary conditions are

$$u_x(0,y,z,t) = 0$$

and

$$u_x(b,y,z,t) = -hu(b,y,z,t).$$

Finally, it should be pointed out that the postulate $\Phi = -K \, du/dn$ also applies to simple diffusion to represent the flux of a substance which is diffusing within a porous solid. In this case the function u denotes *concentration* (the mass of the diffusing substance per unit volume of the solid) and K is the *coefficient of diffusion*. Since the content of the substance in a given region is the integral of u over that region, we can replace c_0 by unity in the derivation in Sec. 6 to see that the concentration u satisfies the *equation of diffusion*

(7) $$u_t = K \, \nabla^2 u.$$

PROBLEMS

1. Let $u(x)$ denote the steady-state temperatures in a slab bounded by the planes $x = 0$ and $x = b$ when those boundaries are kept at fixed temperatures $u = 0$ and $u = u_0$, respectively. Set up the boundary value problem for $u(x)$ and solve it to show that

$$u(x) = \frac{u_0}{b} x, \qquad \Phi = -K \frac{u_0}{b}$$

where Φ is the flux of heat across each plane $x = x_0$ $(0 \le x_0 \le b)$.

2. A slab occupies the region $0 \le x \le b$. There is a constant flux of heat $\Phi = \phi_0$ into the slab through the face $x = 0$. The face $x = b$ is kept at temperature $u = 0$. Set up and solve the boundary value problem for the steady-state temperatures $u(x)$ in the slab.

$$Ans. \quad u(x) = \frac{\phi_0}{K} (b - x).$$

3. Let the slab $0 \le x \le b$ be subjected to surface heat transfer according to Newton's law at its faces $x = 0$ and $x = b$, the surface conductance H being the same on each face. Verify that if the medium $x < 0$ has temperature zero and the medium $x > b$ has the constant temperature T, then the boundary value problem for steady-state temperatures in the slab is

$$u''(x) = 0 \qquad\qquad (0 < x < b),$$

$$Ku'(0) = Hu(0), \qquad -Ku'(b) = H[u(b) - T],$$

where K is the thermal conductivity of the material in the slab. Write $h = H/K$ and derive the expression

$$u(x) = \frac{T}{bh + 2} (hx + 1)$$

for those temperatures.

4. Let heat be generated at a rate q per unit volume per unit time throughout a solid. Point out why expression (5), Sec. 6, should be modified to read

$$Q_0'(t) = -\iint_{S_0} J_{(n)} \, dA + \iiint_{R_0} q \, dV.$$

Assume that q is a constant or a continuous function of x, y, z. Then postulate the continuity of u and its derivatives as before. Complete the derivation of the form (4), Sec. 7, of the heat equation where q is such a function.

5. Let $u(x,t)$ denote temperatures within a solid which is free from sources. Consider a cylindrical element interior to the solid such that the axis of the cylinder is parallel to the x axis and of length Δx. Letting A denote the area of the base of the cylinder, show why

$$A[-Ku_x(x,t) + Ku_x(x + \Delta x, t)] = c_0 A \, \Delta x \, u_t(x,t),$$

approximately, if Δx is small. Thus obtain equation (2), Sec. 7.

6. In Problem 5 let heat be generated at a constant rate q per unit volume per unit time throughout the solid and derive the equation

$$c_0 u_t = K u_{xx} + q.$$

7. Let $u(x,t)$ denote temperatures in a slender wire lying along the x axis when surface heat transfer takes place along the wire into the surrounding medium at fixed temperature T. Assuming that the time rate of transfer per unit length is proportional to $u(x,t) - T$, use the procedure indicated in Problem 5 to derive the equation

$$u_t = k u_{xx} - h(u - T),$$

where h is a positive constant.

8. Suppose that the thermal coefficients c_0 and K are functions of x, y, z, and t. Modify the derivation in Sec. 6 to show that the heat equation takes the form

$$(c_0 u)_t = (K u_x)_x + (K u_y)_y + (K u_z)_z$$

in a domain where all functions and derivatives involved in that equation are continuous.

9. It is sometimes convenient to change the unit of time so that the constant k or a^2 in the heat or wave equation may be considered as unity. Show, for example, that the substitution $\tau = kt$ may be used to write the two-dimensional heat equation (1), Sec. 7, in the form

$$u_\tau = u_{xx} + u_{yy}.$$

8. Laplace's Equation

A function $u(x,y,z)$ that is continuous, together with its partial derivatives of the first and second order, and satisfies Laplace's equation

(1) $$\nabla^2 u = 0,$$

where $\nabla^2 u$ is the laplacian

(2) $$\nabla^2 u = \text{div}(\text{grad } u) = u_{xx} + u_{yy} + u_{zz},$$

is called a *harmonic function*.

We have seen that the steady-state temperatures at points interior to a solid in which no heat is generated are represented by a harmonic function u. In fact, the temperature function u serves as a potential function for the flow of heat in the sense that $-K \text{ grad } u$ represents the current of heat flow by conduction, and so $-K \, du/dn$ represents the flux of heat in the direction of the coordinate n.

The steady-state concentration of a diffusing substance (Sec. 7) is likewise represented by a harmonic function. From equation (5), Sec. 5, we can see that the static transverse displacements $z(x,y)$ of a stretched membrane satisfy Laplace's equation in two dimensions. Here the displacements are the result of displacements perpendicular to the xy plane of parts of the frame that supports the membrane when no external forces are exerted except at the boundary.

Among the many physical examples of harmonic functions, the velocity potential for the steady-state irrotational motion of an incompressible fluid is prominent in hydrodynamics and aerodynamics.†

An important harmonic function in electrical field theory is the electrostatic potential $V(x,y,z)$ in a region of space that is free from electric charges. The potential may be produced by any static distribution of electric charges outside that region. The vector $-\text{grad } V$ at (x,y,z) represents the electrostatic intensity, the electrical force that would be exerted upon a unit positive charge placed at that point. The fact that V is harmonic is a consequence of the inverse-square law of attraction or repulsion between electric charges.

Likewise, gravitational potential is a harmonic function in domains of space not occupied by matter.

The physical problems treated in this book are limited to those for which the differential equations were derived in this chapter. A few other problems in the physical sciences are noted in order to indicate the great variety of applications of partial differential equations. Derivations of the differential equations for such problems can be found in books on hydrodynamics, elasticity, vibrations and sound, electrical field theory, potential theory, and other branches of continuum mechanics.

9. Cylindrical and Spherical Coordinates

The cylindrical coordinates (ρ,ϕ,z), shown in Fig. 4, determine a point P whose rectangular cartesian coordinates are

(1)
$$x = \rho \cos \phi, \qquad y = \rho \sin \phi, \qquad z = z.$$

Thus ρ and ϕ are the polar coordinates in the xy plane of the point Q where Q is the projection of P onto that plane.

The relations (1) can be written

(2)
$$\rho = \sqrt{x^2 + y^2}, \qquad \phi = \tan^{-1}\left(\frac{y}{x}\right), \qquad z = z,$$

† See Churchill, Brown, and Verhey (1974, Chap. 9), listed in the Bibliography.

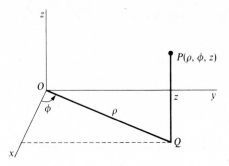

Figure 4

where the quadrant to which the angle ϕ belongs is determined by the signs of x and y, not by the ratio y/x alone.

Let u denote a function of x, y, and z. Then, in view of equations (1), it is also a function of the three independent variables ρ, ϕ, and z. If u and its derivatives of the first and second order with respect to those variables are continuous functions, we can obtain the laplacian of u in terms of them by using the chain rule for differentiating composite functions.

In view of equations (2),

(3)
$$\frac{\partial u}{\partial x} = \frac{\partial u}{\partial \rho}\frac{\partial \rho}{\partial x} + \frac{\partial u}{\partial \phi}\frac{\partial \phi}{\partial x} = \frac{\partial u}{\partial \rho}\frac{x}{\sqrt{x^2+y^2}} - \frac{\partial u}{\partial \phi}\frac{y}{x^2+y^2},$$

where y and z are kept fixed in the differentiation $\partial/\partial x$, ϕ and z are kept fixed in the differentiation $\partial/\partial \rho$, and so on, and ϕ is measured in radians. Similarly,

(4)
$$\frac{\partial u}{\partial y} = \frac{\partial u}{\partial \rho}\frac{y}{\sqrt{x^2+y^2}} + \frac{\partial u}{\partial \phi}\frac{x}{x^2+y^2}.$$

From expression (3) we can write

$$\frac{\partial^2 u}{\partial x^2} = \frac{\partial u}{\partial \rho}\frac{\partial}{\partial x}\left(\frac{x}{\rho}\right) - \frac{\partial u}{\partial \phi}\frac{\partial}{\partial x}\left(\frac{y}{\rho^2}\right) + \frac{x}{\rho}\frac{\partial}{\partial x}\left(\frac{\partial u}{\partial \rho}\right) - \frac{y}{\rho^2}\frac{\partial}{\partial x}\left(\frac{\partial u}{\partial \phi}\right).$$

The chain rule applies to the last two indicated derivatives, giving

$$\frac{\partial}{\partial x}\left(\frac{\partial u}{\partial \rho}\right) = \frac{\partial^2 u}{\partial \rho^2}\frac{x}{\rho} - \frac{\partial^2 u}{\partial \phi\, \partial \rho}\frac{y}{\rho^2},$$

$$\frac{\partial}{\partial x}\left(\frac{\partial u}{\partial \phi}\right) = \frac{\partial^2 u}{\partial \rho\, \partial \phi}\frac{x}{\rho} - \frac{\partial^2 u}{\partial \phi^2}\frac{y}{\rho^2}.$$

Substituting and simplifying, we find that

(5)
$$\frac{\partial^2 u}{\partial x^2} = \frac{y^2}{\rho^3}\frac{\partial u}{\partial \rho} + \frac{2xy}{\rho^4}\frac{\partial u}{\partial \phi} + \frac{x^2}{\rho^2}\frac{\partial^2 u}{\partial \rho^2} - \frac{2xy}{\rho^3}\frac{\partial^2 u}{\partial \rho\, \partial \phi} + \frac{y^2}{\rho^4}\frac{\partial^2 u}{\partial \phi^2}.$$

In like manner, we find from equation (4) that

(6)
$$\frac{\partial^2 u}{\partial y^2} = \frac{x^2}{\rho^3} \frac{\partial u}{\partial \rho} - \frac{2xy}{\rho^4} \frac{\partial u}{\partial \phi} + \frac{y^2}{\rho^2} \frac{\partial^2 u}{\partial \rho^2} + \frac{2xy}{\rho^3} \frac{\partial^2 u}{\partial \rho \partial \phi} + \frac{x^2}{\rho^4} \frac{\partial^2 u}{\partial \phi^2}.$$

According to equations (5) and (6), then,

(7)
$$\frac{\partial^2 u}{\partial x^2} + \frac{\partial^2 u}{\partial y^2} = \frac{\partial^2 u}{\partial \rho^2} + \frac{1}{\rho} \frac{\partial u}{\partial \rho} + \frac{1}{\rho^2} \frac{\partial^2 u}{\partial \phi^2};$$

and so the *laplacian of u in cylindrical coordinates* is

(8)
$$\nabla^2 u = \frac{\partial^2 u}{\partial \rho^2} + \frac{1}{\rho} \frac{\partial u}{\partial \rho} + \frac{1}{\rho^2} \frac{\partial^2 u}{\partial \phi^2} + \frac{\partial^2 u}{\partial z^2}.$$

We may group the first two terms and use the subscript notation for partial derivatives to write this in the form

(9)
$$\nabla^2 u = \frac{1}{\rho} (\rho u_\rho)_\rho + \frac{1}{\rho^2} u_{\phi\phi} + u_{zz}.$$

The spherical coordinates (r,ϕ,θ) of a point (Fig. 5) are related to x, y, and z as follows:

(10) $x = r \sin \theta \cos \phi,$ $y = r \sin \theta \sin \phi,$ $z = r \cos \theta.$

The coordinate ϕ is common to cylindrical and spherical coordinates, and the coordinates in the two systems are related by the equations

(11) $z = r \cos \theta,$ $\rho = r \sin \theta,$ $\phi = \phi.$

Expression (8) can be transformed into spherical coordinates quite readily by means of the proper interchange of letters, *without any further application of the chain rule.* This is accomplished in three steps, described below.

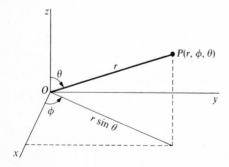

Figure 5

First, we observe that, except for the names of the variables involved, transformation (11) is the same as transformation (1). Since transformation (1) gave us equation (7), it follows that

(12)
$$\frac{\partial^2 u}{\partial z^2} + \frac{\partial^2 u}{\partial \rho^2} = \frac{\partial^2 u}{\partial r^2} + \frac{1}{r}\frac{\partial u}{\partial r} + \frac{1}{r^2}\frac{\partial^2 u}{\partial \theta^2}.$$

Next, we note that when transformation (11) is used, the counterpart of equation (4) is

$$\frac{\partial u}{\partial \rho} = \frac{\rho}{r}\frac{\partial u}{\partial r} + \frac{z}{r^2}\frac{\partial u}{\partial \theta}.$$

With this and the first two of equations (11), we are able to write

(13)
$$\frac{1}{\rho}\frac{\partial u}{\partial \rho} + \frac{1}{\rho^2}\frac{\partial^2 u}{\partial \phi^2} = \frac{1}{r}\frac{\partial u}{\partial r} + \frac{\cot\theta}{r^2}\frac{\partial u}{\partial \theta} + \frac{1}{r^2\sin^2\theta}\frac{\partial^2 u}{\partial \phi^2}.$$

Finally, by grouping the first and last terms in expression (8), and also the second and third terms there, we see that, according to equations (12) and (13), the *laplacian of u in spherical coordinates* is

(14)
$$\nabla^2 u = \frac{\partial^2 u}{\partial r^2} + \frac{2}{r}\frac{\partial u}{\partial r} + \frac{1}{r^2\sin^2\theta}\frac{\partial^2 u}{\partial \phi^2} + \frac{1}{r^2}\frac{\partial^2 u}{\partial \theta^2} + \frac{\cot\theta}{r^2}\frac{\partial u}{\partial \theta}.$$

Other forms of this expression are

(15)
$$\nabla^2 u = \frac{1}{r}(ru)_{rr} + \frac{1}{r^2\sin^2\theta}u_{\phi\phi} + \frac{1}{r^2\sin\theta}(u_\theta\sin\theta)_\theta$$

and

(16)
$$\nabla^2 u = \frac{1}{r^2}(r^2 u_r)_r + \frac{1}{r^2\sin^2\theta}u_{\phi\phi} + \frac{1}{r^2\sin\theta}(u_\theta\sin\theta)_\theta.$$

The two-dimensional laplacian is sometimes needed in the polar coordinates (ρ,ϕ). That form, already given in equation (7), is of course a special case of expression (8) when u is independent of the coordinate z.

PROBLEMS

1. Derive expressions (4) and (6) in Sec. 9.

2. Show that in Sec. 9 expressions (15) and (16) are the same as (14).

3. Obtain equation (7), Sec. 9, by starting with its right-hand side and transforming each term there by means of the chain rule. That is, find

$$\frac{\partial u}{\partial \rho} = \frac{\partial u}{\partial x}\frac{\partial x}{\partial \rho} + \frac{\partial u}{\partial y}\frac{\partial y}{\partial \rho}, \text{ and so on.}$$

4. Let $u(r)$ denote the steady-state temperatures in a solid bounded by two concentric spheres $r = a$ and $r = b$ $(a < b)$ when $u = 0$ on the inner surface $r = a$ and $u = u_0$ on the outer surface $r = b$, where u_0 is a constant. Show why the heat equation for $u(r)$ reduces to

$$\frac{d^2}{dr^2}(ru) = 0,$$

and derive the expression

$$u(r) = \frac{bu_0}{b - a}\left(1 - \frac{a}{r}\right) \qquad (a \leq r \leq b).$$

Sketch the graph of $u(r)$ versus r.

5. In Problem 4 replace the condition on the outer sphere $r = b$ with the condition that there is surface heat transfer into a medium at constant temperature T according to Newton's law. Then obtain the expression

$$u(r) = \frac{hb^2 T}{a + hb(b - a)}\left(1 - \frac{a}{r}\right) \qquad (a \leq r \leq b)$$

for the steady-state temperatures, where h is the ratio of the surface conductance H to the thermal conductivity K of the material.

6. Let $z(\rho)$ represent the static transverse displacements of a membrane stretched between two circles $\rho = 1$ and $\rho = \rho_0$ $(1 < \rho_0)$ in the plane $z = 0$ after the outer support $\rho = \rho_0$ is displaced by a distance $z = z_0$. State why the boundary value problem in $z(\rho)$ can be written

$$\frac{d}{d\rho}\left(\rho \frac{dz}{d\rho}\right) = 0 \qquad (1 < \rho < \rho_0),$$

$$z(1) = 0, \qquad z(\rho_0) = z_0,$$

and obtain the expression

$$z(\rho) = z_0 \frac{\log \rho}{\log \rho_0} \qquad (1 \leq \rho \leq \rho_0).$$

7. Show that the steady-state temperatures $u(\rho)$ in a hollow cylinder $1 \leq \rho \leq \rho_0$, $-\infty < z < \infty$ also satisfy the boundary value problem written in Problem 6 if $u = 0$ on the inner cylindrical surface and $u = z_0$ on the surface $\rho = \rho_0$. Thus show that Problem 6 is a *membrane analogy* for this heat problem. Soap films have been used to display such analogies.

8. A uniform transverse force of F_0 units per unit area acts over a membrane stretched between the two circles $\rho = 1$, $z = 0$ and $\rho = \rho_0$, $z = 0$, where $1 < \rho_0$. From Problem 8, Sec. 5, show that the static transverse displacements $z(\rho)$ satisfy the equation

$$(\rho z')' + f_0 \rho = 0 \qquad (f_0 = F_0/H),$$

and derive the expression

$$z(\rho) = \frac{f_0}{4}(\rho_0^2 - 1)\left(\frac{\log \rho}{\log \rho_0} - \frac{\rho^2 - 1}{\rho_0^2 - 1}\right) \qquad (1 \leq \rho \leq \rho_0).$$

10. Types of Equations and Conditions

The second-order linear partial differential equation

(1) $$Au_{xx} + Bu_{xy} + Cu_{yy} + Du_x + Eu_y + Fu = G,$$

where $A, B, \ldots, G$ are constants or functions of x and y only, is of *elliptic*, *parabolic*, or *hyperbolic* type in a domain of the xy plane if the quantity $B^2 - 4AC$ is negative, zero, or positive, respectively, throughout the domain. The three types require different kinds of boundary conditions to determine a solution.

Poisson's equation $\nabla^2 u = G$ and Laplace's equation $\nabla^2 u = 0$ in $u(x,y)$ are special cases of equation (1) in which $A = C = 1$ and B, D, E, and F vanish. Hence those equations are elliptic in every domain. The heat equation $ku_{xx} - u_y = 0$ is parabolic, and the wave equation $a^2 u_{xx} - u_{yy} = 0$ is hyperbolic.

In the theory of partial differential equations it is shown that equation (1) can be reduced to generalized forms of Laplace's equation or the heat equation or the wave equation, according to whether equation (1) is elliptic, parabolic, or hyperbolic.† The reductions are made by substituting new independent variables for x and y.

Another special case of equation (1) is the *telegraph equation*

$$(2) \qquad v_{xx} = KLv_{tt} + (KR + LS)v_t + RSv.$$

Here $v(x,t)$ represents either the electric potential or current at time t at a point x units from one end of a transmission line or cable which has electrostatic capacity K, self-inductance L, resistance R, and leakage conductance S, all per unit length. The equation is hyperbolic if $KL > 0$. It is parabolic if either K or L is zero.

Let u denote the unknown function in a boundary value problem. A condition that prescribes the values of u itself along a boundary is known as a boundary condition of the *first type*, or a *Dirichlet condition*. A problem of determining a harmonic function interior to a region so that the function assumes prescribed values over the boundaries of that region is a *Dirichlet problem*. In this case the values of the function can be interpreted as steady-state temperatures, and such a physical interpretation leads us to expect that a Dirichlet problem may have a unique solution if the functions considered satisfy certain requirements as to their regularity.

A boundary condition of the *second type*, also called a *Neumann condition*, prescribes the values of the normal derivative du/dn of the function at the boundaries. Among other kinds of boundary conditions are those of the *third type* in which values of $hu + du/dn$ are prescribed at the boundaries, where h is either a constant or a function of the independent variables.

If the partial differential equation in u is of second order with respect to one of the independent variables t and if the values of both u and u_t are prescribed on a boundary $t = 0$, the boundary condition is one of *Cauchy type* with respect to t. When the differential equation is the wave equation

---

† See, for instance, the books on partial differential equations by Courant and Hilbert (vol. 2, 1962), Greenspan (1961), or Weinberger (1965) listed in the Bibliography.

$u_{tt} = a^2 u_{xx}$, such a condition corresponds physically to that of prescribing the initial values of both the transverse displacements u and velocities u_t of a stretched string. Both conditions appear to be needed if the displacements $u(x,t)$ are to be determined.

When the equation is Laplace's equation $u_{xx} = -u_{yy}$ or the heat equation $k u_{xx} = u_t$, however, conditions of Cauchy type with respect to x cannot be imposed without severe restrictions. This again is suggested by interpreting u physically as a temperature function. When the temperatures u of a slab $0 \le x \le b$ are prescribed on the boundary $x = 0$, for example, the flux $K u_x$ to the left through that boundary is ordinarily determined by the values of u there and by other conditions in the problem. Conversely, if the flux $K u_x$ is prescribed at $x = 0$, specific temperatures are needed there to produce that flux.

THE METHOD OF SEPARATION OF VARIABLES

11. Linear Combinations

If c_1 and c_2 are constants and u_1 and u_2 are functions, then the function

$$c_1 u_1 + c_2 u_2$$

is called a *linear combination* of u_1 and u_2. Note that $c_1 u_1, c_2 u_2$, and $u_1 \pm u_2$ are special cases.

Suppose, for example, that u_1 and u_2 are both functions of the independent variables x and y. According to elementary properties of derivatives, a derivative of any linear combination of the two functions can be written as the same linear combination of the individual derivatives. Thus

(1) $$\frac{\partial}{\partial x}(c_1 u_1 + c_2 u_2) = c_1 \frac{\partial u_1}{\partial x} + c_2 \frac{\partial u_2}{\partial x},$$

provided $\partial u_1 / \partial x$ and $\partial u_2 / \partial x$ exist.

A linear space of functions, or *function space*, is a class of functions, with a common domain of definition, such that each linear combination of any two functions in that class remains in it; that is, if u_1 and u_2 are in the class, then so is $c_1 u_1 + c_2 u_2$.

A *linear operator* on a function space is an operator L that transforms each function u of that space into a function Lu and has the property that, for each pair of functions u_1 and u_2,

(2) $$L(c_1 u_1 + c_2 u_2) = c_1 L u_1 + c_2 L u_2$$

whenever c_1 and c_2 are constants. In particular, then,

$$L(u_1 + u_2) = L u_1 + L u_2, \qquad L(c_1 u_1) = c_1 L u_1, \qquad L(0) = 0.$$

If u_3 is a third function in the space,

$$L(c_1 u_1 + c_2 u_2 + c_3 u_3) = L(c_1 u_1 + c_2 u_2) + L(c_3 u_3)$$

$$= c_1 L u_1 + c_2 L u_2 + c_3 L u_3.$$

23

Proceeding by induction, we find that L transforms linear combinations of m functions in this manner:

$$(3) \qquad L\left(\sum_{n=1}^{m} c_n u_n\right) = \sum_{n=1}^{m} c_n L u_n.$$

In view of equation (1), $\partial/\partial x$ is a linear operator on functions of x and y that have derivatives of the first order with respect to x. It is naturally classified as a linear *differential* operator. The operator L which multiplies each function $u(x,y)$ by a fixed function $f(x,y)$ is an example of a still more elementary linear operator. Let us now point out why such composites as $\partial^2/\partial x^2$, $f\,\partial/\partial x$, and $\partial^2/\partial x^2 + f\,\partial/\partial x$ are also linear operators.

If linear operators L and M, distinct or not, are such that M transforms each function u of some function space into a function Mu to which L applies and if u_1 and u_2 are functions of that space, it follows from equation (2) that

$$(4) \qquad LM(c_1 u_1 + c_2 u_2) = L(c_1 Mu_1 + c_2 Mu_2) = c_1 LMu_1 + c_2 LMu_2.$$

That is, the *product* LM of linear operators is a linear operator. If L and M both represent $\partial/\partial x$, it follows that $\partial^2/\partial x^2$ is a linear operator on the space of all functions u whose derivatives u_x and u_{xx} exist. Similarly, if $L = f(x,y)$ and $M = \partial/\partial x$, it follows that $f\,\partial/\partial x$ is a linear operator.

The *sum* of two linear operators is defined by the equation

$$(5) \qquad (L + M)u = Lu + Mu.$$

If we replace u here by $c_1 u_1 + c_2 u_2$, we can see that the sum $L + M$ is a linear operator and hence that the sum of any finite number of linear operators is linear. Thus $\partial^2/\partial x^2 + f\,\partial/\partial x$ is a linear operator.

Each term of a linear homogeneous differential equation in u consists of a product of a function of the independent variables by one of the derivatives of u or by u itself. Hence every linear homogeneous differential equation has the form

$$(6) \qquad Lu = 0$$

where L is a linear differential operator. For example, if

$$(7) \qquad L = A\frac{\partial^2}{\partial x^2} + B\frac{\partial^2}{\partial y\,\partial x} + C\frac{\partial^2}{\partial y^2} + D\frac{\partial}{\partial x} + E\frac{\partial}{\partial y} + F,$$

where the letters A through F denote functions of x and y only, equation (6) is the linear homogeneous partial differential equation

$$(8) \qquad Au_{xx} + Bu_{xy} + Cu_{yy} + Du_x + Eu_y + Fu = 0$$

in $u(x,y)$.

Linear homogeneous boundary conditions also have the form (6). Then the variables appearing as arguments of u and as arguments of functions that serve as coefficients in the linear operator L are restricted so that they represent points on the boundaries.

Now let u_n denote functions that satisfy equation (6); that is, $Lu_n = 0$ $(n = 1, 2, \ldots, m)$. It follows from equation (3) that each linear combination of those functions also satisfies equation (6). We state that *principle of superposition* of solutions, which is fundamental to one of the most powerful methods of solving linear boundary value problems, as follows.

Theorem 1. *If each of m functions* u_1, u_2, $\ldots$, u_m *satisfies a linear homogeneous differential equation* $Lu = 0$, *then every linear combination*

$$(9) \qquad u = c_1 u_1 + c_2 u_2 + \cdots + c_m u_m,$$

where the c's are arbitrary constants, satisfies that differential equation. If each of the m functions satisfies a linear homogeneous boundary condition $Lu = 0$, *then every linear combination* (9) *satisfies that boundary condition.*

12. Examples

The principle of superposition is useful in ordinary differential equations. For example, from the two solutions $y = e^x$ and $y = e^{-x}$ of the linear homogeneous equation $y'' - y = 0$, the general solution $y = c_1 e^x + c_2 e^{-x}$ can be written.

To illustrate the application of Theorem 1 to partial differential equations, consider the linear homogeneous heat equation

$$(1) \qquad u_t(x,t) = u_{xx}(x,t)$$

and the two linear homogeneous boundary conditions

$$(2) \qquad u_x(0,t) = 0, \qquad u_x(1,t) = 0.$$

It is easy to verify that each of the functions

$$u_0 = 1, \qquad u_n = \exp\left(-n^2\pi^2 t\right) \cos n\pi x \qquad (n = 1, 2, \ldots)$$

satisfies equation (1) and conditions (2). Thus every linear combination

$$(3) \qquad u = c_0 + \sum_{n=1}^{m} c_n \exp\left(-n^2\pi^2 t\right) \cos n\pi x$$

of those functions satisfies equation (1) and conditions (2).

A principle of superposition for *non*homogeneous linear differential equations should be noted. Suppose that L is a given linear differential operator and that

$$(4) \qquad Lu_1 = f_1 \qquad \text{and} \qquad Lu_2 = f_2,$$

where f_1 and f_2 are functions of the independent variables only. The function $u = u_1 + u_2$ clearly satisfies the linear differential equation

$$(5) \qquad\qquad Lu = f_1 + f_2,$$

where L is the same operator as in equations (4).

A special case of this principle is the fact that if

$$(6) \qquad\qquad Lu_1 = 0$$

and

$$(7) \qquad\qquad Lu_2 = f,$$

then the function

$$(8) \qquad\qquad u = u_1 + u_2$$

satisfies the equation

$$(9) \qquad\qquad Lu = f.$$

It is also true that if functions u and u_2 satisfy equations (9) and (7), respectively, the function $u_1 = u - u_2$ satisfies equation (6). Thus the totality of solutions u of equation (9) consists of all sums of the form (8) where u_1 denotes solutions of the corresponding homogeneous equation (6) and u_2 is any particular solution of equation (9) itself.

The above discussion of nonhomogeneous linear differential equations is also valid when L is used to represent linear boundary conditions.

13. Series of Solutions

In order to extend our results on linear combinations of solutions to an infinite set of functions u_1, u_2, ..., we must deal with convergence and differentiability of infinite series of functions.

Let the constants c_n and the functions u_n be such that the infinite series with terms $c_n u_n$ converges throughout some domain of the independent variables. The sum of that series is a function

$$(1) \qquad\qquad u = \sum_{n=1}^{\infty} c_n u_n.$$

Let x represent one of the independent variables. The series is *differentiable*, or *termwise differentiable*, with respect to x if the derivatives $\partial u_n / \partial x$ and $\partial u / \partial x$ exist and the series of functions $c_n \, \partial u_n / \partial x$ converges to $\partial u / \partial x$:

$$(2) \qquad\qquad \frac{\partial u}{\partial x} = \sum_{n=1}^{\infty} c_n \frac{\partial u_n}{\partial x}.$$

Note that a series must necessarily be convergent to be differentiable. Sufficient conditions for differentiability will be noted in Sec. 14.

If, in addition, series (2) is differentiable with respect to x, then series (1) is differentiable twice with respect to x, and so on for other derivatives.

Let L be a linear operator consisting of a function f of the independent variables times a derivative, or a sum of a finite number of such terms. We now show that if series (1) is differentiable for all the derivatives involved in L and if each of the functions u_n in series (1) satisfies the linear homogeneous differential equation $Lu_n = 0$, then so does u; that is, $Lu = 0$.

To accomplish this, we first note that according to the definition of the sum of an infinite series,

$$(3) \qquad f\frac{\partial u}{\partial x} = f \lim_{m \to \infty} \sum_{n=1}^{m} c_n \frac{\partial u_n}{\partial x} = \lim_{m \to \infty} f \frac{\partial}{\partial x} \sum_{n=1}^{m} c_n u_n$$

when series (1) is differentiable with respect to x. Here the operator $\partial/\partial x$ can be replaced by other derivatives if the series is so differentiable. Then, by adding corresponding sides of equations similar to equation (3), we find that

$$(4) \qquad Lu = \lim_{m \to \infty} L\left(\sum_{n=1}^{m} c_n u_n \right).$$

The sum on the right-hand side of equation (4) is a linear combination of the functions $u_1, u_2, \ldots, u_m$; and since $Lu_n = 0$ $(n = 1, 2, \ldots)$, Theorem 1 allows us to write

$$L\left(\sum_{n=1}^{m} c_n u_n \right) = 0$$

for every m. Hence from equation (4) we have the desired result $Lu = 0$.

A linear homogeneous boundary condition may also be represented by an equation $Lu = 0$. In that case we may require the function Lu to satisfy a condition of continuity at points on the boundary in order that its values there will represent limiting values as those points are approached from the interior of the domain.

The following generalization of Theorem 1 is now established.

Theorem 2. *If each function of an infinite set $u_1, u_2, \ldots$ satisfies a linear homogeneous differential equation or boundary condition $Lu = 0$, then the function*

$$(5) \qquad u = \sum_{n=1}^{\infty} c_n u_n$$

also satisfies $Lu = 0$, provided the constants c_n and the functions u_n are such that series (5) converges and is differentiable for all derivatives involved in L and provided the required continuity condition at the boundary is satisfied by Lu when $Lu = 0$ is a boundary condition.

14. Uniform Convergence and Example

We recall some facts about uniformly convergent series of functions.†

Let $S(x)$ denote the sum of an infinite series of functions $f_n(x)$, where the series is convergent for all x in some interval $a \leq x \leq b$. That is,

$$(1) \qquad S(x) = \sum_{n=1}^{\infty} f_n(x) = \lim_{m \to \infty} S_m(x) \qquad (a \leq x \leq b)$$

where $S_m(x)$ is the partial sum consisting of the first m terms of the series. The series converges *uniformly* with respect to x if the absolute value of its remainder $R_m(x) = S(x) - S_m(x)$ can be made arbitrarily small for all x in the interval by taking m sufficiently large; that is, for each positive number ε, there exists an integer m_ε, *independent of* x, such that

$$(2) \qquad |S(x) - S_m(x)| < \varepsilon \qquad \text{whenever } m > m_\varepsilon \qquad (a \leq x \leq b).$$

A useful sufficient condition for uniform convergence is given by the *Weierstrass M-test*: if there is a convergent series

$$\sum_{n=1}^{\infty} M_n$$

of positive constants such that for each n

$$(3) \qquad |f_n(x)| \leq M_n \qquad (a \leq x \leq b),$$

then series (1) is uniformly convergent on the stated interval.

Because $S(x) = S_m(x) + R_m(x)$ and $|R_m(x)|$ is uniformly small when m is sufficiently large, the following properties of uniformly convergent series can be established.

If the functions $f_n(x)$ are continuous and if series (1) is uniformly convergent, then the sum $S(x)$ of that series is a continuous function. Also, the series can be integrated term by term over the interval to give the integral of $S(x)$ over the interval.

If the functions f_n and their derivatives f'_n are continuous, if series (1) converges, and if the series whose terms are $f'_n(x)$ is uniformly convergent, then series (1) is differentiable with respect to x.

Corresponding results hold for series of functions of two or more independent variables.

Let us use the above results and Theorem 2 to write a more general solution for the example on the heat equation introduced in Sec. 12. To be specific, we now require the function u to satisfy the equation

$$(4) \qquad u_t(x,t) = u_{xx}(x,t)$$

---

† See, for instance, Kaplan (1973, pp. 407 ff) or Taylor and Mann (1972, pp. 644 ff), listed in the Bibliography.

in the domain $0 < x < 1$, $t > t_0$ whenever $t_0 > 0$ and these two Neumann conditions at the boundaries $x = 0$ and $x = 1$:

(5) $$u_x(0,t) = 0, \qquad u_x(1,t) = 0.$$

Also, $u_x(x,t)$ is to be a continuous function of x on the closed interval $0 \le x \le 1$ for each fixed t ($t \ge t_0$). Then $u_x(0,t)$ is the same as the limit of $u_x(x,t)$ as $x \to 0$ from the interior of the interval, and similarly for $u_x(1,t)$; that is, conditions (5) then imply that

$$\lim_{\substack{h \to 0 \\ h > 0}} u_x(h,t) = 0, \qquad \lim_{\substack{h \to 0 \\ h > 0}} u_x(1 - h, t) = 0 \qquad (t \ge t_0 > 0).$$

In Sec. 12 we noted that each function of the infinite sequence

$$u_0 = 1, \qquad u_n = \exp\left(-n^2\pi^2 t\right) \cos n\pi x \qquad (n = 1, 2, \ldots)$$

satisfies the linear homogeneous equations (4) and (5). Those functions and their partial derivatives are continuous everywhere. According to Theorem 2, the function

(6) $$u(x,t) = c_0 + \sum_{n=1}^{\infty} c_n \exp\left(-n^2\pi^2 t\right) \cos n\pi x$$

also satisfies equations (4) and (5) if the constant coefficients c_n in the infinite series can be restricted so that the series is differentiable twice with respect to x and once with respect to t and so that u_x is continuous in x ($0 \le x \le 1$). We now show that those conditions are satisfied when the sequence of constants c_n is bounded; thus we assume that a number B, independent of n, exists such that

(7) $$|c_n| \le B \qquad (n = 0, 1, 2, \ldots).$$

The terms $c_n u_n$ of series (6) satisfy the condition

(8) $$\left| c_n \exp\left(-n^2\pi^2 t\right) \cos n\pi x \right| \le B \exp\left(-n^2\pi^2 t_0\right) \qquad (t \ge t_0),$$

and the ratio test shows the convergence of the series whose terms are the constants $B \exp\left(-n^2\pi^2 t_0\right)$ when $t_0 > 0$. Series (6) is therefore absolutely convergent by the comparison test and also uniformly convergent with respect to x and t according to the Weierstrass M-test. In like manner we see that

$$\left| \frac{\partial}{\partial x}(c_n u_n) \right| \le B\pi n \exp\left(-n^2\pi^2 t_0\right).$$

The series of constants $n^k \exp\left(-n^2\pi^2 t_0\right)$, where k is any integer, converges according to the ratio test. Hence the series whose terms are $(c_n u_n)_x$ is uniformly convergent with respect to x and, in fact, with respect to x and t together. Series (6) is therefore differentiable with respect to x, and the sum

$u_x(x,t)$ of the differentiated series is continuous with respect to x for all x when $t \geq t_0$.

The uniform convergence of the series of derivatives $(c_n u_n)_{xx}$, which are the same as $(c_n u_n)_t$, is also evident. Thus series (6) is differentiable twice with respect to x and once with respect to t. Its sum u therefore satisfies equations (4) and (5), and it involves an infinite set of constants c_n which are arbitrary except for the boundedness condition (7).

We shall see later on that those constants can be determined so that u will satisfy a nonhomogeneous initial condition

$$(9) \qquad\qquad u(x,0) = f(x) \qquad\qquad (0 < x < 1).$$

Then, physically, u will represent temperatures in a slab $0 \leq x \leq 1$ with insulated faces, in view of conditions (5), and with initial temperatures given by condition (9). Note that condition (9) is satisfied by the function (6) if c_n $(n = 0, 1, 2, \ldots)$ can be found such that f has the representation

$$(10) \qquad\qquad f(x) = c_0 + \sum_{n=1}^{\infty} c_n \cos n\pi x \qquad\qquad (0 < x < 1).$$

PROBLEMS

1. Verify that each of the products $u_n = \cos nx \sinh ny$ $(n = 1, 2, \ldots)$, as well as the function $u_0 = y$, satisfies Laplace's equation $u_{xx} + u_{yy} = 0$ and the three boundary conditions

$$u_x(0,y) = u_x(\pi,y) = u(x,0) = 0.$$

Then apply Theorem 1 to show that the linear combination

$$u = c_0 y + \sum_{n=1}^{m} c_n \cos nx \sinh ny$$

also satisfies that differential equation and those boundary conditions. Find values of the coefficients c_n $(n = 0, 1, 2, \ldots)$ for which that linear combination also satisfies the nonhomogeneous boundary condition

$$u(x,2) = 4 + 3 \cos x - \cos 2x.$$

$$\textit{Ans.} \quad c_0 = 2, \, c_1 = \frac{3}{\sinh 2}, \, c_2 = \frac{-1}{\sinh 4}, \, c_n = 0 \text{ when } n \geq 3.$$

2. Show that each of the functions

$$u_n = \exp\left[-(n - \tfrac{1}{2})^2 \pi^2 t\right] \sin\left[(n - \tfrac{1}{2})\pi x\right] \qquad\qquad (n = 1, 2, \ldots)$$

satisfies the heat equation

$$u_t(x,t) = u_{xx}(x,t)$$

and the boundary conditions

$$u(0,t) = 0, \qquad u_x(1,t) = 0.$$

Then apply Theorem 1 to show that the function

$$u = \sum_{n=1}^{m} c_n \exp\left[-(n - \tfrac{1}{2})^2 \pi^2 t\right] \sin\left[(n - \tfrac{1}{2})\pi x\right]$$

also satisfies those three equations. Find values of the constants c_n such that this function also satisfies the nonhomogeneous boundary condition

$$u(x,0) = 2 \sin \frac{\pi x}{2} - \sin \frac{5\pi x}{2}.$$

Ans. $c_1 = 2$, $c_2 = 0$, $c_3 = -1$, $c_n = 0$ when $n \geq 4$.

3. Verify that each of the functions

$$u_{mn} = \sin mx \cos ny \exp\left(-z\sqrt{m^2 + n^2}\right) \qquad (m = 1, 2, \ldots; n = 0, 1, 2, \ldots)$$

satisfies Laplace's partial differential equation

$$u_{xx}(x,y,z) + u_{yy}(x,y,z) + u_{zz}(x,y,z) = 0$$

and the boundary conditions

$$u(0,y,z) = u(\pi,y,z) = u_y(x,0,z) = u_y(x,\pi,z) = 0.$$

Use superposition to obtain a function that satisfies not only Laplace's equation and those boundary conditions, but also the condition

$$u_z(x,y,0) = (-6 + 5 \cos 4y) \sin 3x.$$

Ans. $u = (2e^{-3z} - e^{-5z} \cos 4y) \sin 3x.$

4. Let u and v denote two functions that satisfy the one-dimensional heat equation in x and t, where the thermal diffusivity k need not be constant; that is, $u_t = ku_{xx}$ and $v_t = kv_{xx}$, where k may be a function of x and t. Multiply each side of those two equations by constants c_1 and c_2, respectively, and add to show that the linear combination $c_1 u + c_2 v$ also satisfies the heat equation. This illustrates a variation in the proof of Theorem 1.

5. Show that if an operator L has the two properties

$$L(u_1 + u_2) = Lu_1 + Lu_2, \qquad L(cu) = cLu,$$

for all functions u_1, u_2, and u and for every constant c, then L is linear; that is, show that it has property (2), Sec. 11.

6. Let u_1 and u_2 satisfy a linear *non*homogeneous differential equation $Lu = f$, where f is a function of the independent variables only. Prove that the linear combination $c_1 u_1 + c_2 u_2$ fails to satisfy that equation when $c_1 + c_2 \neq 1$.

7. Show that each of the functions $y_1 = 1/x$ and $y_2 = 1/(1 + x)$ satisfies the *nonlinear* differential equation $y' + y^2 = 0$. Then show that if c is a constant $(c \neq 0, c \neq 1)$, neither cy_1 nor cy_2 satisfies that equation. Also show that the function $y_1 + y_2$ does not satisfy it.

8. Use special cases of linear operators, such as $L = x$ and $M = \partial/\partial x$, to show that the operators LM and ML are not always the same.

9. Assuming that c_n is a bounded sequence of constants and $y_0 > 0$, prove that the function

$$u(x,y) = \sum_{n=1}^{\infty} c_n e^{-ny} \sin nx \qquad (y \geq y_0)$$

is twice differentiable with respect to x and y and satisfies Laplace's equation in the domain $y > y_0$.

10. Prove that if $n^4|c_n| \leq B$ $(n = 1, 2, \ldots)$, where B is a constant, the function

$$y(x,t) = \sum_{n=1}^{\infty} c_n \sin nx \cos nt$$

satisfies the wave equation $y_{tt} = y_{xx}$ for all x and t.

15. Separation of Variables

Let us find an expression for the transverse displacements $y(x,t)$ of a string stretched between the points $(0,0)$ and $(c,0)$ if the string is initially displaced into a position $y = f(x)$ and released at rest from that position.

At this stage some of our steps in solving the problem must be *formal*, or manipulative. Their validity cannot be established until we have developed further theory.

We assume that no external forces act along the string. Then the function y must satisfy the wave equation (Sec. 3)

$$(1) \qquad\qquad y_{tt}(x,t) = a^2 y_{xx}(x,t) \qquad\qquad (0 < x < c, t > 0).$$

It must also satisfy these boundary conditions:

$$(2) \qquad\qquad y(0,t) = 0, \qquad y(c,t) = 0, \qquad y_t(x,0) = 0,$$

$$(3) \qquad\qquad y(x,0) = f(x) \qquad\qquad (0 \leq x \leq c),$$

where the prescribed displacement function f is continuous on the interval $0 \leq x \leq c$ and $f(0) = f(c) = 0$.

In determining nontrivial $(y \not\equiv 0)$ solutions of *all homogeneous equations* (1) and (2) in the above boundary value problem, using ordinary differential equations, we seek functions of the form

$$(4) \qquad\qquad y(x,t) = X(x)T(t)$$

which satisfy those equations. Note that X is a function of x alone and T a function of t alone. Note, too, that X and T must be nontrivial $(X \not\equiv 0, T \not\equiv 0)$.

If $y = XT$ satisfies equation (1), then

$$X(x)T''(t) = a^2 X''(x)T(t);$$

and we can divide by $a^2 XT$ to separate the variables:

$$(5) \qquad\qquad \frac{X''(x)}{X(x)} = \frac{T''(t)}{a^2 T(t)}.$$

Since the left-hand side here is a function of x alone, it does not vary with t. However, it is equal to a function of t alone, and so it cannot vary with x.

Hence both sides must have some constant value, which we write as $-\lambda$, in common; that is,

(6) $$X''(x) = -\lambda X(x), \qquad T''(t) = -\lambda a^2 T(t).$$

While it may seem more natural to denote the *separation constant* by λ, rather than $-\lambda$, the choice is of course a minor matter of notation. It is only for convenience later on (Secs. 31 through 33) that we have written $-\lambda$.

If XT is to satisfy the first of conditions (2), then $X(0)T(t)$ must vanish for all t $(t > 0)$. With our requirement that $T \not\equiv 0$, it follows that $X(0) = 0$. Likewise, the last two of conditions (2) are satisfied by XT if $X(c) = 0$ and $T'(0) = 0$.

Thus XT satisfies equations (1) and (2) when X and T satisfy these two homogeneous problems:

(7) $$X''(x) + \lambda X(x) = 0, \qquad X(0) = 0, \qquad X(c) = 0,$$

(8) $$T''(t) + \lambda a^2 T(t) = 0, \qquad T'(0) = 0,$$

where the parameter λ has the *same* value in both problems. To find nontrivial solutions of this pair of problems, we first note that problem (8) has only one boundary condition and therefore many solutions for each value of λ. Since problem (7) has two boundary conditions, it may have nontrivial solutions for particular values of λ. As we shall see in the next chapter, problem (7) is a so-called Sturm-Liouville problem, and λ *must be real-valued* if the problem is to have a nontrivial solution.

If $\lambda = 0$, the differential equation in problem (7) becomes $X'' = 0$ and its general solution is $X(x) = Ax + B$. Since $B = 0$ and $Ac + B = 0$ if $X(0) = 0$ and $X(c) = 0$, it follows that $A = B = 0$; therefore this problem has just the trivial solution $X(x) \equiv 0$ when $\lambda = 0$.

If $\lambda > 0$, we may write $\lambda = \alpha^2$ $(\alpha > 0)$. The differential equation in problem (7) then takes the form $X'' + \alpha^2 X = 0$, its general solution being

$$X(x) = C_1 \sin \alpha x + C_2 \cos \alpha x.$$

The condition $X(0) = 0$ implies that $C_2 = 0$; and if the condition $X(c) = 0$ is to hold, $C_1 \sin \alpha c = 0$. In order for there to be a nontrivial solution, then, α must satisfy the equation $\sin \alpha c = 0$. That is,

(9) $$\alpha = \frac{n\pi}{c} \qquad (n = 1, 2, \ldots).$$

Thus, except for the constant factor C_1,

(10) $$X(x) = \sin \frac{n\pi x}{c} \qquad (n = 1, 2, \ldots).$$

The numbers $\lambda = \alpha^2 = n^2\pi^2/c^2$ for which problem (7) has nontrivial solutions are called *eigenvalues* of that problem, and the functions (10) are the corresponding *eigenfunctions*.

When $\lambda < 0$, let us write $\lambda = -\beta^2$ $(\beta > 0)$. Then

$$X(x) = D_1 \sinh \beta x$$

is the solution of $X'' - \beta^2 X = 0$ that satisfies the condition $X(0) = 0$. Since $\sinh \beta c \neq 0$, $D_1 = 0$ if $X(c) = 0$. Thus problem (7) has no negative eigenvalues.

When $\lambda = n^2\pi^2/c^2$, problem (7) is a distinct problem for each different positive integer n. For a fixed integer n it has the solution (10), and problem (8) becomes

$$T''(t) + \left(\frac{n\pi a}{c}\right)^2 T(t) = 0, \qquad T'(0) = 0.$$

Except for a constant factor, then,

$$T(t) = \cos\frac{n\pi a t}{c}.$$

Therefore each function of the infinite set

$$(11) \qquad y_n(x,t) = \sin\frac{n\pi x}{c}\cos\frac{n\pi a t}{c} \qquad (n = 1, 2, \ldots)$$

satisfies all the homogeneous equations (1) and (2).

The procedure here illustrates a fundamental method of obtaining nontrivial solutions of the homogeneous equations in boundary value problems. It is called the *method of separation of variables.*

A linear combination of a finite number of the functions (11) also satisfies the homogeneous equations, according to Theorem 1. But when $t = 0$, it reduces to a linear combination of a finite number of the functions $\sin(n\pi x/c)$. Thus it will not satisfy the nonhomogeneous condition (3) unless $f(x)$ has that special character.

According to Theorem 2, the function

$$(12) \qquad y(x,t) = \sum_{n=1}^{\infty} b_n \sin\frac{n\pi x}{c}\cos\frac{n\pi a t}{c}$$

also satisfies the homogeneous equations (1) and (2), provided the coefficients b_n can be restricted so that the infinite series is suitably convergent and differentiable. That function will satisfy the remaining condition (3) if f can be represented in the form

$$(13) \qquad f(x) = \sum_{n=1}^{\infty} b_n \sin\frac{n\pi x}{c} \qquad (0 \leq x \leq c).$$

In the following section we shall indicate why the coefficients b_n should have the values

(14) $$b_n = \frac{2}{c} \int_0^c f(x) \sin \frac{n\pi x}{c} \, dx \qquad (n = 1, 2, \ldots)$$

if representation (13) is to be valid.

The formal solution of our boundary value problem in the displacements of the string is, then, expression (12) with coefficients (14).

16. The Fourier Sine Series

Recall from trigonometry the identities

$$2 \sin A \sin B = \cos (A - B) - \cos (A + B),$$

$$2 \sin^2 A = 1 - \cos 2A.$$

If c denotes any positive number and m and n are *distinct* positive integers, it is then easy to see that

$$\int_0^c \sin \frac{m\pi x}{c} \sin \frac{n\pi x}{c} \, dx = \frac{1}{2} \int_0^c \left[\cos \frac{(m - n)\pi x}{c} - \cos \frac{(m + n)\pi x}{c} \right] dx = 0;$$

also,

$$\int_0^c \sin^2 \frac{n\pi x}{c} \, dx = \frac{1}{2} \int_0^c \left[1 - \cos \frac{2n\pi x}{c} \right] dx = \frac{c}{2}.$$

That is, when m and n are positive integers,

(1) $$\int_0^c \sin \frac{m\pi x}{c} \sin \frac{n\pi x}{c} \, dx = \begin{cases} 0 & \text{if } m \neq n, \\ \dfrac{c}{2} & \text{if } m = n. \end{cases}$$

Using terminology to be developed in the next chapter, we thus find that the set of functions $\sin (n\pi x/c)$ $(n = 1, 2, \ldots)$ is *orthogonal* on the interval $(0, c)$; that is, the integral over that interval of the product of any two distinct functions of the set is zero.

In Sec. 15 we needed to determine the coefficients b_n so that a series of those sine functions would converge to a prescribed function $f(x)$ on the interval $0 \leq x \leq c$. *Assuming* that an expansion

(2) $$f(x) = b_1 \sin \frac{\pi x}{c} + b_2 \sin \frac{2\pi x}{c} + \cdots + b_n \sin \frac{n\pi x}{c} + \cdots$$

is possible and that the series can be integrated term by term after being multiplied by $\sin(n\pi x/c)$, for any fixed value of the positive integer n, we can use the orthogonality property (1) to find a formula for b_n.

Let all terms in equation (2) be multiplied by $\sin(n\pi x/c)$ and integrated from $x = 0$ to $x = c$. The first term on the right becomes

$$b_1 \int_0^c \sin\frac{\pi x}{c}\sin\frac{n\pi x}{c}\,dx.$$

This is zero unless $n = 1$, according to the orthogonality property. Likewise, all other terms on the right, except the nth one, become zero. In view of property (1), then,

$$\int_0^c f(x)\sin\frac{n\pi x}{c}\,dx = b_n\int_0^c \sin^2\frac{n\pi x}{c}\,dx = \frac{c}{2}b_n.$$

Hence the coefficients in representation (2) have the values

$$(3) \qquad b_n = \frac{2}{c}\int_0^c f(x)\sin\frac{n\pi x}{c}\,dx \qquad (n = 1, 2, \ldots).$$

The sine series (2) with coefficients (3) is called the *Fourier sine series* corresponding to the function f on the interval $0 < x < c$. Using the tilde symbol $\sim$ to indicate correspondence, we can write

$$(4) \qquad f(x) \sim \frac{2}{c}\sum_{n=1}^{\infty}\sin\frac{n\pi x}{c}\int_0^c f(s)\sin\frac{n\pi s}{c}\,ds \qquad (0 < x < c).$$

In Chap. 4 we shall establish general conditions on f under which this series converges to the values $f(x)$, so that the correspondence (4) becomes an equality.

17. A Plucked String

To treat a special case of the boundary value problem considered in Sec. 15, let the string be stretched between the points (0,0) and (2,0) and suppose that its mid-point is raised to a height h above the x axis. The string is then released from rest in that position, consisting of two line segments (Fig. 6).

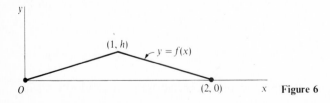

(1, h)

$y = f(x)$

0

(2, 0)

x **Figure 6**

In this case the function f, which describes the initial position of the string, is given by the equations

(1)
$$f(x) = \begin{cases} hx & \text{when } 0 \leq x \leq 1, \\ -hx + 2h & \text{when } 1 \leq x \leq 2. \end{cases}$$

The coefficients b_n in the Fourier sine series corresponding to that function on the interval $0 < x < 2$ can be written

$$b_n = \int_0^2 f(x) \sin \frac{n\pi x}{2} dx$$

$$= h \int_0^1 x \sin \frac{n\pi x}{2} dx + h \int_1^2 (-x + 2) \sin \frac{n\pi x}{2} dx.$$

After integrating and simplifying, we find that

(2)
$$b_n = \frac{8h}{\pi^2 n^2} \sin \frac{n\pi}{2}.$$

Our formal solution (12), Sec. 15, for the displacements of the string then becomes

(3)
$$y(x,t) = \frac{8h}{\pi^2} \sum_{n=1}^{\infty} \frac{1}{n^2} \sin \frac{n\pi}{2} \sin \frac{n\pi x}{2} \cos \frac{n\pi at}{2}.$$

Note that $\sin (n\pi/2) = 0$ when n is even. Series (3) can therefore be written more efficiently by summing only the terms that occur when n is odd. This is accomplished by replacing n by $2n - 1$ wherever n appears on the right of the summation symbol. The result is

(4)
$$y(x,t) = \frac{8h}{\pi^2} \sum_{n=1}^{\infty} \frac{(-1)^{n+1}}{(2n-1)^2} \sin \frac{(2n-1)\pi x}{2} \cos \frac{(2n-1)\pi at}{2},$$

where we have also used the fact that

$$\sin \frac{(2n-1)\pi}{2} = \sin \left(n\pi - \frac{\pi}{2} \right) = -\cos n\pi \sin \frac{\pi}{2} = (-1)^{n+1}.$$

The infinite series in equations (3) and (4) is not differentiable twice with respect to x or t. But in Sec. 51 we shall show how the series can be represented in a form simple enough to see that the sum $y(x,t)$ of the series is indeed a solution of the boundary value problem.

PROBLEMS

1. As pointed out in Sec. 12, the linear homogeneous differential $u_t = u_{xx}$ and boundary conditions $u_x(0,t) = 0$, $u_x(1,t) = 0$ are all satisfied by these functions of type $X(x)T(t)$:

$$u_0 = 1, \qquad u_n = \exp(-n^2\pi^2 t) \cos n\pi x \qquad (n = 1, 2, \ldots).$$

Use the method of separation of variables (Sec. 15) to derive that set of solutions. [The solution $u_0 = 1$ corresponds to the eigenvalue $\lambda = 0$ for the problem $X'' + \lambda X = 0$, $X'(0) = X'(1) = 0$.]

2. Use the method of separation of variables to obtain the set of functions $u_n = \exp\left[-(n - \tfrac{1}{2})^2\pi^2 t\right] \sin\left[(n - \tfrac{1}{2})\pi x\right]$ $(n = 1, 2, \ldots)$ used in Problem 2, Sec. 14.

3. In Problem 1, Sec. 14, it was pointed out that each function of the set

$$u_0 = y, \qquad u_n = \cos nx \sinh ny \qquad\qquad (n = 1, 2, \ldots)$$

satisfies Laplace's equation and the homogeneous boundary conditions

$$u_x(0,y) = u_x(\pi,y) = u(x,0) = 0.$$

Use the method of separation of variables to derive that set.

4. Suppose that equation (5), Sec. 15, had been written in the form

$$a^2 \frac{X''(x)}{X(x)} = \frac{T''(t)}{T(t)}.$$

Set each side here equal to $-\lambda$ and show that the desired result (11), Sec. 15, still follows.

5. Show that the Fourier sine series corresponding to the function $f(x) = 3 \sin 4\pi x$ on the interval $0 < x < 1$ reduces to $3 \sin 4\pi x$, the function itself.

6. Show that the Fourier sine series corresponding to the function $f(x) = 1$ on the interval $0 < x < \pi$ is

$$\frac{4}{\pi} \sum_{n=1}^{\infty} \frac{\sin (2n - 1)x}{2n - 1} = \frac{4}{\pi}\left(\sin x + \frac{1}{3}\sin 3x + \cdots\right).$$

7. Find the Fourier sine series corresponding to the function $f(x) = x$ on the interval $0 < x < 1$.

$$Ans. \quad \frac{2}{\pi} \sum_{n=1}^{\infty} \frac{(-1)^{n+1}}{n} \sin n\pi x.$$

8. Give an example to show that linear combinations of functions of the type $X(x)T(t)$ are not necessarily functions of that restricted type.

18. Ordinary Differential Equations

In Sec. 15 we found that the homogeneous boundary value problem

$$(1) \qquad\qquad X''(x) + \lambda X(x) = 0, \qquad X(0) = 0, \qquad X(c) = 0$$

has many solutions $X(x) = C_1 \sin (n\pi x/c)$, one for each value of C_1, when $\lambda = n^2\pi^2/c^2$ where n is a fixed positive integer. When $\lambda \neq n^2\pi^2/c^2$, the problem has the unique solution $X(x) \equiv 0$. This illustrates the fact that uniqueness of solutions of boundary value problems may depend on the coefficients in the differential equation and the kind of boundary conditions, not solely on the order of the equation and the number of boundary conditions. This is also true of the existence of solutions, as we shall point out below.

The theory of ordinary differential equations ensures the existence and uniqueness of solutions of certain types of *initial value problems*, problems in which all boundary data are given at one point. We state, without proof, a theorem for linear equations of the second order.

Theorem 3. *Let A, B, and C denote continuous functions of x on an interval $a \leq x \leq b$. If x_0 is a fixed point in that interval and y_0 and y_0' are prescribed constants, then there is one and only one function y, continuous together with its derivative y' when $a \leq x \leq b$, which satisfies the differential equation*

$$\text{(2)} \qquad y''(x) + A(x)y'(x) + B(x)y(x) = C(x) \qquad (a < x < b)$$

and the two initial conditions

$$\text{(3)} \qquad y(x_0) = y_0, \qquad y'(x_0) = y_0'.$$

Proofs of this and corresponding theorems for equations of other orders, by methods of successive approximations, will be found in books on the theory of ordinary differential equations.† According to equation (2), $y'' = C - Ay' - By$; thus y'' is also continuous. That the equation has a general solution involving two arbitrary constants follows from the fact that arbitrary values can be assigned to the constants y_0 and y_0'.

As an example, the equation

$$\text{(4)} \qquad x''(t) + k^2 x(t) = 0,$$

where k is a constant and $k \neq 0$, is satisfied by

$$\text{(5)} \qquad x(t) = C_1 \sin kt + C_2 \cos kt$$

where C_1 and C_2 are constants. On each interval of the t axis that contains the point $t = 0$ the function (5) is *the* solution of equation (4) that satisfies the conditions $x(0) = C_2$ and $x'(0) = kC_1$.

Suppose that $x(t)$ satisfies the two boundary conditions

$$\text{(6)} \qquad x(0) = 0, \qquad x(1) = 1$$

along with equation (4) and the requirement that x and x' be continuous on the interval $0 \leq t \leq 1$. Since $x(0) = 0$, it is necessary that

$$x(t) = C_1 \sin kt;$$

and since $x(1) = 1$, a value of C_1 must exist such that

$$\text{(7)} \qquad C_1 \sin k = 1.$$

---

† See pp. 73 ff of the book by Ince (1956) listed in the Bibliography.

This last equation determines C_1 provided $\sin k \neq 0$. Thus

(8)
$$x(t) = \frac{\sin kt}{\sin k} \qquad (k \neq \pm n\pi, \; n = 1, 2, \ldots).$$

But when $k = \pm n\pi$, the two-point boundary value problem consisting of equations (4) and (6) *has no solution* because there is no value of C_1, or $x'(0)/k$, that satisfies equation (7).

It is interesting to note a physical interpretation of the case $k = \pi$. Since $x'' = -\pi^2 x$, the function x represents the displacement of a unit mass along the x axis under a force $-\pi^2 x$ proportional to the displacement from the origin $x = 0$. The mass is initially at the origin and is required to have the displacement $x = 1$ at time $t = 1$. But the number π is the natural frequency k of this system; consequently, the mass returns to the origin at the instants $t = 1, 2, \ldots$, regardless of its initial velocity. Therefore $x(1) = 0$, and the condition $x(1) = 1$ cannot be satisfied.

19. General Solutions of Partial Differential Equations

Some boundary value problems in partial differential equations can be solved by a method corresponding to the one usually used to solve such problems in ordinary differential equations, the method of first finding a general solution of the differential equation.

Example 1. Let us solve the boundary value problem

(1)
$$u_{xx}(x,y) = 0, \qquad u(0,y) = y^2, \qquad u(1,y) = 1.$$

Successive integrations of the equation $u_{xx} = 0$ with respect to x, with y kept fixed, give the equations $u_x = f(y)$ and

(2)
$$u = xf(y) + g(y),$$

where f and g are arbitrary functions. The boundary conditions in problem (1) require that

$$g(y) = y^2, \qquad f(y) + g(y) = 1;$$

thus $f(y) = 1 - y^2$, and the solution of the problem is

(3)
$$u(x,y) = x(1 - y^2) + y^2.$$

Example 2. We next solve the wave equation

(4)
$$y_{tt}(x,t) = a^2 y_{xx}(x,t)$$

in the domain $-\infty < x < \infty$, $t > 0$, subject to the boundary conditions

(5) $$y(x,0) = f(x), \qquad y_t(x,0) = 0 \qquad (-\infty < x < \infty),$$

in terms of the constant a and the function f.

The differential equation (4) can be simplified by introducing new independent variables r and s:

$$r = x + at, \qquad s = x - at.$$

According to the chain rule for differentiation,

$$y_t = y_r r_t + y_s s_t = a y_r - a y_s.$$

Continuing with that rule, we find that

$$y_{tt} = a^2(y_{rr} - 2y_{rs} + y_{ss}), \qquad y_{xx} = y_{rr} + 2y_{rs} + y_{ss}.$$

Equation (4) therefore reduces to the form

$$y_{rs} = 0,$$

which can be solved by successive integrations to give $y_r = g'(r)$ and

$$y = g(r) + h(s)$$

where g and h are arbitrary differentiable functions.

A general solution of the wave equation (4) is therefore

(6) $$y = g(x + at) + h(x - at).$$

In this example the boundary conditions are simple enough that we can determine the functions g and h. To be precise, the function (6) satisfies conditions (5) when

$$g(x) + h(x) = f(x) \qquad \text{and} \qquad ag'(x) - ah'(x) = 0.$$

Thus $g(x) - h(x) = c$, where c is a constant, and it follows that

$$2g(x) = f(x) + c \qquad \text{and} \qquad 2h(x) = f(x) - c.$$

Consequently,

(7) $$y(x,t) = \frac{1}{2}[f(x + at) + f(x - at)].$$

The solution (7) of the boundary value problem consisting of equations (4) and (5) is known as *d'Alembert's solution*. It is easily verified under the assumption that $f'(x)$ and $f''(x)$ exist for all x.

The method illustrated in the two examples here has severe limitations. The general solutions (2) and (6), solutions involving arbitrary functions, were obtained by successive integrations, a procedure that applies to relatively few types of partial differential equations. But, even in the exceptional cases where such general solutions can be found, the determination of the arbitrary functions directly from the boundary conditions is often too difficult.

20. On Other Methods

We have stressed a process based on superposition for solving linear boundary value problems in partial differential equations. The process, illustrated in Sec. 15, consists first of finding solutions of all homogeneous equations in the problem by the method of separation of variables. A generalized linear combination, or superposition, of those solutions which will also satisfy a nonhomogeneous boundary condition is then sought.

That process used in Sec. 15, which is sometimes called the *Fourier method*, is a classical and powerful one. The treatment of its theory and applications is the object of this book. Limitations of the method will be noted also.

There are other important methods of solving linear boundary value problems. The procedures of using Laplace transforms, Fourier transforms, or other integral transforms, all included in the subject of operational mathematics, are especially effective.† The classical method of conformal mapping in the theory of functions of a complex variable applies to a prominent class of problems involving Laplace's equation in two dimensions.‡ There are still other ways of reducing or solving such problems, including applications of so-called Green's functions and numerical or computational methods.

Even when a problem yields to more than one method, however, different methods sometimes produce different forms of the solution; and each form may have its own desirable features. On the other hand, some problems require successive applications of two or more methods. Others, including some linear problems of fairly simple types, have defied all known exact methods. The development of new methods is an activity in present-day mathematical research.

---

† See Churchill (1972), listed in the Bibliography.
‡ See Churchill, Brown, and Verhey (1974), listed in the Bibliography.

PROBLEMS

Use general solutions of the partial differential equations to solve the boundary value problems in Problems 1 through 4:

1. $u_{xx}(x,y) = 6xy$, $u(0,y) = y$, $u_x(1,y) = 0$.

$$\textit{Ans.} \quad u = (x^3 - 3x + 1)y.$$

2. $u_{xy}(x,y) = 2x$, $u(0,y) = 0$, $u(x,0) = x^2$.

$$\textit{Ans.} \quad u = x^2(1 + y).$$

3. $y_{tt}(x,t) = a^2 y_{xx}(x,t)$, $y(x,0) = 0$, $y_t(x,0) = 1/(1 + x^2)$.

$$\textit{Ans.} \quad y = \frac{1}{2a}\left[\tan^{-1}(x + at) - \tan^{-1}(x - at)\right].$$

4. $y_{tt}(x,t) = a^2 y_{xx}(x,t)$, $y(x,0) = 0$, $y_t(x,0) = g(x)$.

$$\textit{Ans.} \quad y = \frac{1}{2a}\int_{x-at}^{x+at} g(s)\, ds.$$

5. Use superposition of the solution (7), Sec. 19, and the solution of Problem 4 to write the solution

$$y = \frac{1}{2}[f(x + at) + f(x - at)] + \frac{1}{2a}\int_{x-at}^{x+at} g(s)\, ds$$

of the boundary value problem

$$y_{tt}(x,t) = a^2 y_{xx}(x,t), \qquad y(x,0) = f(x), \qquad y_t(x,0) = g(x).$$

6. In Example 2, Sec. 19, $y(x,t)$ represents transverse displacements of a stretched string of infinite length, initially released at rest from a position $y = f(x)$ $(-\infty < x < \infty)$. From the solution $y = \frac{1}{2}f(x + at) + \frac{1}{2}f(x - at)$ show how the instantaneous position of the string at time t can be described as the curve obtained by adding ordinates of two curves, one obtained by translating the curve $y = \frac{1}{2}f(x)$ to the left through the distance at, the other by translating it to the right through the same distance. As t varies, the curve $y = \frac{1}{2}f(x)$ moves as a wave with velocity a. Show graphically some instantaneous positions of the string when $f(x)$ is zero except on a small interval about the origin.

7. Derive the general solution

$$u = x + e^{-y}g(x) + h(y)$$

of the partial differential equation $u_{xy} + u_x = 1$.

8. The boundary value problem

$$y_{tt}(x,t) = a^2 y_{xx}(x,t) \qquad\qquad (x > 0,\ t > 0),$$

$$y(x,0) = e^{-x}, \qquad y_t(x,0) = 0 \qquad\qquad (x > 0),$$

$$y_x(0,t) = y(0,t), \qquad \lim_{x \to \infty} y(x,t) = 0 \qquad\qquad (t > 0)$$

can be solved with the aid of Laplace transforms. Its solution is given by the equations

$$y = \begin{cases} e^{-x}\cosh at & \text{when } x \geq at, \\ e^{-x}\cosh at + \sinh(x - at) + (x - at)\exp(x - at) & \text{when } x \leq at. \end{cases}$$

Show that this is a special case of the general solution (6), Sec. 19, and hence that y satisfies the wave equation in the domain $x > 0$, $t > 0$ except at points on the line $x = at$, where the derivatives of y do not exist. Point out that when $t = 0$ and $x > 0$, $y = e^{-x}$ since $x > at$. Show that y satisfies the rest of the boundary conditions. [Even though the general solution of the partial differential equation here is known, this fairly simple boundary value problem is poorly adapted to the method (Sec. 19) of using the general solution.]

9. Consider the partial differential equation

$$Ay_{xx} + By_{xt} + Cy_{tt} = 0 \qquad\qquad (A \neq 0, C \neq 0),$$

where the coefficients A, B, and C are constants.

(a) Use the transformation

$$r = x + \alpha t, \qquad s = x + \beta t,$$

where α and β are *distinct* constants, to obtain a new differential equation in y with independent variables r and s.

(b) Assuming that the given differential equation is of hyperbolic type (Sec. 10), so that $B^2 - 4AC > 0$, find values α_0 and β_0 of α and β, respectively, such that the transformed equation is

$$y_{rs} = 0.$$

Then show that the general solution of the given differential equation is

$$y = g(x + \alpha_0 t) + h(x + \beta_0 t)$$

where g and h are arbitrary differentiable functions.

(c) By applying the solution obtained in part (b) to the wave equation $-a^2 y_{xx} + y_{tt} = 0$, show how the solution (6), Sec. 19, of that equation follows as a special case.

$$Ans. \quad (b) \ \alpha_0, \beta_0 = \frac{-B \pm \sqrt{B^2 - 4AC}}{2C}.$$

21. Historical Development

Mathematical sciences experienced a burst of activity following the invention of calculus by Newton (1642–1727) and Leibnitz (1646–1716). Among topics in mathematical physics that attracted the attention of great scientists during that period were boundary value problems in vibrations of strings stretched between fixed points and vibrations of bars or columns of air, all associated with mathematical theories of musical vibrations. Early contributors to the theory of vibrating strings included the English mathematician Brook Taylor (1685–1731), the Swiss mathematicians Daniel Bernoulli (1700–1782) and Leonhard Euler (1707–1783), and Jean d'Alembert (1717–1783) in France.

By the 1750's d'Alembert, Bernoulli, and Euler had advanced the theory of vibrating strings to the stage where the partial differential equation $y_{tt} = a^2 y_{xx}$ was known and a solution of the boundary value problem for strings had been found from the general solution of that equation. Also, the concept of fundamental modes of vibration led those men to notions of superposition of solutions, to a solution of the form (12), Sec. 15, where a series of trigonometric functions appears, and thus to the matter of representing an arbitrary function by a trigonometric series. Later on, Euler gave the formulas for the coefficients in the series. But the general concept of a function had not been clarified, and a lengthy controversy took place over the question of representing arbitrary functions on a bounded interval by series of sine functions. This question of representation was finally settled by the German mathematician P. G. Lejeune Dirichlet (1805–1859) about 70 years later.

The French mathematical physicist Jean Baptiste Joseph Fourier (1768–1830) presented many instructive examples of expansions in trigonometric series in connection with boundary value problems in the conduction of heat. His book "Théorie analytique de la chaleur," published in 1822, is a classic on the theory of heat conduction. It was actually the third version of a monograph that he originally submitted to the Institut de France on December 21, 1807.† He effectively illustrated the basic procedures of separation of variables and superposition, and his work did much toward arousing interest in trigonometric series representations.

But Fourier's contributions to the representation problem did not include conditions of validity of representations; he was interested in applications and methods. As noted above, Dirichlet was the first to give such conditions. In 1829 he firmly established general conditions on a function sufficient to ensure the convergence of its Fourier series to values of the function.‡

Representation theory has been refined and greatly extended since Dirichlet's time. It is still growing.

---

† Freeman's early translation of Fourier's book into English was reprinted by Dover, New York, in 1955. The original 1807 monograph itself remained unpublished until 1972, when the critical edition by Grattan-Guinness that is listed in the Bibliography appeared.

‡ For supplementary reading on the history of Fourier series, see the articles by Langer (1947) and Van Vleck (1914) listed in the Bibliography.

CHAPTER
THREE

ORTHOGONAL SETS OF FUNCTIONS

22. The Inner Product of Two Vectors

The concept of an orthogonal set of functions is analogous to the concept of an orthogonal, or mutually perpendicular, set of vectors in ordinary three-dimensional space. Fundamental properties of the set of functions are suggested by corresponding properties of the set of vectors. In fact, many aspects of orthogonal functions and three-dimensional vectors can be treated simultaneously within the context of generalized vector spaces. We shall limit ourselves, however, to analogies which will motivate the terminology and results needed in our brief introduction to orthogonal functions. Our notation for three-dimensional vectors in this and the following sections is chosen in order to point up those analogies.

Let f denote a vector in three-dimensional space whose rectangular components are the three real numbers a_1, a_2, and a_3. It can be thought of as the radius vector of the point having those three numbers as rectangular cartesian coordinates. When f is the zero vector, the vector whose components are all zero, the radius vector reduces of course to a single point, the origin. If a vector g has components b_1, b_2, and b_3, the *inner*, or scalar, *product* of f and g is given by the equation

$$(1) \qquad (f,g) = \sum_{k=1}^{3} a_k b_k.$$

The *norm*, or length, of f can then be written

$$(2) \qquad \|f\| = (f,f)^{1/2},$$

the number (f,f) being nonnegative.

Although we have introduced the inner product first, and then the norm, the norm is usually defined first by writing

$$(3) \qquad \|f\| = \left(\sum_{k=1}^{3} a_k^2 \right)^{1/2},$$

which is the same as equation (2). When neither f nor g is the zero vector, the angle θ $(0 \leq \theta \leq \pi)$ between the radius vectors representing them is well-defined; and the inner product (f,g) is introduced in terms of norms by means of the equation

(4) $$(f,g) = \|f\| \|g\| \cos \theta.$$

Expression (1) is then obtained from equation (4). If either f or g is the zero vector, the angle θ in equation (4) is undefined; but it is understood that $(f,g) = 0$ then.

That approach has considerable advantage from a geometric point of view. Equation (3) is simply a statement that the norm of f is the length of the directed line segment representing it. Also, the length

(5) $$\|f - g\| = \left[\sum_{k=1}^{3} (a_k - b_k)^2 \right]^{1/2}$$

of the vector $f - g$, whose components are $a_1 - b_1, a_2 - b_2$, and $a_3 - b_3$, can be thought of as the distance between the tips of the vectors f and g, interpreted as radius vectors. Note, too, that if $\|g\| = 1$, then, according to equation (4), (f,g) is the projection of f onto the direction of g. As we shall see, the analogous norms and inner products of *functions* do not have such immediate geometric interpretations, and it will be more convenient to start with inner products when we develop the analogy in Sec. 25.

Finally, we note that when f and g are *orthogonal*, or perpendicular, to each other, the value of $\cos \theta$ in equation (4) is zero; that is,

(6) $$(f,g) = 0.$$

If $\|f\| = 1$, then f is a unit vector, also called a *normalized* vector. In the next section we combine the concepts of orthogonal and normalized vectors.

23. Orthonormal Sets of Vectors

Let ψ_n $(n = 1, 2, 3)$ denote nonzero vectors in three-dimensional space which form an orthogonal set; that is, they are mutually orthogonal, so that $(\psi_m, \psi_n) = 0$ when $m \neq n$. A set of *unit* vectors $\phi_n (n = 1, 2, 3)$ having the same directions can be formed by dividing each vector ψ_n by its length:

$$\phi_n = \frac{\psi_n}{\|\psi_n\|} \qquad (n = 1, 2, 3).$$

This orthogonal set of normalized vectors ϕ_n is called an *orthonormal set*. Such a set is described by means of inner products by writing

(1) $$(\phi_m, \phi_n) = \delta_{mn} \qquad (m, n = 1, 2, 3)$$

where δ_{mn} is *Kronecker's* δ:

$$\delta_{mn} = \begin{cases} 0 & \text{if } m \neq n, \\ 1 & \text{if } m = n. \end{cases}$$

Condition (1) states that each of the vectors ϕ_1, ϕ_2, ϕ_3 is perpendicular to the other two and that each has unit length.

The symbol $\{\phi_n\}$ will be used to denote such orthonormal sets. A simple example is the set consisting of unit vectors along the three coordinate axes of the rectangular cartesian coordinate system.

Every vector f in three-dimensional space can be expressed as a linear combination of the vectors ϕ_1, ϕ_2, and ϕ_3. That is, three numbers $c_1, c_2,$ and c_3 can always be found such that

$$(2) \qquad f = c_1\phi_1 + c_2\phi_2 + c_3\phi_3$$

when the vector f is given. To find the number c_1, take the inner product of both sides of the vector equation (2) by ϕ_1. This gives

$$(f,\phi_1) = (c_1\phi_1 + c_2\phi_2 + c_3\phi_3, \phi_1)$$
$$= c_1(\phi_1,\phi_1) + c_2(\phi_2,\phi_1) + c_3(\phi_3,\phi_1) = c_1$$

since $(\phi_1,\phi_1) = 1$ and $(\phi_2,\phi_1) = (\phi_3,\phi_1) = 0$, according to condition (1). Similarly, c_2 and c_3 are found by taking the inner products of both sides of equation (2) by ϕ_2 and ϕ_3, respectively. Thus

$$(3) \qquad c_n = (f,\phi_n) \qquad (n = 1, 2, 3),$$

c_n being the projection of f onto ϕ_n. The representation (2) can then be written

$$(4) \qquad f = (f,\phi_1)\phi_1 + (f,\phi_2)\phi_2 + (f,\phi_3)\phi_3 = \sum_{n=1}^{3} (f,\phi_n)\phi_n.$$

The representation (2), or (4), is an expansion of the arbitrary vector f into a finite series of the orthonormal reference vectors. Those orthogonal reference vectors were assumed to be normalized only as a matter of convenience in order to obtain the simple formulas (3) for the coefficients in the expansion.

The definitions and results just given can be extended immediately to vectors in N-dimensional space, where a vector f has N components instead of 3. It is natural to define the inner product of two such vectors f and g to be the number

$$(5) \qquad (f,g) = \sum_{k=1}^{N} a_k b_k$$

when f and g have components $a_1, a_2, \ldots, a_N$ and $b_1, b_2, \ldots, b_N$, respectively.

A generalization of another sort is also possible. The units of length on the rectangular coordinate axes, with respect to which the components of vectors are measured, may vary from one axis to another. In such a case an inner product of two vectors f and g in three-dimensional space has the form

$$(f,g) = \sum_{k=1}^{3} p_k a_k b_k,$$

where the positive weight numbers p_1, p_2, and p_3 depend upon the units of length used along the three axes.

24. Piecewise Continuous Functions

Before introducing inner products and norms of functions, we need to specify the function space (Sec. 11) for which this is to be done.

Let a function f be continuous at all points of a bounded interval $a \leq x \leq b$ except possibly for some finite set of points a, x_1, x_2, ..., x_{n-1}, b where $a < x_1 < x_2 < \cdots < x_{n-1} < b$. Then f is continuous on each of the open intervals

$$a < x < x_1, \qquad x_1 < x < x_2, \qquad \ldots, \qquad x_{n-1} < x < b.$$

It is not necessarily continuous or even defined at their end points. But if in each of those subintervals f has finite limits as x approaches the end points from the interior, f is said to be *piecewise continuous* on the interval (a,b). To be precise, the one-sided limits $f(a+)$, $f(x_1-)$, $f(x_1+)$, ..., $f(b-)$ are required to exist, where such limits from the right and left at a point x_0 are defined, respectively, as follows:

(1) $\qquad f(x_0+) = \lim_{\substack{h \to 0 \\ h > 0}} f(x_0 + h), \qquad f(x_0-) = \lim_{\substack{h \to 0 \\ h > 0}} f(x_0 - h).$

The step function illustrated in Fig. 7 is piecewise continuous on the interval $(0,4)$, although it is not continuous there. Note that a function is piecewise continuous on (a,b) if it is continuous on the *closed* interval

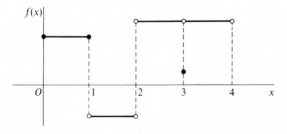

Figure 7

$a \leq x \leq b$. Continuity on the *open* interval $a < x < b$ does not, however, imply piecewise continuity there. The function $f(x) = 1/x$ defined on the interval $0 < x < 1$ is, for example, continuous but not piecewise continuous on that interval since $f(0+)$ fails to exist.

The integral of a piecewise continuous function exists. It is the sum of the integrals over the subintervals:

$$(2) \qquad \int_a^b f(x)\, dx = \int_a^{x_1} f\, dx + \int_{x_1}^{x_2} f\, dx + \cdots + \int_{x_{n-1}}^b f\, dx.$$

The first integral on the right exists because it is the integral of a continuous function over the interval $a \leq x \leq x_1$ if we merely assign the values $f(a+)$ and $f(x_1-)$ to f at a and x_1, respectively. Likewise, the other integrals exist as integrals of continuous functions.

If two functions f_1 and f_2 are each piecewise continuous on an interval (a,b), there is a subdivision of the interval such that both functions are continuous on each closed subinterval when the functions are given their limiting values from the interior at the two end points. Hence a linear combination $c_1 f_1 + c_2 f_2$, or the product $f_1 f_2$, has that continuity on each subinterval and is therefore piecewise continuous on (a,b). Consequently, the integral of the functions $c_1 f_1 + c_2 f_2$, $f_1 f_2$, and $[f_1(x)]^2$ all exist on the interval (a,b). Also, one-sided limits of those functions exist as the corresponding combinations of one-sided limits of f_1 and f_2.

Except when otherwise stated, in this book *we shall restrict our attention to functions that are piecewise continuous* on all bounded intervals under consideration. When it is stated that a function is piecewise continuous on an interval, it is to be understood that the interval is bounded; and the notion of piecewise continuity clearly applies regardless of whether the interval is open or closed.

Since, as noted above, any linear combination of functions that are piecewise continuous also has that property, we may use the terminology of Sec. 11 and refer to the class of all piecewise continuous functions defined on the interval (a,b) as a function space; we denote it by $C_p(a,b)$. It is analogous to three-dimensional space, where linear combinations of vectors are well-defined vectors in that space. In the next section we shall extend the analogy by developing the concept of orthogonal functions.

Other function spaces can be used in the theory of orthogonal functions. The subspace consisting of all continuous functions on the interval $a \leq x \leq b$ is simpler but more restricted than the space $C_p(a,b)$. The space of all integrable functions f on (a,b) whose products, including squares $[f(x)]^2$, are integrable is used in the general theory of functional analysis where a more general type of integral, known as the Lebesgue integral, is also used. Our special case involves more elementary concepts in mathematical analysis.

25. Orthogonality of Functions

We now introduce an inner product which can be applied to any two functions f and g in the function space $C_p(a,b)$ defined in the previous section. In order to simplify our initial remarks, we assume for the moment that f and g are actually *continuous* on the interval $a \le x \le b$.

Dividing that interval into subintervals of equal length $\Delta x = (b - a)/N$, where N is any positive integer, and letting x_k denote any point in the kth subinterval, we recall from elementary calculus that when N is large,

$$\int_a^b f(x)g(x)\, dx \doteq \sum_{k=1}^N f(x_k)g(x_k)\, \Delta x,$$

the symbol $\doteq$ here denoting approximate equality. That is,

$$(1) \qquad \int_a^b f(x)g(x)\, dx \doteq \sum_{k=1}^N a_k b_k$$

where

$$a_k = f(x_k)\sqrt{\Delta x} \qquad \text{and} \qquad b_k = g(x_k)\sqrt{\Delta x}.$$

In view of equation (5), Sec. 23, the left-hand side of expression (1) is then approximately the inner product of two vectors in N-dimensional space, when N is large. The approximate equality becomes exact in the limit as the number of components considered tends to infinity.†

This suggests defining the *inner product of the functions f and g* as the number

$$(2) \qquad (f,g) = \int_a^b f(x)g(x)\, dx.$$

The inner product here is, of course, also well-defined when f and g are allowed to be piecewise continuous on (a,b).

The function space $C_p(a,b)$ with inner product (2) is analogous to ordinary three-dimensional space. Indeed, the following counterparts of familiar properties of vectors in three-dimensional space hold for any functions f, g, and h in $C_p(a,b)$:

$$(3) \qquad\qquad (f,g) = (g,f),$$

$$(4) \qquad\qquad (f,g + h) = (f,g) + (f,h),$$

$$(5) \qquad\qquad (cf,g) = c(f,g),$$

---

† See pp. 210 ff of the book by Lanczos (1966), listed in the Bibliography, for an elaboration of this idea.

where c is any real number, and

$$(6) \qquad\qquad (f,f) \geq 0.$$

The analogy is carried further with the introduction of the *norm*

$$(7) \qquad\qquad \|f\| = (f,f)^{1/2}$$

of a function f in $C_p(a,b)$. It is evident from equation (2) that the norm of f can be written

$$(8) \qquad\qquad \|f\| = \left\{ \int_a^b [f(x)]^2 \, dx \right\}^{1/2}.$$

Two functions f and g in $C_p(a,b)$ are *orthogonal* when

$$(f,g) = 0,$$

or

$$(9) \qquad\qquad \int_a^b f(x)g(x) \, dx = 0.$$

Also, if $\|f\| = 1$, the function f is said to be *normalized*.

We have carried our analogy too far to preserve the original meaning of our geometric terminology. The norm of a function f has no interpretation as a length associated with f. Geometrically, it is simply the square root of the area under the graph of $[f(x)]^2$. The orthogonality of two functions f and g signifies nothing about perpendicularity, but instead that the product fg assumes both negative and positive values on the interval in such a manner that equation (9) holds. The so-called *distance between two functions* f *and* g,

$$(10) \qquad\qquad \|f - g\| = \left\{ \int_a^b [f(x) - g(x)]^2 \, dx \right\}^{1/2},$$

is a measure of the mean distance between their graphs.

A set of functions $\{\psi_n(x)\}$ $(n = 1, 2, \ldots)$ is orthogonal on an interval (a,b) if $(\psi_m, \psi_n) = 0$ whenever $m \neq n$. If none of the functions ψ_n have zero norms, each function ψ_n can be normalized by dividing it by the positive constant $\|\psi_n\|$. The new set $\{\phi_n(x)\}$ so formed, where

$$(11) \qquad\qquad \phi_n(x) = \frac{\psi_n(x)}{\|\psi_n\|} \qquad\qquad (n = 1, 2, \ldots),$$

is *orthonormal* on the interval; that is,

$$(12) \qquad\qquad (\phi_m, \phi_n) = \delta_{mn} \qquad\qquad (m,n = 1, 2, \ldots),$$

where δ_{mn} is Kronecker's δ (Sec. 23). Written in full, the characterization (12) of an orthonormal set becomes

$$(13) \qquad\qquad \int_a^b \phi_m(x)\phi_n(x) \, dx = \begin{cases} 0 & \text{if } m \neq n, \\ 1 & \text{if } m = n. \end{cases}$$

The interval (a,b) over which the functions and their inner products are defined is called the *fundamental interval.* The same terminology is used when the interval under consideration is closed, such as when the inner product is defined on the space of functions that are continuous on the entire interval $a \leq x \leq b$.

An example of an orthogonal set of functions was cited in Sec. 16, where it was noted that the set

(14) $$\left\{ \sin \frac{n\pi x}{c} \right\} \qquad (n = 1, 2, \ldots)$$

is orthogonal on the interval $(0,c)$. According to that section, the norm of each function is $\sqrt{c/2}$; so the corresponding orthonormal set on $(0,c)$ is

(15) $$\left\{ \sqrt{\frac{2}{c}} \sin \frac{n\pi x}{c} \right\} \qquad (n = 1, 2, \ldots).$$

The set (14) is also orthogonal on the interval $(-c,c)$; in that case the normalizing factor is $1/\sqrt{c}$.

PROBLEMS

1. Show that the set

$$\left\{ \frac{1}{\sqrt{c}}, \sqrt{\frac{2}{c}} \cos \frac{n\pi x}{c} \right\} \qquad (n = 1, 2, \ldots)$$

is orthonormal on the interval $(0,c)$.

Suggestion: Use the trigonometric identities (compare Sec. 16)

$$2 \cos A \cos B = \cos (A - B) + \cos (A + B),$$

$$2 \cos^2 A = 1 + \cos 2A.$$

2. (*a*) Use the fact that the set (15), Sec. 25, is orthonormal on the interval $(0,c)$ to show that the set

$$\left\{ \frac{1}{\sqrt{c}} \sin \frac{n\pi x}{c} \right\} \qquad (n = 1, 2, \ldots)$$

is orthonormal on the interval $(-c,c)$.

(*b*) Use the fact that the set in Problem 1 is orthonormal on the interval $(0,c)$ to show that the set

$$\left\{ \frac{1}{\sqrt{2c}}, \frac{1}{\sqrt{c}} \cos \frac{n\pi x}{c} \right\} \qquad (n = 1, 2, \ldots)$$

is orthonormal on the interval $(-c,c)$.

(*c*) Using the results in parts (*a*) and (*b*), show that the set

$$\left\{ \frac{1}{\sqrt{2c}}, \frac{1}{\sqrt{c}} \cos \frac{m\pi x}{c}, \frac{1}{\sqrt{c}} \sin \frac{n\pi x}{c} \right\} \qquad (m,n = 1, 2, \ldots)$$

is orthonormal on the interval $(-c,c)$.

Suggestion: Observe that if $f(-x) = f(x)$ for all x in the interval $(-c,c)$, its graph $y = f(x)$ for that interval is symmetric with respect to the y axis and

$$\int_{-c}^{c} f(x) \, dx = 2 \int_{0}^{c} f(x) \, dx,$$

provided f is integrable from $x = 0$ to $x = c$. Likewise, if $f(-x) = -f(x)$, the graph is symmetric with respect to the origin and

$$\int_{-c}^{c} f(x) \, dx = 0.$$

3. Show that the functions $\psi_1(x) = 1$ and $\psi_2(x) = x$ are orthogonal on the interval $(-1,1)$, and determine constants A and B such that the function $\psi_3(x) = 1 + Ax + Bx^2$ is orthogonal to both ψ_1 and ψ_2 on that interval.

Ans. $A = 0, B = -3$.

4. Two continuous functions $f(x)$ and $\psi_1(x)$ are linearly independent on an interval (a,b); that is, one is not a constant times the other. Determine the linear combination $f + A\psi_1$ of those functions which is orthogonal to ψ_1 on the interval, and thus obtain the orthogonal pair ψ_1, ψ_2 where

$$\psi_2(x) = f(x) - \frac{(f,\psi_1)}{\|\psi_1\|^2} \psi_1(x).$$

Also, give the geometric interpretation of this expression for ψ_2 when f, ψ_1, and ψ_2 represent vectors in three-dimensional space.

5. If a function f is continuous on a closed bounded interval, it is bounded there; that is, a positive number M exists such that $|f(x)| \le M$ for all points x in the interval. Using this fact, state why a piecewise continuous function is bounded on the set of points in its interval at which it is defined.

6. Write $f(x) = 1/\sqrt{x}$ and (*a*) state why f is not piecewise continuous on the interval $(0,1)$; also, (*b*) show that

$$\int_{0}^{1} f(x) \, dx$$

exists as an improper integral but that $[f(x)]^2$ is not integrable over the interval $(0,1)$.

7. Suppose that $f(x) = 0$ except at a finite number of points on an interval $a \le x \le b$. State why f is a piecewise continuous function with zero norm ($\|f\| = 0$) on that interval.

8. Given that the integral of a nonnegative continuous function has a positive value if the function has a positive value somewhere in the interval of integration, show that if a function f is piecewise continuous and $\|f\| = 0$ on an interval (a,b) of the x axis, then $f(x) = 0$ everywhere, except possibly for a finite number of points, in the interval.

9. Verify that

$$\frac{1}{2} \int_{a}^{b} \int_{a}^{b} [f(x)g(y) - g(x)f(y)]^2 \, dx \, dy = \|f\|^2 \|g\|^2 - (f,g)^2,$$

assuming that f and g are piecewise continuous. Thus establish the *Schwarz inequality*

$$|(f,g)| \le \|f\| \, \|g\|,$$

which is also valid when f and g denote vectors in three-dimensional space. In that case it is known as Cauchy's inequality and is an immediate consequence of definition (4), Sec. 22, of the inner product of two vectors.

10. Prove that if f and g are piecewise continuous functions on a fundamental interval (a,b) and if either has norm zero, say $\|f\| = 0$, then $(f,g) = 0$ (use Problem 9).

11. Prove that if f and g are functions in the space $C_p(a,b)$, then

$$\|f + g\| \le \|f\| + \|g\|.$$

If f and g denote, instead, vectors in three-dimensional space, this is the familiar triangle inequality, which states that the length of one side of a triangle is less than or equal to the sum of the lengths of the other two sides.

Suggestion: Start the proof by showing that

$$\|f + g\|^2 = \|f\|^2 + 2(f,g) + \|g\|^2,$$

and then use the Schwarz inequality (Problem 9).

26. Generalized Fourier Series

Let f be an arbitrarily given function in $C_p(a,b)$, the space of piecewise continuous functions defined on the interval (a,b). When an orthonormal set of functions $\phi_n(x)$ $(n = 1, 2, \ldots)$ in $C_p(a,b)$ is specified, it *may* be possible to represent f by a linear combination of those functions, generalized to an infinite series which converges to $f(x)$ at all but possibly a finite number of points x in the fundamental interval (a,b):

$$(1) \qquad f(x) = c_1 \phi_1(x) + c_2 \phi_2(x) + \cdots + c_n \phi_n(x) + \cdots \quad (a < x < b).$$

This corresponds to the representation (2), Sec. 23, of any vector in three-dimensional space in terms of the three vectors of an orthonormal set. We recall from Sec. 23 that for three-dimensional space only three vectors ϕ_n are required in any orthonormal reference set, three being the number of components of a vector in that space. However, as suggested by the discussion at the beginning of Sec. 25 leading up to the inner product (2) in that section, the number of functions $\phi_n(x)$ in the above orthonormal set will in general need to be countably infinite; that is, there is a one-to-one correspondence between those functions and the set of all positive integers.

If the series in equation (1) does converge to $f(x)$ and if, after we multiply all terms in the equation by $\phi_n(x)$, the resulting series is integrable, we can obtain the coefficients c_n as inner products by the process used for vectors. Upon multiplying through by $\phi_n(x)$ and integrating over the interval (a,b), we see that

$$(f,\phi_n) = c_1(\phi_1,\phi_n) + c_2(\phi_2,\phi_n) + \cdots + c_n(\phi_n,\phi_n) + \cdots.$$

Since $(\phi_m,\phi_n) = \delta_{mn}$, it follows that

$$(2) \qquad\qquad (f,\phi_n) = c_n \qquad\qquad (n = 1, 2, \ldots).$$

The numbers c_n are called the *Fourier constants* for f corresponding to the orthonormal set $\{\phi_n(x)\}$; they can be written

$$(3) \qquad c_n = \int_a^b f(x)\phi_n(x)\,dx \qquad (n = 1, 2, \ldots).$$

The series in equation (1) with those coefficients is the *generalized Fourier series* corresponding to the function f, and we write

$$(4) \qquad f(x) \sim \sum_{n=1}^{\infty} c_n \phi_n(x) = \sum_{n=1}^{\infty} \phi_n(x) \int_a^b f(s)\phi_n(s)\,ds \qquad (a < x < b).$$

When $\{\phi_n(x)\}$ is the orthonormal set of functions (Sec. 25)

$$\phi_n(x) = \sqrt{\frac{2}{c}} \sin \frac{n\pi x}{c} \qquad (n = 1, 2, \ldots)$$

in the function space $C_p(0,c)$, for example, the correspondence (4) becomes

$$f(x) \sim \frac{2}{c} \sum_{n=1}^{\infty} \sin \frac{n\pi x}{c} \int_0^c f(s) \sin \frac{n\pi s}{c}\,ds \qquad (0 < x < c),$$

which was obtained earlier in Sec. 16.

The correspondence (4) between $f(x)$ and its series will not always be an equality, even at all but a finite number of points. We may anticipate this limitation by considering the case of vectors in three-dimensional space. In that case, if only two vectors ϕ_1 and ϕ_2 make up the orthonormal set, any vector not in the plane of those two fails to have a representation of the form $c_1\phi_1 + c_2\phi_2$. The orthonormal reference system here is not complete in the sense that there are vectors in three-dimensional space which are perpendicular to both ϕ_1 and ϕ_2.

Likewise in the correspondence (4), if $f(x)$ is orthogonal to every function $\phi_n(x)$ in the orthonormal set, then every term in the series is zero; and so the series does not represent f, unless $\|f\| = 0$ (see Problem 8, Sec. 25).

An orthonormal set $\{\phi_n(x)\}$ is *complete* in the function space being considered if there is no function in that space, with positive norm, which is orthogonal to each of the functions $\phi_n(x)$. We have just noted that the set must necessarily be complete if the correspondence (4) is to be an equality for each function f of the space.

27. Approximation in the Mean

In three-dimensional space, a linear combination

$$K = \gamma_1 \phi_1 + \gamma_2 \phi_2$$

of two of the vectors of an orthonormal set $\{\phi_1, \phi_2, \phi_3\}$ is a vector in the plane

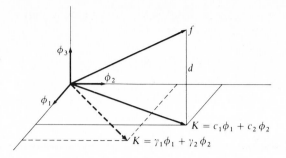

Figure 8

of ϕ_1 and ϕ_2. To make K the best approximation to a given vector f in three dimensions, in the sense that the distance $d = \| f - K \|$ between the tips of f and K is to be as small as possible when f and K are thought of as radius vectors, we can see geometrically that K must be the projection of f onto the plane of ϕ_1 and ϕ_2 (Fig. 8). Consequently, γ_1 has the value $(f,\phi_1) = c_1$, the projection of f onto ϕ_1, and $\gamma_2 = (f,\phi_2) = c_2$. Similarly, we see that the coefficients $c_n = (f,\phi_n)$ are those for which a linear combination of any one, two, or three of the vectors ϕ_n best approximates f.

There is a corresponding characterization of the Fourier constants c_n for a function $f(x)$.

Consider functions in the space of piecewise continuous functions on an interval (a,b). Let $\phi_1(x),\, \phi_2(x),\, \ldots,\, \phi_m(x)$ denote m functions of an orthonormal set $\{\phi_n(x)\}$ $(n = 1,\, 2,\, \ldots)$ on that interval, and let $K_m(x)$ be a linear combination of them:

(1) $$K_m(x) = \gamma_1 \phi_1(x) + \gamma_2 \phi_2(x) + \cdots + \gamma_m \phi_m(x).$$

We shall determine the constants γ_n so that K_m is the *best approximation in the mean* to a given function f in the sense that the value of the integral

(2) $$E = \int_a^b [f(x) - K_m(x)]^2 \, dx,$$

a measure of the error, is to be as small as possible. This is approximation of f by *least squares*. We note that E is the square of the distance (Sec. 25) $\| f - K_m \|$ between the functions f and K_m.

Let c_n be the Fourier constants $c_n = (f,\phi_n)$ for f; then

$$E = \int_a^b [f(x) - \gamma_1 \phi_1(x) - \gamma_2 \phi_2(x) - \cdots - \gamma_m \phi_m(x)]^2 \, dx$$

$$= \int_a^b [f(x)]^2 \, dx$$

$$+ \gamma_1^2 + \gamma_2^2 + \cdots + \gamma_m^2 - 2\gamma_1 c_1 - 2\gamma_2 c_2 - \cdots - 2\gamma_m c_m.$$

We add and subtract $c_1^2, c_2^2, \ldots, c_m^2$ to complete the squares in this last line; thus

$$(3) \qquad E = \int_a^b [f(x)]^2 \, dx - c_1^2 - c_2^2 - \cdots - c_m^2$$

$$+ (\gamma_1 - c_1)^2 + (\gamma_2 - c_2)^2 + \cdots + (\gamma_m - c_m)^2.$$

It is clear from equation (2) that $E \geq 0$, and so it follows from equation (3) that E has its least value when $\gamma_1 = c_1, \gamma_2 = c_2, \ldots, \gamma_m = c_m$. The result can be stated as follows.

Theorem 1. *If $c_1, c_2, \ldots, c_m$ are the Fourier constants of a function $f(x)$ with respect to the functions $\phi_1(x), \phi_2(x), \ldots, \phi_m(x)$ of an orthonormal set, then, of all possible linear combinations of those m functions, the combination*

$$c_1 \phi_1(x) + c_2 \phi_2(x) + \cdots + c_m \phi_m(x)$$

is the best approximation in the mean to $f(x)$ on the fundamental interval.

Write $\gamma_n = c_n$ in equation (3). Then, since $E \geq 0$,

$$(4) \qquad c_1^2 + c_2^2 + \cdots + c_m^2 \leq \int_a^b [f(x)]^2 \, dx = \|f\|^2;$$

this is known as *Bessel's inequality*. The term $\|f\|^2$ on the right is independent of m; and, as the number m of elements from the set $\{\phi_n(x)\}$ ($n = 1, 2, \ldots$) increases, the sums on the left form a sequence which is bounded and nondecreasing. That sequence therefore converges to a limit not greater than $\|f\|^2$. That is, the infinite series of squares of the Fourier constants for each function f in our function space converges, and

$$(5) \qquad \sum_{n=1}^{\infty} c_n^2 \leq \int_a^b [f(x)]^2 \, dx.$$

The convergence of this series implies that the general term tends to zero as n becomes infinite. Hence *the Fourier constants always approach zero as $n \to \infty$*:

$$(6) \qquad \lim_{n \to \infty} c_n = \lim_{n \to \infty} (f, \phi_n) = 0.$$

Note that since the set

$$\left\{ \sqrt{\frac{2}{\pi}} \sin nx \right\} \qquad (n = 1, 2, \ldots)$$

is orthonormal on $(0, \pi)$ (Sec. 25), it follows that

$$(7) \qquad \lim_{n \to \infty} \int_0^\pi f(x) \sin nx \, dx = 0.$$

The set

$$\left\{ \frac{1}{\sqrt{\pi}}, \sqrt{\frac{2}{\pi}} \cos nx \right\} \qquad (n = 1, 2, \ldots)$$

is also orthonormal on $(0,\pi)$, according to Problem 1, Sec. 25; and consequently,

(8) $$\lim_{n \to \infty} \int_0^\pi f(x) \cos nx \, dx = 0.$$

28. Closed and Complete Sets

A sequence S_m $(m = 1, 2, \ldots)$ of functions defined on a fundamental interval (a,b) is said to *converge in the mean*, or in the norm, to a function f over that interval if $\| f - S_m \| \to 0$ as $m \to \infty$; that is, it converges in the mean to f if

(1) $$\lim_{m \to \infty} \int_a^b [f(x) - S_m(x)]^2 \, dx = 0.$$

Condition (1) is also written

$$\text{l.i.m.}_{m \to \infty} S_m(x) = f(x),$$

where the abbreviation l.i.m. stands for *limit in the mean*.

Suppose now that the functions S_m are the partial sums of a generalized Fourier series corresponding to f on the fundamental interval (a,b):

(2) $$S_m(x) = \sum_{n=1}^{m} c_n \phi_n(x).$$

This is the linear combination $K_m(x)$ in Sec. 27 when $\gamma_n = c_n$.

If condition (1) is satisfied by each function f in our function space, we say that the orthonormal set $\{\phi_n(x)\}$ is *closed* in the sense of mean convergence. Thus each function f can be approximated arbitrarily closely in the mean by some linear combination of functions $\phi_n(x)$ of a closed set, namely the linear combination (2) when m is large enough.†

By expanding the integrand in equation (1) and keeping the definition of c_n in mind, we can write that equation in the form

(3) $$\lim_{m \to \infty} \left\{ \int_a^b [f(x)]^2 \, dx - 2 \sum_{n=1}^{m} c_n^2 + \sum_{n=1}^{m} c_n^2 \right\} = 0.$$

---

† In the mathematical literature, the terms closed and complete are sometimes applied to sets which we have called complete (Sec. 26) and closed, respectively.

Hence for a closed set $\{\phi_n(x)\}$ it is true that

$$(4) \qquad \sum_{n=1}^{\infty} c_n^2 = \int_a^b [f(x)]^2 \, dx.$$

This is known as *Parseval's equation*. When written in the form

$$(5) \qquad \sum_{n=1}^{\infty} (f,\phi_n)^2 = \|f\|^2,$$

it identifies the sum of the squares of the components of f, with respect to the generalized reference set $\{\phi_n(x)\}$, as the square of the norm of f.

Conversely, if each function f of the space satisfies Parseval's equation, the set $\{\phi_n(x)\}$ is closed in the sense of mean convergence. This is true because equations (3) and (4) are merely alternative forms of equation (1).

Suppose that a function $\theta(x)$ in the space, where $\|\theta\| \neq 0$, is orthogonal to each function $\phi_n(x)$ of a closed orthonormal set. Then equation (5), with f replaced by θ, gives the contradiction $\|\theta\| = 0$. Thus the set is complete (Sec. 26), and we have the following theorem.

Theorem 2. *If an orthonormal set $\{\phi_n(x)\}$ is closed, it is complete.*

This bare introduction to the theory of orthogonal functions, based on convergence in the mean, will not be continued here. *Convergence in the mean does not ensure pointwise convergence;* that is, the statement (1) is not the same as the statement

$$(6) \qquad \lim_{m \to \infty} S_m(x) = f(x)$$

for each point x of the interval (a,b), even if some points may be excepted.†
We shall be concerned with pointwise convergence.

An orthonormal set is *closed in the sense of pointwise convergence*, let us say, if its generalized Fourier series for each function f of our function space converges pointwise to $f(x)$, except possibly at a finite number of points, on the fundamental interval. At the end of Sec. 26 we indicated why a set that is closed in that sense must be complete. Thus Theorem 2 is true for such closed sets in the space of piecewise continuous functions defined on the fundamental interval.

The direct application of Theorem 2 is limited, however, to denying that a set is closed when there is a function with positive norm which is orthogonal to each function of the set. It is our representation theorems which will show that certain sets *are* closed, and hence complete, in specified function spaces.

---

† A simple example of a sequence of functions which converges in the mean to zero but which *diverges at each point* of the interval is given by Franklin (1964, p. 408), listed in the Bibliography.

PROBLEMS

1. According to Problem 1, Sec. 25, the set $\{\sqrt{2/c}\, \cos{(n\pi x/c)}\}$ $(n = 1, 2, \ldots)$ is orthonormal on the interval $0 \leq x \leq c$. Show that without the inclusion of some constant function, corresponding to the case $n = 0$, the set is not complete even in the space of continuous functions with continuous derivatives on that interval. We shall see later (Sec. 44) why the larger set *is* complete in the space.

2. From Problem 2(*a*), Sec. 25, we know that the set $\{\sin{n\pi x}\}$ $(n = 1, 2, \ldots)$ is orthonormal on the interval $-1 \leq x \leq 1$. Show that it is not complete even in the space of continuous functions on that interval.

3. Suppose that a function f has a representation

$$f(x) = \sum_{n=1}^{\infty} A_n \psi_n(x)$$

on a fundamental interval (a,b) on which the set $\{\psi_n(x)\}$ is orthogonal but not normalized $(0 < \|\psi_n\| \neq 1)$. Use inner products to show formally that

$$A_n = \frac{(f,\psi_n)}{\|\psi_n\|^2}.$$

Also, point out how the above series with these coefficients is the generalized Fourier series (4), Sec. 26, where $\phi_n(x) = \psi_n(x)/\|\psi_n\|$.

4. Consider continuous functions on a fundamental interval $a \leq x \leq b$. The norm of such a function vanishes if and only if the function vanishes at every point of the interval. Show that if two functions f and g have the same set of Fourier constants with respect to a *complete* orthonormal set $\{\phi_n(x)\}$ $[c_n = (f,\phi_n) = (g,\phi_n), n = 1, 2, \ldots]$, then the functions must be identical; that is, $f(x) = g(x)$ for all x in the interval $a \leq x \leq b$. This is a uniqueness property for functions with Fourier constants c_n.
Suggestion: Write $h(x) = f(x) - g(x)$ and show that $h(x) \equiv 0$.

5. Consider continuous functions on a fundamental interval $a \leq x \leq b$ and let $\{\phi_n(x)\}$ be a *complete* orthogonal set. Prove that if the corresponding generalized Fourier series for a function f converges uniformly over the interval, its sum $S(x)$ is identical to $f(x)$. (See Problem 4.) Also, point out why the conclusion holds if the condition of completeness is replaced by the condition that the set is closed in the sense of mean convergence.

6. Show that the sequence of functions S_m $(m = 1, 2, \ldots)$ where $S_m(x) = 0$ when $0 \leq x \leq 1/m$, $S_m(x) = \sqrt{m}$ when $1/m < x < 2/m$, and $S_m(x) = 0$ when $2/m \leq x \leq 1$ converges to zero at each point x of the interval $0 \leq x \leq 1$ but that the sequence does not converge in the mean to the function $f(x) = 0$ over $(0,1)$.

7. Consider the sequence of functions S_m $(m = 1, 2, \ldots)$ defined on the interval $0 \leq x \leq 1$ as follows:

$$S_m(x) = \begin{cases} 0 & \text{when } x = 1, \tfrac{1}{2}, \ldots, \tfrac{1}{m}, \\ 1 & \text{when } x \neq 1, \tfrac{1}{2}, \ldots, \tfrac{1}{m}. \end{cases}$$

Show that it converges in the mean to the function $f(x) = 1$ over $(0,1)$ but that, for each positive integer N, $S_m(1/N) \to 0$ as $m \to \infty$.
Suggestion: Observe that $S_m(1/N) = 0$ when $m \geq N$.

29. Complex-valued Functions

We shall have a few occasions to use complex-valued functions of a real variable, functions of the type

$$(1) \qquad w(t) = u(t) + iv(t)$$

where u and v are real-valued functions of a real variable t.†
    The derivative of w is defined in a natural way as

$$(2) \qquad w'(t) = u'(t) + iv'(t),$$

provided u and v are differentiable. Similarly,

$$(3) \qquad \int_a^b w(t)\, dt = \int_a^b u(t)\, dt + i \int_a^b v(t)\, dt.$$

    The operation $\bar{w} = u - iv$ of taking the complex conjugate of w commutes with the operations of differentiation and integration; thus, for differentiation, $[\overline{w(t)}]' = \overline{w'(t)}$. From definition (2) one can show that the elementary rules of differentiation for real-valued functions, such as the formula for the derivative of a product, apply to these complex-valued functions. Also, from definitions (2) and (3) it follows that if $u(t)$ and $v(t)$ are continuous on the interval $a \le t \le b$ and if $W(t) = U(t) + iV(t)$ is a function such that $W'(t) = w(t)$, then

$$\int_a^b w(t)\, dt = W(b) - W(a).$$

    The exponential function e^z, or exp z, is defined by

$$(4) \qquad e^z = e^x(\cos y + i \sin y),$$

where $z = x + iy$, x and y are real, and the angle y means y radians. From this definition it can be seen that e^z satisfies the usual laws of exponents. Also, if x and y are functions of a real variable t, then z and e^z are functions of type (1); and, with the aid of equation (2), one can show that

$$\frac{d}{dt} e^z = e^z \frac{dz}{dt}.$$

Thus, for example, the function $w = \exp(ict)$ satisfies the differential equation $w''(t) + c^2 w(t) = 0$ when c is a complex number.
    From the definitions

$$(5) \qquad \sin z = \frac{1}{2i}(e^{iz} - e^{-iz}), \qquad \cos z = \frac{1}{2}(e^{iz} + e^{-iz})$$

---

† See Churchill, Brown, and Verhey (1974), listed in the Bibliography, for a more complete treatment of points raised in this section.

and definition (4), we find that

(6) $|\sin z|^2 = \sin^2 x + \sinh^2 y,$ $|\cos z|^2 = \cos^2 x + \sinh^2 y;$

and, when $z = z(t)$,

$$\frac{d}{dt}\sin z = \cos z \frac{dz}{dt}, \qquad \frac{d}{dt}\cos z = -\sin z \frac{dz}{dt}.$$

The hyperbolic functions

(7) $$\sinh z = \frac{1}{2}(e^z - e^{-z}), \qquad \cosh z = \frac{1}{2}(e^z + e^{-z})$$

are clearly related to the trigonometric functions as follows:

$$\sin iz = i \sinh z, \qquad \cos iz = \cosh z,$$
$$\sinh iz = i \sin z, \qquad \cosh iz = \cos z.$$

The identity

(8) $$\sum_{n=1}^{m} z^n = \frac{z(1 - z^m)}{1 - z} \qquad\qquad (z \neq 1)$$

gives the sum of a finite number of terms in a geometric progression of complex numbers, and its validity is clear upon observing that

$$(1 - z)\sum_{n=1}^{m} z^n = \sum_{n=1}^{m}(z^n - z^{n+1}) = z - z^{m+1}.$$

It enables us to write, for example,

$$2\sum_{n=1}^{m}\cos n\theta = \sum_{n=1}^{m}(e^{i\theta})^n + \sum_{n=1}^{m}(e^{-i\theta})^n$$

$$= \frac{e^{i\theta}(1 - e^{im\theta})}{1 - e^{i\theta}}\frac{e^{-i\theta/2}}{e^{-i\theta/2}} + \frac{e^{-i\theta}(1 - e^{-im\theta})}{1 - e^{-i\theta}}\frac{e^{i\theta/2}}{e^{i\theta/2}}$$

$$= -1 + \frac{\exp\left[i(m + \frac{1}{2})\theta\right] - \exp\left[-i(m + \frac{1}{2})\theta\right]}{\exp\left(i\theta/2\right) - \exp\left(-i\theta/2\right)},$$

or

(9) $$2\sum_{n=1}^{m}\cos n\theta = -1 + \frac{\sin\left[(m + \frac{1}{2})\theta\right]}{\sin\left(\theta/2\right)},$$

where θ is any angle such that $\theta \neq 0, \pm 2\pi, \pm 4\pi, \ldots$. Identity (9) is *Lagrange's trigonometric identity,* which will be used later on (Sec. 40) in the theory of Fourier Series.

30. Other Types of Orthogonality

Extensions of the concept of orthogonal sets of functions should be noted.

(*a*) A set $\{\psi_n(x)\}$ is orthogonal on an interval (a,b) with respect to a *weight function* $p(x)$, which is positive when $a < x < b$, if

$$(1) \qquad \int_a^b p(x)\psi_m(x)\psi_n(x)\,dx = 0 \qquad\qquad \text{when } m \neq n.$$

The integral here represents the inner product (ψ_m,ψ_n) with respect to the weight function, corresponding to the inner product of vectors when weight numbers are used (Sec. 23). The set is normalized by dividing $\psi_n(x)$ by $\|\psi_n\|$, where

$$\|\psi_n\|^2 = (\psi_n,\psi_n) = \int_a^b p(x)[\psi_n(x)]^2\,dx$$

and where it is assumed that $\|\psi_n\| \neq 0$.

This type of orthogonality is reduced to the ordinary type by using the products $\sqrt{p(x)}\,\psi_n(x)$ as the functions of the set.

Weight functions other than unity will arise in orthogonal sets of Bessel functions and in other sets generated by Sturm-Liouville problems. A further example is the set of *Tchebysheff polynomials*

$$(2) \qquad\qquad T_n(x) = \cos\,(n\,\cos^{-1}\,x) \qquad\qquad (n = 0,\,1,\,2,\,\ldots),$$

which is orthogonal on the interval $(-1,1)$ with respect to the weight function

$$(3) \qquad\qquad\qquad p(x) = \frac{1}{\sqrt{1 - x^2}}.$$

(*b*) A set $\{w_n(t)\}$ of complex-valued functions (Sec. 29) of a real variable t is orthogonal in the *hermitian* sense on an interval (a,b) if

$$(4) \qquad\qquad \int_a^b w_m(t)\overline{w}_n(t)\,dt = 0 \qquad\qquad (m \neq n).$$

The integral here is the hermitian inner product (w_m,w_n). The square of the norm of w_n is real and nonnegative since

$$(5) \qquad\qquad \|w_n\|^2 = \int_a^b w_n\overline{w}_n\,dt = \int_a^b (u_n^2 + v_n^2)\,dt$$

if $w_n = u_n + iv_n$ where u_n and v_n are real-valued functions of t.

Certain complex-valued exponential functions furnish the most prominent examples of such sets. The functions

$$(6) \qquad\qquad\qquad e^{int} = \cos\,nt + i\,\sin\,nt \qquad\qquad (n = 0,\,\pm 1,\,\pm 2,\,\ldots),$$

for example, constitute a set with hermitian orthogonality on the interval $(-\pi,\pi)$. The proof is left to the problems.

(c) For sets of functions of two independent variables the fundamental interval is replaced by a region in the xy plane, and the integrations are made over that region. Similar extensions apply to functions of three or more variables.

PROBLEMS

1. Verify that solutions of the differential equation

$$w''(t) + c^2 w(t) = 0,$$

in which c is any nonzero complex constant, may be written in each of the forms

$$w = C_1 \exp(ict) + C_2 \exp(-ict)$$

$$= C_3 \sin ct + C_4 \cos ct$$

$$= \tfrac{1}{2}C_5\{\exp[i(ct + C_6)] + \exp[-i(ct + C_6)]\}$$

$$= C_5 \cos(ct + C_6),$$

where $C_1, C_2, \ldots, C_6$ are arbitrary complex constants.

2. Use equations (6), Sec. 29, to show (a) that the zeros of the functions $\sin z$ and $\cos z$ are all real; (b) that $|\sin z|$ and $|\cos z|$ are unbounded as $|y| \to \infty$.

3. *Euler's formula*

$$e^{i\theta} = \cos\theta + i\sin\theta,$$

where θ is any angle, is a special case of equation (4), Sec. 29. Use it to write

$$\int_0^\pi e^x \cos x \, dx + i \int_0^\pi e^x \sin x \, dx = \int_0^\pi e^{(1+i)x} \, dx,$$

and then evaluate the two integrals on the left-hand side here by evaluating the single integral on the right.

$$Ans. \quad -\frac{1 + e^\pi}{2}, \ \frac{1 + e^\pi}{2}.$$

4. Derive the identity

$$2 \sum_{n=1}^m \cos(2n - 1)\theta = \frac{\sin 2m\theta}{\sin\theta} \qquad (\theta \neq 0, \pm\pi, \pm2\pi, \ldots)$$

by noting that its left-hand side can be written

$$e^{-i\theta} \sum_{n=1}^m (e^{i2\theta})^n + e^{i\theta} \sum_{n=1}^m (e^{-i2\theta})^n$$

and then using equation (8), Sec. 29.

5. Show that

$$\sum_{n=1}^\infty a^n \cos n\theta = \frac{a \cos\theta - a^2}{1 - 2a \cos\theta + a^2} \qquad (-1 < a < 1)$$

by using equation (8), Sec. 29, to find an expression for the sum of the first m terms of the series on the left here and then letting m tend to infinity. Do this by first writing

$$2 \sum_{n=1}^{m} a^n \cos n\theta = \sum_{n=1}^{m} (ae^{i\theta})^n + \sum_{n=1}^{m} (ae^{-i\theta})^n.$$

6. Establish the hermitian orthogonality of the exponential functions (6), Sec. 30, on the interval $(-\pi,\pi)$. Also show that the norm of each function is $\sqrt{2\pi}$.

7. Show that the set

$$\left| \frac{1}{\sqrt{b-a}} \exp\left(i\frac{2n\pi t}{b-a} \right) \right| \qquad\qquad (n = 0, \pm 1, \pm 2, \ldots)$$

is orthonormal in the hermitian sense on the interval (a,b).

8. Substitute $\theta = \cos^{-1} x$ to establish the orthogonality of the Tchebysheff polynomials (2), Sec. 30, with weight function (3) in that section.

9. Write $x = \cos \theta$ in expression (2), Sec. 30, for the functions $T_n(x)$. Then, in *de Moivre's formula*

$$(\cos \theta + i \sin \theta)^n = \cos n\theta + i \sin n\theta$$

use the binomial expansion and equate real parts to show that $\cos n\theta$ is a polynomial of degree n in $\cos \theta$ and hence that $T_n(x)$ is actually a polynomial of degree n in x.

10. Let R be the square region $0 < x < \pi$, $0 < y < \pi$ in the xy plane. Show that if $\phi_{mn}(x,y) = \sin mx \sin ny$ $(m,n = 1, 2, \ldots)$, the set $\{\phi_{mn}(x,y)\}$ is orthogonal over R in the sense that

$$\iint_R \phi_{mn}(x,y)\phi_{jk}(x,y)\, dA = 0 \qquad\qquad \text{if } m \neq j \text{ or } n \neq k.$$

31. Sturm-Liouville Problems

In Sec. 15 the method of separation of variables, followed by superposition of functions of the type $X(x)T(t)$ that satisfy the homogeneous equations, was used to write a formal solution of a boundary value problem in the displacements $y(x,t)$ of a stretched string. The process required the function X to be a solution of the homogeneous problem

(1) $$X''(x) + \lambda X(x) = 0, \qquad X(0) = 0, \qquad X(c) = 0.$$

For a discrete set of values of the parameter, $\lambda = n^2\pi^2/c^2$ $(n \doteq 1, 2, \ldots)$, we found that problem (1) has nontrivial solutions $X = \sin (n\pi x/c)$ and (Sec. 16) that those functions are orthogonal on the interval $(0,c)$.

When applied to more general boundary value problems in partial differential equations, the process frequently leads to a homogeneous ordinary differential equation of the type

(2) $$X''(x) + R(x)X'(x) + [Q(x) + \lambda P(x)]X(x) = 0,$$

involving a parameter λ in the manner indicated, and to a pair of homogeneous boundary conditions of the type

$$(3) \qquad a_1 X(a) + a_2 X'(a) = 0, \qquad b_1 X(b) + b_2 X'(b) = 0.$$

The functions P, Q, and R and the constants a, b, a_1, a_2, b_1, and b_2 are prescribed by the problem in partial differential equations; λ and $X(x)$ are to be determined.

When its terms are multiplied by a function

$$r(x) = \exp\left[\int R(x)\, dx\right],$$

which is an integrating factor for $X'' + RX'$, equation (2) takes the form

$$(4) \qquad \frac{d}{dx}\left[r(x)\frac{dX}{dx}\right] + [q(x) + \lambda p(x)]X = 0.$$

Other types of homogeneous two-point boundary value problems involving a parameter may arise instead of the problem consisting of equation (4) with boundary conditions (3). But that problem, called a *Sturm-Liouville problem*, is of fundamental importance.†

Under rather general conditions on the functions p, q, and r it can be shown that there is always a countable infinity of values λ_1, λ_2, ... of the parameter λ for each of which the above Sturm-Liouville problem has a solution that is not identically zero. The numbers λ_n are the *eigenvalues*, or characteristic numbers, of the problem; and the corresponding solutions $X_n(x)$ are the *eigenfunctions*, or characteristic functions. Note that $CX_n(x)$ is also an eigenfunction, where C is any constant other than zero. In the special case (1), $\lambda_n = n^2\pi^2/c^2$ and $X_n(x) = \sin(n\pi x/c)$.

The orthogonality of the eigenfunctions with weight function p, on the interval (a,b), is established in the following section. With respect to the set of normalized eigenfunctions

$$\phi_n(x) = \frac{X_n(x)}{\|X_n\|}, \quad \text{where } \|X_n\|^2 = \int_a^b pX_n^2\, dx,$$

the generalized Fourier series for a function f is

$$(5) \qquad \sum_{n=1}^{\infty} c_n \phi_n(x) \qquad \text{where} \qquad c_n = \int_a^b pf\phi_n\, dx.$$

For prominent special cases of the Sturm-Liouville problem we shall establish the convergence of this series to $f(x)$ on the interval (a,b).

---

† Papers by J. C. F. Sturm and J. Liouville giving the first extensive development of the theory of this problem appeared in vols. 1–3 of *Journal de mathématique*, 1836–1838.

The convergence of the series in the general case is not treated in this book.† For the general case, proofs that series (5) converges to $f(x)$ usually employ either the theory of residues of functions of a complex variable or a comparison of the series with an ordinary Fourier series that represents f. Proofs are complicated by the fact that explicit solutions of the Sturm-Liouville equation with arbitrary coefficients cannot be written. Properties of solutions are found, by interesting and useful devices, from the differential equation itself.

The *adjoint* of a second-order linear differential operator M, where

$$(6) \qquad M[X(x)] = A(x)X''(x) + B(x)X'(x) + C(x)X(x),$$

is the operator M^* such that

$$(7) \qquad M^*[X(x)] = [A(x)X(x)]'' - [B(x)X(x)]' + C(x)X(x).$$

The operator L defined by the equation

$$(8) \qquad L[X(x)] = [r(x)X'(x)]' + q(x)X(x)$$

is *self-adjoint;* that is, L and L^* are the same since

$$rX'' + r'X' + qX = (rX)'' - (r'X)' + qX.$$

Form (4) of the Sturm-Liouville equation turns out to be especially useful because it is in self-adjoint form:

$$(9) \qquad L[X(x)] + \lambda p(x)X(x) = 0.$$

Some properties of L will be noted explicitly in the problems.

32. Orthogonality of the Eigenfunctions

A few results in the general theory can be established here. They will be useful in the chapters to follow. In special cases the eigenvalues and eigenfunctions will be found, and so their existence will not be in doubt.

We assume, *unless otherwise stated*, that the coefficients in the Sturm-Liouville problem

$$(1) \qquad (rX')' + (q + \lambda p)X = 0,$$

$$(2) \qquad a_1 X(a) + a_2 X'(a) = 0, \qquad b_1 X(b) + b_2 X'(b) = 0$$

satisfy these conditions: *p, q, r, and r' are real-valued continuous functions of*

---

† The general case is treated in Churchill (1972, Chap. 9). For other treatments of Sturm-Liouville theory, some quite extensive, see the book by Ince (1956), the one by Coddington and Levinson (1955), and the two volumes by Titchmarsh (1962, 1958). These are all listed in the Bibliography.

the real variable x $(a \leq x \leq b)$ and are independent of λ; also, $p(x) > 0$ and $r(x) > 0$ when $a < x < b$, and the constants a_1, a_2, b_1, and b_2 are real and independent of λ. It is usually understood that a_1 and a_2 are not both zero, and the same is true of the constants b_1 and b_2. Eigenfunctions X_n are to satisfy the regularity conditions usually required of solutions of differential equations of the second order (Sec. 18); namely, X_n and X'_n are to be continuous on the interval $a \leq x \leq b$.

Theorem 3. Let λ_m and λ_n be any two distinct eigenvalues of the Sturm-Liouville problem (1) and (2), with corresponding eigenfunctions X_m and X_n. Then X_m and X_n are orthogonal with weight function p on the interval (a,b). The orthogonality also holds in the following cases: .

(a) when $r(a) = 0$ and the first of boundary conditions (2) is dropped from the problem;

(b) when $r(b) = 0$ and the second of conditions (2) is dropped;

(c) when $r(a) = r(b)$ and conditions (2) are replaced by the conditions

$$(3) \qquad\qquad X(a) = X(b), \qquad X'(a) = X'(b).$$

In cases (a) and (b) the Sturm-Liouville problem is said to be *singular*. This terminology is used when either $p(x)$ or $r(x)$ vanishes at an end point or $q(x)$ is discontinuous there. It also applies when the interval is replaced by an unbounded one. Note that the dropping of the first of conditions (2) in case (a) is the same as letting both a_1 and a_2 be zero; a similar remark regarding b_1 and b_2 applies to the dropping of the second of those conditions in case (b). Conditions (3) in case (c) are called *periodic boundary conditions*. They commonly arise when x represents the angle ϕ in cylindrical coordinates, on the interval $(-\pi,\pi)$.

To prove the theorem, we first note that

$$(rX'_m)' + qX_m = -\lambda_m p X_m, \qquad (rX'_n)' + qX_n = -\lambda_n p X_n$$

since each eigenfunction satisfies equation (1) when λ is the corresponding eigenvalue. We multiply each side of these two equations by X_n and X_m, respectively, and subtract to get the relation

$$(4) \quad (\lambda_m - \lambda_n)pX_mX_n = X_m(rX'_n)' - X_n(rX'_m)' = \frac{d}{dx}[r(X_mX'_n - X'_mX_n)].$$

While the final reduction here to an exact derivative is elementary, it is made possible by the special nature of the Sturm-Liouville operator L defined in equation (8), Sec. 31. Details regarding this point are left to the problems.

Our continuity conditions now permit us to write

$$(5) \qquad\qquad (\lambda_m - \lambda_n) \int_a^b pX_mX_n \, dx = [r(x)\, \Delta(x)]_a^b$$

where $\Delta(x)$ is the determinant

$$(6) \qquad \Delta(x) = \begin{vmatrix} X_m(x) & X'_m(x) \\ X_n(x) & X'_n(x) \end{vmatrix};$$

that is,

$$(7) \qquad (\lambda_m - \lambda_n) \int_a^b pX_m X_n \, dx = r(b)\, \Delta(b) - r(a)\, \Delta(a).$$

The first of boundary conditions (2) requires that

$$a_1 X_m(a) + a_2 X'_m(a) = 0, \qquad a_1 X_n(a) + a_2 X'_n(a) = 0;$$

and for these simultaneous equations in a_1 and a_2 to be satisfied by numbers a_1 and a_2, not both zero, it is necessary that the determinant $\Delta(a)$ be zero. Similarly, from the second boundary condition, where b_1 and b_2 are not both zero, we see that $\Delta(b) = 0$. Then, according to equation (7),

$$(8) \qquad (\lambda_m - \lambda_n) \int_a^b pX_m X_n \, dx = 0$$

and, since $\lambda_m \neq \lambda_n$, the desired orthogonality property follows:

$$(9) \qquad \int_a^b p(x)X_m(x)X_n(x) \, dx = 0 \qquad\qquad (\lambda_m \neq \lambda_n).$$

If $r(a) = 0$, property (9) follows from equation (7) even when $\Delta(a) \neq 0$, or when $a_1 = a_2 = 0$, in which case the first of boundary conditions (2) disappears. Similarly, if $r(b) = 0$, the second of those conditions is not needed.

When $r(a) = r(b)$ and the periodic conditions (3) are used in place of conditions (2), then

$$r(b)\, \Delta(b) = r(a)\, \Delta(a)$$

and, again, equation (9) follows. This completes the proof of Theorem 3.

Suppose that X is an eigenfunction corresponding to an eigenvalue $\lambda = \alpha + i\beta$, where α and β are real numbers. Then $X(x)$, which may be complex-valued, satisfies equations (1) and (2). Taking complex conjugates of all terms in those equations and recalling which coefficients are real-valued, we see that (Sec. 29)

$$(r\bar{X}')' + (q + \bar{\lambda}p)\bar{X} = 0,$$

$$a_1 \bar{X}(a) + a_2 \bar{X}'(a) = 0, \qquad b_1 \bar{X}(b) + b_2 \bar{X}'(b) = 0.$$

Thus $\bar{X}$ is an eigenfunction corresponding to the eigenvalue $\bar{\lambda}$; and, according to equation (8),

$$(\lambda - \bar{\lambda}) \int_a^b p(x)X(x)\bar{X}(x) \, dx = 0.$$

But $p(x) > 0$ when $a < x < b$, and $X\bar{X} = |X|^2$. So the integral here has a positive value. Since $\lambda - \bar{\lambda} = 2i\beta$, it follows that $\beta = 0$; that is, λ is real.

The argument also applies to cases (a), (b), and (c) in Theorem 3.

Theorem 4. *For the Sturm-Liouville problem* (1) *and* (2), *and its modifications* (a), (b), *and* (c) *cited in Theorem 3, each eigenvalue is real.*

33. Uniqueness of Eigenfunctions

Let X and Y denote two eigenfunctions of the Sturm-Liouville problem (1) and (2), Sec. 32, corresponding to the same real eigenvalue λ. Suppose that r does not vanish at one end point of the interval, say at $x = a$, where the value of r is positive.

The linear combination

$$(1) \qquad\qquad W(x) = Y'(a)X(x) - X'(a)Y(x)$$

satisfies the linear homogeneous differential equation

$$(2) \qquad\qquad (rW')' + (q + \lambda p)W = 0;$$

also $W'(a) = 0$. Since X and Y satisfy the equations

$$(3) \qquad a_1 X(a) + a_2 X'(a) = 0, \qquad a_1 Y(a) + a_2 Y'(a) = 0,$$

where a_1 and a_2 are not both zero, and since $W(a)$ is the determinant of that pair of equations in a_1 and a_2, then $W(a) = 0$. According to our uniqueness theorem for solutions of linear differential equations (Sec. 18), $W(x) \equiv 0$ is the only solution of equation (2) for which $W(a) = W'(a) = 0$. Thus

$$(4) \qquad\qquad Y'(a)X(x) - X'(a)Y(x) = 0.$$

Unless $Y'(a) = X'(a) = 0$, equation (4) states that the functions X and Y are linearly dependent; that is, one is a constant times the other. Recall that zero is not an eigenfunction. If $Y'(a) = X'(a) = 0$, then $a_1 = 0$ in equations (3). In that case we use the linear combination

$$(5) \qquad\qquad Z(x) = Y(a)X(x) - X(a)Y(x)$$

to prove in like manner that $Z(x) \equiv 0$ and hence that X and Y are linearly dependent.

Suppose that $X = u + iv$ is a complex-valued eigenfunction corresponding to a real eigenvalue λ. When we substitute $u + iv$ for X in the Sturm-Liouville problem and separate real and imaginary parts, we see that u and v are each eigenfunctions corresponding to λ. Hence $v = ku$ and $X = (1 + ik)u$, where k is a constant. That is, X is real except possibly for an imaginary constant factor.

We collect our results as follows.

Theorem 5. *Under the additional condition that either* $r(a) > 0$ *or* $r(b) > 0$, *the Sturm-Liouville problem* (1) *and* (2) *in Sec.* 32 *cannot have two linearly independent eigenfunctions that correspond to the same eigenvalue; also, each eigenfunction can be made real-valued by multiplying it by an appropriate nonzero constant.*

That Theorem 5 does not apply when the conditions (2), Sec. 32, are replaced by periodic boundary conditions is shown by this important special case:

$$(6) \qquad X'' + \lambda X = 0, \qquad X(-\pi) = X(\pi), \qquad X'(-\pi) = X'(\pi).$$

When $\lambda > 0$, or $\lambda = \alpha^2$ $(\alpha > 0)$, the solution of the differential equation here is

$$X(x) = C_1 \sin \alpha x + C_2 \cos \alpha x.$$

If it is to satisfy both boundary conditions, we find that

$$(7) \qquad C_1 \sin \alpha\pi = 0 \qquad \text{and} \qquad C_2 \sin \alpha\pi = 0.$$

Since C_1 and C_2 must not both vanish if X is to be an eigenfunction, it follows that $\lambda = n^2$ $(n = 1, 2, \ldots)$, while the constants C_1 and C_2 are otherwise arbitrary.

In particular, for any constant B_n, *the two functions*

$$(8) \qquad X_n(x) = \sin nx, \qquad Y_n(x) = B_n \sin nx + \cos nx$$

are linearly independent eigenfunctions corresponding to the same eigenvalue $\lambda = n^2$. Other linear combinations of $\sin nx$ and $\cos nx$ can be used to form such pairs.

When two functions are linearly independent, there is always a linear combination of the two which is orthogonal to one of the functions (Problem 4, Sec. 25). If $B_n = 0$ in the second of equations (8), Y_n is orthogonal to X_n on the interval $(-\pi, \pi)$. [See Problem 2(c), Sec. 25.]

A constant is the only eigenfunction of problem (6) corresponding to the eigenvalue $\lambda = 0$, and there are no negative eigenvalues. According to Theorem 3, eigenfunctions corresponding to distinct eigenvalues are orthogonal. Here $p(x) = 1$. Consequently, the set

$$(9) \qquad \{1, \cos x, \cos 2x, \ldots, \sin x, \sin 2x, \ldots\},$$

a set of eigenfunctions of problem (6), is orthogonal on the interval $(-\pi, \pi)$. Note that this includes the orthogonality of $\sin nx$ and $\cos nx$, each of which corresponds to the same eigenvalue $\lambda = n^2$.

34. Another Example

Graphical techniques can often be used to demonstrate the existence of an infinite number of eigenvalues. For an illustration, we consider here the problem

$$(1) \qquad\qquad X''(x) + \lambda X(x) = 0,$$

$$(2) \qquad\qquad X'(0) = 0, \qquad hX(1) + X'(1) = 0$$

where h is a positive constant.

The case $\lambda = 0$ leads to the trivial solution $X(x) \equiv 0$, and zero is therefore not an eigenvalue.

When $\lambda > 0$, or $\lambda = \alpha^2$ $(\alpha > 0)$, the general solution of the differential equation together with the first boundary condition is $X(x) = C_2 \cos \alpha x$. Imposing the second boundary condition, we find that in order for C_2 to be nonzero α must be a (positive) root of the equation

$$(3) \qquad\qquad \tan \alpha = \frac{h}{\alpha}.$$

Figure 9, where the graphs of $y = \tan \alpha$ and $y = h/\alpha$ are plotted, shows that equation (3) has an infinite number of consecutive positive roots $\alpha_1, \alpha_2, \ldots$; they are the positive values of α for which those graphs intersect.† The

† Roots of this and a related equation, for certain values of h, are tabulated, for example, on pp. 224–225 of the handbook edited by Abramowitz and Stegun (1965) which is listed in the Bibliography.

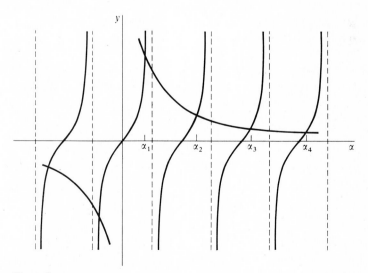

Figure 9

corresponding eigenvalues are then $\lambda_n = \alpha_n^2$ $(n = 1, 2, \ldots)$, with eigenfunctions $X_n(x) = \cos \alpha_n x$.

Finally, when $\lambda < 0$, or $\lambda = -\beta^2$ $(\beta > 0)$, it is straightforward to show that in order for a nontrivial solution to occur β must satisfy the equation

(4)
$$\tanh \beta = -\frac{h}{\beta}.$$

This equation, however, has no real roots since the graphs of the functions $y = \tanh \beta$ and $y = -h/\beta$ do not intersect.

In summary, then, the complete solution of the given Sturm-Liouville problem is

$$\lambda_n = \alpha_n^2, \quad X_n(x) = \cos \alpha_n x \qquad (n = 1, 2, \ldots)$$

where $\alpha_1, \alpha_2, \ldots$ are the consecutive positive roots of equation (3).

Keeping in mind the fact that $\alpha_n = (h \cos \alpha_n)/\sin \alpha_n$, we can readily show that

(5)
$$\int_0^1 X_n^2(x)\,dx = \frac{h + \sin^2 \alpha_n}{2h} \qquad (n = 1, 2, \ldots).$$

Thus the normalized eigenfunctions are

(6)
$$\phi_n(x) = \left(\frac{2h}{h + \sin^2 \alpha_n}\right)^{1/2} \cos \alpha_n x \qquad (n = 1, 2, \ldots).$$

PROBLEMS

Find all eigenvalues and eigenfunctions of the Sturm-Liouville problems in Problems 1 through 7 below. Also state what the interval and weight function p are in the orthogonality relation ensured by Theorem 3.

1. $X'' + \lambda X = 0$, $X(0) = 0$, $X'(\pi/2) = 0$.

Ans. $X_n = \sin (2n - 1)x \qquad (n = 1, 2, \ldots)$.

2. $X'' + \lambda X = 0$, $X'(0) = 0$, $X'(c) = 0$.

Ans. $X_0 = 1$, $X_n = \cos \dfrac{n\pi x}{c} \qquad (n = 1, 2, \ldots)$.

3. $X'' + \lambda X = 0$, $X'(0) = 0$, $X(c) = 0$.

Ans. $X_n = \cos \dfrac{(2n - 1)\pi x}{2c} \qquad (n = 1, 2, \ldots)$.

4. $X'' + \lambda X = 0$, $X'(-\pi) = 0$, $X'(\pi) = 0$.
Suggestion: Make the substitution $x = s - \pi$ and transform this problem into a special case of Problem 2.

Ans. $X_0 = 1$, $X_n = \cos \dfrac{n(\pi + x)}{2} \qquad (n = 1, 2, \ldots)$.

5. $X'' + \lambda X = 0$, $X(0) = 0$, $hX(1) + X'(1) = 0$ where h is a positive constant.

Ans. $X_n = \sin \alpha_n x$ $(n = 1,2,\ldots)$ where α_n are the consecutive positive roots of the equation $\tan \alpha = -\alpha/h$.

6. $X'' + \lambda X = 0$, $X(0) = 0$, $X(1) - X'(1) = 0$.

Ans. $X_0 = x$, $X_n = \sin \alpha_n x$ $(n = 1, 2, \ldots)$ where α_n are the consecutive positive roots of the equation $\tan \alpha = \alpha$.

7. $X'' + \lambda X = 0$, $X(0) = 0$, $hX(1) + X'(1) = 0$ where h is a constant and $h < -1$.

Ans. $X_0 = \sinh \alpha_0 x$ where α_0 is the positive root of the equation $\tanh \alpha = -\alpha/h$ and $X_n = \sin \alpha_n x$ $(n = 1, 2, \ldots)$ where α_n are the consecutive positive roots of the equation $\tan \alpha = -\alpha/h$.

8. Give the steps that lead to equations (7), Sec. 33. Also, show that $X(x) = 1$ is the eigenfunction of the Sturm-Liouville problem (6), Sec. 33, corresponding to the eigenvalue $\lambda = 0$ and that there are no negative eigenvalues.

9. In the Sturm-Liouville problem (6), Sec. 33,

(a) point out why it is true that if n is a positive integer and α (radians) is a constant angle such that $\sin \alpha \neq 0$, then the two linearly independent functions $\sin nx$ and $\sin (nx + \alpha)$ are eigenfunctions corresponding to the same eigenvalue $\lambda = n^2$;

(b) verify that

$$\sin (nx + \alpha) \cos \alpha - \cos (nx + \alpha) \sin \alpha = \sin nx$$

and hence that $\cos (nx + \alpha)$ is a linear combination of the two eigenfunctions in part (a);

(c) show that $\cos (nx + \alpha)$ is orthogonal to $\sin (nx + \alpha)$ and, with the aid of parts (a) and (b), show that the functions

$$1, \cos (x + \alpha), \cos (2x + \alpha), \ldots, \sin (x + \alpha), \sin (2x + \alpha), \ldots$$

constitute an orthogonal set of eigenfunctions on the interval $(-\pi, \pi)$.

10. For the eigenvalue problem

$$X'' + \lambda X = 0, \qquad X(-c) = X(c), \qquad X'(-c) = X'(c),$$

obtain this set of eigenfunctions, which is orthogonal on the interval $(-c, c)$ (see Sec. 33):

$$\left\{ 1, \cos \frac{m\pi x}{c}, \sin \frac{n\pi x}{c} \right\} \qquad (m, n = 1, 2, \ldots).$$

11. Use the function Z defined by equation (5), Sec. 33, to prove that the functions X and Y there are linearly dependent when $a_1 = 0$.

12. Show that the eigenvalues and normalized eigenfunctions of the Sturm-Liouville problem

$$X'' + \lambda X = 0, \qquad X'(0) = 0, \qquad hX(c) + X'(c) = 0 \qquad (h > 0),$$

which is a generalization of the one solved in Sec. 34, are

$$\lambda_n = \alpha_n^2, \qquad \phi_n(x) = \left(\frac{2h}{hc + \sin^2 \alpha_n c} \right)^{1/2} \cos \alpha_n x \qquad (n = 1, 2, \ldots)$$

where α_n are the consecutive positive roots of the equation $\tan \alpha c = h/\alpha$. Do this (a) by using the method of Sec. 34 and (b) by substituting $s = x/c$ and writing the new Sturm-Liouville problem

$$\frac{d^2 X}{ds^2} + \mu X = 0, \qquad X'(0) = 0, \qquad kX(1) + X'(1) = 0 \qquad (\mu = \lambda c^2, k = hc)$$

in the independent variable s and then using the results already obtained in Sec. 34.

13. If A, B, and C are constants, the differential equation

$$Ax^2 X'' + BxX' + CX = 0$$

is called a *Cauchy-Euler equation.*

(a) Show that with the substitution $x = e^s$ this differential equation can be put into the form

$$A\frac{d^2 X}{ds^2} + (B - A)\frac{dX}{ds} + CX = 0.$$

(b) Use the result in part (a) and the solution of the Sturm-Liouville problem (1), Sec. 31, to solve the related problem

$$(xX')' + \lambda\left(\frac{1}{x}\right)X = 0, \qquad X(1) = 0, \qquad X(b) = 0,$$

where b is a constant and $b > 1$. Normalize the eigenfunctions.

Ans. $\lambda_n = \left(\frac{n\pi}{\log b}\right)^2,$ $\phi_n(x) = \sqrt{\frac{2}{\log b}} \sin\left(n\pi \frac{\log x}{\log b}\right)$ $(n = 1, 2, \ldots).$

14. Solve the Sturm-Liouville problem

$$(x^3 X')' + \lambda x X = 0, \qquad X(1) = 0, \qquad X(e) = 0$$

by noting that the differential equation can be written as a Cauchy-Euler equation and applying the transformation $x = e^s$ [see Problem 13(a)]. State the weight function for the eigenfunctions.

Ans. $X_n = \frac{1}{x}\sin(n\pi \log x)$ $(n = 1, 2, \ldots);$ $p(x) = x.$

15. Solve the Sturm-Liouville problem

$$(xX')' + \lambda\left(\frac{1}{x}\right)X = 0, \qquad X'(1) = 0, \qquad X(b) = 0 \qquad (b > 1)$$

by noting that the differential equation can be written in Cauchy-Euler form (see Problem 13), using the transformation $x = e^s$, and then referring to the solution of Problem 3. Normalize the eigenfunctions.

Ans. $\lambda_n = \left[\frac{(2n - 1)\pi}{2 \log b}\right]^2,$ $\phi_n(x) = \sqrt{\frac{2}{\log b}} \cos\left[\frac{(2n - 1)\pi}{2} \frac{\log x}{\log b}\right]$ $(n = 1, 2, \ldots).$

16. Note that the differential equation in the Sturm-Liouville problem

$$(xX')' + \lambda\left(\frac{1}{x}\right)X = 0, \qquad X'(1) = 0, \qquad hX(b) + X'(b) = 0 \qquad (h > 0, b > 1)$$

can be written as a Cauchy-Euler equation (Problem 13). Then, with the aid of the transformation $x = e^s$ and using the solution of Problem 12, find the eigenvalues and normalized eigenfunctions.

Ans. $\lambda_n = \alpha_n^2,$ $\phi_n(x) = \left[\frac{2bh}{bh \log b + \sin^2(\alpha_n \log b)}\right]^{1/2} \cos(\alpha_n \log x)$ $(n = 1, 2, \ldots)$

where α_n are the consecutive positive roots of the equation $\tan(\alpha \log b) = bh/\alpha.$

17. (a) Prove that the self-adjoint Sturm-Liouville operator L defined by equation (8), Sec. 31, satisfies *Lagrange's identity*

$$X L[Y] - Y L[X] = \frac{d}{dx} [r(X Y' - X' Y)]$$

for each pair of functions X and Y, assuming that all derivatives involved exist.

(b) Show that this identity can also be written

$$X (r Y')' - Y(r X')' = \frac{d}{dx} [r(X Y' - X' Y)].$$

18. Let the operator L and the functions X and Y be those in Problem 17. With the aid of Lagrange's identity, obtained in that problem, show that if X and Y satisfy the Sturm-Liouville boundary conditions (2) or (3) in Sec. 32, then

$$(X, L[Y]) = (Y, L[X]),$$

where these inner products on the interval (a,b) are with weight function unity.

19. Show that if N is the operator d^4/dx^4, then

$$X N[Y] - Y N[X] = \frac{d}{dx} (X Y''' - Y X''' - X' Y'' + Y' X'').$$

Thus show that if X_1 and X_2 are eigenfunctions of the fourth-order eigenvalue problem

$$N[X] + \lambda X = 0, \quad X(0) = X''(0) = 0, \quad X(c) = X''(c) = 0,$$

corresponding to distinct eigenvalues λ_1 and λ_2, then X_1 is orthogonal to X_2 on the interval $(0,c)$.

FOUR

FOURIER SERIES

35. The Basic Series

The trigonometric series

(1)
$$\frac{1}{2}a_0 + \sum_{n=1}^{\infty}(a_n \cos nx + b_n \sin nx)$$

is a *Fourier series* if its coefficients are given by the formulas

(2)
$$a_n = \frac{1}{\pi}\int_{-\pi}^{\pi} f(x) \cos nx \, dx \qquad (n = 0, 1, 2, \ldots),$$

$$b_n = \frac{1}{\pi}\int_{-\pi}^{\pi} f(x) \sin nx \, dx \qquad (n = 1, 2, \ldots)$$

where f is some function defined on the interval $(-\pi,\pi)$.

In Sec. 33 we saw that the functions $1, \cos x, \cos 2x, \ldots, \sin x, \sin 2x, \ldots$ constitute a set of eigenfunctions, orthogonal on the interval $(-\pi,\pi)$, for the eigenvalue problem

(3)
$$X''(x) + \lambda X(x) = 0,$$

$$X(-\pi) = X(\pi), \qquad X'(-\pi) = X'(\pi).$$

The normalized orthogonal set is [Problem 2(c), Sec. 25]

(4)
$$\left\{ \frac{1}{\sqrt{2\pi}}, \frac{\cos mx}{\sqrt{\pi}}, \frac{\sin nx}{\sqrt{\pi}} \right\} \qquad (m, n = 1, 2, \ldots);$$

and the generalized Fourier series (4), Sec. 26, corresponding to a function f with respect to that orthonormal set, is

$$\frac{1}{\sqrt{2\pi}}\int_{-\pi}^{\pi} f(s)\frac{1}{\sqrt{2\pi}}\,ds$$

$$+ \sum_{n=1}^{\infty}\left[\frac{\cos nx}{\sqrt{\pi}}\int_{-\pi}^{\pi} f(s)\frac{\cos ns}{\sqrt{\pi}}\,ds + \frac{\sin nx}{\sqrt{\pi}}\int_{-\pi}^{\pi} f(s)\frac{\sin ns}{\sqrt{\pi}}\,ds\right].$$

This is series (1) with coefficients (2).

The correspondence can be written

(5) $f(x) \sim \dfrac{1}{2\pi} \displaystyle\int_{-\pi}^{\pi} f(s)\, ds$

$\qquad + \dfrac{1}{\pi} \displaystyle\sum_{n=1}^{\infty} \left[\cos nx \int_{-\pi}^{\pi} f(s) \cos ns\, ds + \sin nx \int_{-\pi}^{\pi} f(s) \sin ns\, ds \right].$

A more compact form is

(6) $\qquad f(x) \sim \dfrac{1}{2\pi} \displaystyle\int_{-\pi}^{\pi} f(s)\, ds + \dfrac{1}{\pi} \displaystyle\sum_{n=1}^{\infty} \int_{-\pi}^{\pi} f(s) \cos\left[n(s - x) \right] ds.$

Note that *the constant term* in the series, or the term $\frac{1}{2}a_0$ in series (1), *is the mean value of $f(x)$ over the interval* $(-\pi,\pi)$.

Each term of series (1) is periodic in x with period 2π. Consequently, when the series converges to $f(x)$ on the fundamental interval $(-\pi,\pi)$, it represents a *periodic extension* of f; that is, it converges to a periodic function of period 2π which coincides with f on the fundamental interval. Note, too, that if f is defined as a periodic function with period 2π, so that $f(x + 2\pi) = f(x)$, series (1) represents f outside the interval $(-\pi,\pi)$ when the representation is valid on it.

Thus *the Fourier series* (1) *serves either of two important purposes:* (a) to represent a function defined on the interval $(-\pi,\pi)$, for values of x in that interval, or (b) to represent a periodic function, with period 2π, for all values of x. Clearly, it cannot represent a function for all x if the function is not periodic.

In this chapter we shall establish the convergence of the *basic* Fourier series (1) to $f(x)$ under fairly general conditions on f. Representations by Fourier series for fundamental intervals other than $(-\pi,\pi)$ will follow easily.

36. Example

Let us write the Fourier series corresponding to the function f defined on the interval $-\pi < x < \pi$ as follows:

(1) $\qquad\qquad f(x) = \begin{cases} 0 & \text{when } -\pi < x \le 0, \\ x & \text{when } \quad 0 < x < \pi. \end{cases}$

The graph of the function is indicated by bold line segments in Fig. 10.

Note that f is piecewise continuous on $(-\pi,\pi)$ and the existence of the integrals in formulas (2), Sec. 35, is therefore ensured. The Fourier coefficients obtained from those formulas are

$$ a_0 = \frac{1}{\pi} \int_{-\pi}^{\pi} f(x)\, dx = \frac{1}{\pi} \left(\int_{-\pi}^{0} 0\, dx + \int_{0}^{\pi} x\, dx \right) = \frac{\pi}{2} $$

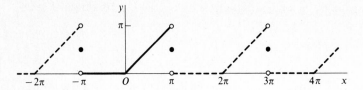

Figure 10

and

$$a_n = \frac{1}{\pi} \int_0^\pi x \cos nx \, dx = \frac{nx \sin nx + \cos nx}{\pi n^2} \bigg|_0^\pi = \frac{(-1)^n - 1}{\pi n^2},$$

$$b_n = \frac{1}{\pi} \int_0^\pi x \sin nx \, dx = \frac{-nx \cos nx + \sin nx}{\pi n^2} \bigg|_0^\pi = -\frac{(-1)^n}{n}$$

where $n = 1, 2, \ldots$. Therefore, on the interval $(-\pi, \pi)$,

$$(2) \qquad f(x) \sim \frac{\pi}{4} + \sum_{n=1}^\infty \left[\frac{(-1)^n - 1}{\pi n^2} \cos nx - \frac{(-1)^n}{n} \sin nx \right].$$

We shall soon see (Sec. 42) why this Fourier series *converges to* $f(x)$ when $-\pi < x < \pi$. Then it will follow that the series also represents the periodic extension of the function f indicated by the dotted line segments in Fig. 10. The periodic extension has jumps at the points $x = \pm\pi, \pm 3\pi, \ldots$. Our theory will show that the sum of the series at any such point must be $\pi/2$, the mean value of the one-sided limits at the point. With the dotted line segments and the solid dots included, Fig. 10 shows the graph of the function that series (2) represents for all x.

As an illustration of the convergence of the series to the function (1) on the interval $-\pi < x < \pi$, a few of its terms may be summed by addition of ordinates. It will be found, for instance, that the graph of the function

$$y = \frac{\pi}{4} - \frac{2}{\pi} \cos x + \sin x - \frac{1}{2} \sin 2x$$

is a wavy approximation to the graph of $y = f(x)$ shown in Fig. 10.

37. Fourier Cosine and Sine Series

If $f(-x) = f(x)$ for all values of x for which $f(x)$ is defined, f is an *even* function. The graph of $y = f(x)$ is symmetric with respect to the y axis and, if f is integrable from $x = 0$ to $x = c$,

$$(1) \qquad \int_{-c}^c f(x) \, dx = 2 \int_0^c f(x) \, dx.$$

An *odd* function f is one such that $f(-x) = -f(x)$. Its graph is symmetric with respect to the origin, and

(2)
$$\int_{-c}^{c} f(x)\, dx = 0.$$

As already indicated in Problem 2, Sec. 25, properties (1) and (2) are often useful in dealing with Fourier series.

The functions 1, x^2, $\cos nx$, and $x \sin nx$ are even, while x, x^3, $\sin nx$, and $x^2 \sin nx$ are odd functions. Although most functions are neither even nor odd, each function defined on an interval $(-c,c)$ is expressed as a sum of an even and an odd function by means of the identity

(3)
$$f(x) = \frac{1}{2}[f(x) + f(-x)] + \frac{1}{2}[f(x) - f(-x)].$$

When f is an even function on the interval $(-\pi,\pi)$, the products $f(x) \cos nx$ $(n = 0, 1, 2, \ldots)$ are even. The products $f(x) \sin nx$ $(n = 1, 2, \ldots)$ are, on the other hand, odd. Consequently, when the function f in integrals (2), Sec. 35, is even, the Fourier coefficients have the values

(4)
$$a_n = \frac{2}{\pi} \int_0^{\pi} f(x) \cos nx\, dx \qquad (n = 0, 1, 2, \ldots),$$

$b_n = 0$ $(n = 1, 2, \ldots)$; and the correspondence (5), Sec. 35, reduces to

(5)
$$f(x) \sim \frac{1}{\pi} \int_0^{\pi} f(s)\, ds + \frac{2}{\pi} \sum_{n=1}^{\infty} \cos nx \int_0^{\pi} f(s) \cos ns\, ds.$$

This *Fourier cosine series* is the generalized Fourier series for f with respect to the set $\{1/\sqrt{\pi}, \sqrt{2/\pi} \cos nx\}$ $(n = 1, 2, \ldots)$ of normalized eigenfunctions of the Sturm-Liouville problem

(6)
$$X''(x) + \lambda X(x) = 0, \qquad X'(0) = 0, \qquad X'(\pi) = 0,$$

the fundamental interval being $(0,\pi)$. (See Problems 2, Sec. 34, and 1, Sec. 25.)

Since the eigenfunctions are even, series (5) represents an even function on the interval $(-\pi,\pi)$ if it converges when $0 < x < \pi$. When f is defined only on the interval $(0,\pi)$, the series is the Fourier series for the even periodic extension of f, with period 2π.

When f is an odd function, $a_n = 0$ $(n = 0, 1, 2, \ldots)$ and

(7)
$$b_n = \frac{2}{\pi} \int_0^{\pi} f(x) \sin nx\, dx \qquad (n = 1, 2, \ldots);$$

thus

(8)
$$f(x) \sim \frac{2}{\pi} \sum_{n=1}^{\infty} \sin nx \int_0^{\pi} f(s) \sin ns\, ds.$$

This *Fourier sine series* (compare Sec. 16) will serve to represent odd functions defined on the interval $(-\pi,\pi)$ or odd periodic functions of period 2π. It will also serve to represent functions defined only on the interval $(0,\pi)$. In fact, it is the generalized Fourier series for f with respect to the orthonormal set $\{\sqrt{2/\pi} \sin nx\}$ $(n = 1, 2, \ldots)$, which is the set of normalized (Sec. 25) eigenfunctions of the Sturm-Liouville problem

$$(9) \qquad X''(x) + \lambda X(x) = 0, \qquad X(0) = 0, \qquad X(\pi) = 0$$

that arose in Sec. 15 with c instead of π.

38. Further Examples

To write the cosine series (5), Sec. 37, for the function $f(x) = \sin x$ on the interval $(0,\pi)$, we observe that

$$a_0 = \frac{2}{\pi} \int_0^\pi \sin x \, dx = \frac{4}{\pi}, \qquad a_1 = \frac{2}{\pi} \int_0^\pi \sin x \cos x \, dx = 0$$

and, when $n = 2, 3, \ldots$, that

$$a_n = \frac{2}{\pi} \int_0^\pi \sin x \cos nx \, dx$$

$$= \frac{1}{\pi} \int_0^\pi [\sin (1 + n)x + \sin (1 - n)x] \, dx$$

$$= \frac{1}{\pi} \left[-\frac{\cos (1 + n)x}{1 + n} - \frac{\cos (1 - n)x}{1 - n} \right]_0^\pi = \frac{2}{\pi} \frac{1 + (-1)^n}{1 - n^2}.$$

Thus

$$\sin x \sim \frac{2}{\pi} + \frac{2}{\pi} \sum_{n=2}^\infty \frac{1 + (-1)^n}{1 - n^2} \cos nx \qquad (0 < x < \pi).$$

Observe that in this particular series the terms occurring when n is an odd integer vanish; that is,

$$a_{2n+1} = 0 \qquad \text{and} \qquad a_{2n} = \frac{4}{\pi} \frac{1}{1 - 4n^2} \qquad (n = 1, 2, \ldots).$$

Hence we may replace n by $2n$ after the summation symbol and write (compare Sec. 17)

$$(1) \qquad \sin x \sim \frac{2}{\pi} - \frac{4}{\pi} \sum_{n=1}^\infty \frac{\cos 2nx}{4n^2 - 1} \qquad (0 < x < \pi).$$

Let us assume that the correspondence here is an equality for each value of x in the interval $0 \le x \le \pi$, as we shall see later on. Then for all values of x

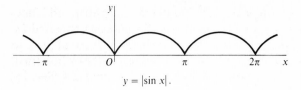

$$y = |\sin x|.$$

Figure 11

outside that interval the series converges to the even periodic extension, with period 2π, of $\sin x$ $(0 \le x \le \pi)$. That extension, shown in Fig. 11, is the function $y = |\sin x|$.

Note that since $\sin x$ is orthogonal to $\sin nx$ when $n = 2, 3, \ldots$, on the interval $(0,\pi)$, the Fourier sine series (8), Sec. 37, for the function $f(x) = \sin x$ on $(0,\pi)$ consists of a single term, namely $\sin x$.

In the evaluation of integrals representing Fourier coefficients, it is sometimes necessary to apply integration by parts more than once. We now give an example where this can be accomplished by means of a single formula due to L. Kronecker (1823–1891).† Let $p(x)$ be a polynomial of degree m and suppose that $f(x)$ is continuous. Then, except for an arbitrary additive constant,

$$(2) \qquad \int p(x)f(x)\,dx = pF_1 - p'F_2 + p''F_3 - \cdots + (-1)^m p^{(m)} F_{m+1}$$

where p is successively differentiated, F_1 denotes an indefinite integral of f, F_2 an indefinite integral of F_1, and so on, and alternating signs are affixed to the terms. Note that the differentiation of p begins with the *second* term, while the integration of f begins with the *first* term. The formula is readily proved by differentiating its right-hand side to obtain $p(x)f(x)$.

To illustrate the usefulness of formula (2), let us write the Fourier sine series (8), Sec. 37, for the particular function $f(x) = x^3$ on the interval $(0,\pi)$. With the aid of that formula, we find that

$$b_n = \frac{2}{\pi} \int_0^\pi x^3 \sin nx \, dx$$

$$= \frac{2}{\pi} \left[(x^3)\left(-\frac{\cos nx}{n} \right) - (3x^2)\left(-\frac{\sin nx}{n^2} \right) + (6x)\left(\frac{\cos nx}{n^3} \right) - (6)\left(\frac{\sin nx}{n^4} \right) \right]_0^\pi$$

$$= 2(-1)^{n+1} \frac{n^2\pi^2 - 6}{n^3} \qquad\qquad (n = 1, 2, \ldots).$$

---

† Kronecker actually treated the problem more extensively in papers which originally appeared in the *Berlin Sitzungsberichte*, 1885 and 1889.

Hence

$$(3) \qquad x^3 \sim 2 \sum_{n=1}^{\infty} (-1)^{n+1} \frac{n^2\pi^2 - 6}{n^3} \sin nx \qquad (0 < x < \pi).$$

Since x^3 is an odd function, the series here is also the basic Fourier series for $f(x) = x^3$ on the interval $(-\pi,\pi)$.

PROBLEMS

Write the Fourier series on the interval $(-\pi,\pi)$ for the functions described in Problems 1 through 6:

1. $f(x) = x$ when $-\pi < x < \pi$. Note that the sum of the series is zero when $x = \pm\pi$.

$$Ans. \quad 2 \sum_{n=1}^{\infty} \frac{(-1)^{n+1}}{n} \sin nx.$$

2. $f(x) = \begin{cases} -\pi \text{ when } -\pi < x < 0, \\ 0 \text{ when } 0 < x < \pi. \end{cases}$

$$Ans. \quad -\frac{\pi}{2} + 2 \sum_{n=1}^{\infty} \frac{\sin (2n-1)x}{2n-1}.$$

3. $f(x)$ is the function such that the graph of $y = f(x)$ consists of the two line segments shown in Fig. 12.

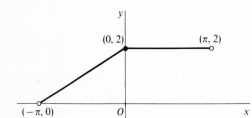

$(0, 2)$ $(\pi, 2)$

$(-\pi, 0)$ O x **Figure 12**

$$Ans. \quad \frac{3}{2} + 2 \sum_{n=1}^{\infty} \left[\frac{1-(-1)^n}{(n\pi)^2} \cos nx + \frac{(-1)^{n+1}}{n\pi} \sin nx \right].$$

4. $f(x) = 0$ when $-\pi \le x \le 0, f(x) = \sin x$ when $0 < x \le \pi$. Also, given that the series converges to $f(x)$ when $-\pi \le x \le \pi$, show graphically the function represented by the series for all x.

Suggestion: To find the series, write the function in the form

$$f(x) = \frac{\sin x + |\sin x|}{2} \qquad (-\pi \le x \le \pi)$$

and then use the Fourier cosine series for $\sin x$ found in Sec. 38.

$$Ans. \quad \frac{1}{\pi} + \frac{1}{2}\sin x - \frac{2}{\pi} \sum_{n=1}^{\infty} \frac{\cos 2nx}{4n^2 - 1}.$$

5. $f(x) = e^{ax}$ $(a \neq 0)$ when $-\pi < x < \pi$.

Suggestion: Using Euler's formula (see Problem 3, Sec. 30), write

$$a_n + ib_n = \frac{1}{\pi} \int_{-\pi}^{\pi} f(x)\, e^{inx}\, dx \qquad\qquad (n = 1, 2, \ldots)$$

and evaluate the integral here.

$$Ans. \quad \frac{2 \sinh a\pi}{\pi} \left[\frac{1}{2a} + \sum_{n=1}^{\infty} \frac{(-1)^n}{a^2 + n^2} (a \cos nx - n \sin nx) \right].$$

6. $f(x) = \sinh x$ when $-\pi < x < \pi$.

Suggestion: Use the series found in Problem 5.

$$Ans. \quad \frac{2 \sinh \pi}{\pi} \sum_{n=1}^{\infty} (-1)^{n+1} \frac{n}{1 + n^2} \sin nx.$$

7. Write the Fourier cosine series for the function $f(x) = x$ $(0 \leq x \leq \pi)$ on the interval $(0, \pi)$. Given that the series converges to $f(x)$ when $0 \leq x \leq \pi$, show graphically the function represented by it for all x.

$$Ans. \quad \frac{\pi}{2} - \frac{4}{\pi} \sum_{n=1}^{\infty} \frac{\cos (2n-1)x}{(2n-1)^2}.$$

8. Prove that the cosine series found in Problem 7 converges uniformly (Sec. 14) with respect to x for all x and hence that the series really does represent an even periodic function which is everywhere continuous.

Write (*a*) the Fourier cosine series and (*b*) the Fourier sine series for the functions defined on the interval $(0, \pi)$ in Problems 9 through 11:

9. $f(x) = 1$ $(0 < x < \pi)$.

$$Ans. \quad (a)\ 1;\ (b)\ \frac{4}{\pi} \sum_{n=1}^{\infty} \frac{\sin (2n-1)x}{2n-1}.$$

10. $f(x) = \pi - x$ $(0 < x < \pi)$.

$$Ans. \quad (a)\ \frac{\pi}{2} + \frac{4}{\pi} \sum_{n=1}^{\infty} \frac{\cos (2n-1)x}{(2n-1)^2};\ (b)\ 2 \sum_{n=1}^{\infty} \frac{\sin nx}{n}.$$

11. $f(x) = 1$ when $0 < x < \pi/2$, $f(x) = 0$ when $\pi/2 < x < \pi$.

$$Ans. \quad (b)\ \frac{2}{\pi} \sum_{n=1}^{\infty} \left(1 - \cos \frac{n\pi}{2}\right) \frac{\sin nx}{n}.$$

Write the Fourier cosine series for the functions defined on the interval $(0, \pi)$ in Problems 12 through 14 below. Note that they are also the Fourier series for those functions defined on the interval $(-\pi, \pi)$.

12. $f(x) = x^2$ $(0 < x < \pi)$.

$$Ans. \quad \frac{\pi^2}{3} + 4 \sum_{n=1}^{\infty} \frac{(-1)^n}{n^2} \cos nx.$$

13. $f(x) = x^4$ $(0 < x < \pi)$.

$$Ans. \quad \frac{\pi^4}{5} + 8 \sum_{n=1}^{\infty} (-1)^n \frac{n^2\pi^2 - 6}{n^4} \cos nx.$$

14. $f(x) = \cos ax$ $(0 < x < \pi)$, where $a \neq 0, \pm 1, \pm 2, \dots$.

Suggestion: The identity $\cos ax = (e^{iax} + e^{-iax})/2$ (see Sec. 29) and the series obtained in Problem 5 can be used here.

$$Ans. \quad \frac{2a \sin a\pi}{\pi}\left[\frac{1}{2a^2} + \sum_{n=1}^{\infty} \frac{(-1)^{n+1}}{n^2 - a^2} \cos nx\right].$$

15. Using the series in Problems 1 and 12, find the Fourier series for the function

$$f(x) = x + \frac{1}{4}x^2 \qquad (-\pi < x < \pi).$$

$$Ans. \quad \frac{\pi^2}{12} + \sum_{n=1}^{\infty} (-1)^n \left(\frac{\cos nx}{n^2} - \frac{2 \sin nx}{n}\right).$$

16. Prove Kronecker's formula (2), Sec. 38.

39. One-sided Derivatives

Let f denote a function whose right-hand limit $f(x_0 +)$ exists at a point x_0 (see Sec. 24). The *right-hand derivative*, or derivative from the right, of f at x_0 is defined as follows:

$$(1) \qquad f'_R(x_0) = \lim_{\substack{h \to 0 \\ h > 0}} \frac{f(x_0 + h) - f(x_0+)}{h},$$

provided the limit here exists. Note that although $f(x_0)$ need not exist, $f(x_0 +)$ must exist if $f'_R(x_0)$ does. When the ordinary, or two-sided, derivative $f'(x_0)$ exists, it is obvious that $f'_R(x_0) = f'(x_0)$.

Similarly, if $f(x_0 -)$ exists, the *left-hand derivative* of f at x_0 is given by the equation

$$(2) \qquad f'_L(x_0) = \lim_{\substack{h \to 0 \\ h < 0}} \frac{f(x_0 + h) - f(x_0-)}{h}$$

when this limit exists; and $f'_L(x_0) = f'(x_0)$ when $f'(x_0)$ exists. A useful alternative form of definition (2),

$$(3) \qquad f'_L(x_0) = \lim_{\substack{h \to 0 \\ h > 0}} \frac{f(x_0-) - f(x_0 - h)}{h},$$

is obtained by replacing h by $-h$ there.

To illustrate, consider first the continuous function

$$f(x) = \begin{cases} x^2 & \text{when } x \leq 0, \\ \sin x & \text{when } x > 0. \end{cases}$$

We find that $f'_R(0) = 1$ and $f'_L(0) = 0$, as the graph of f indicates; $f'(0)$ does not exist.

For the step function

$$f(x) = \begin{cases} 0 & \text{when } x < 0, \\ 1 & \text{when } x > 0, \end{cases}$$

$f'(0)$ does not exist; but both $f'_R(0)$ and $f'_L(0)$ exist and have the common value zero.

The function $f(x) = \sqrt{x}$ $(x \geq 0)$ is an example of a function that has no right-hand derivative at $x = 0$ although it is continuous there.

A number of properties of ordinary derivatives are also valid for one-sided derivatives. If, for example, each of two functions f and g has a right-hand derivative at a point x_0, then so does their product. A direct proof is left to the problems. But a proof can be based on the corresponding property of ordinary derivatives in the following way. We use $f(x_0 +)$ and $g(x_0 +)$ as the values of f and g at x_0; we also define those functions when $x \leq x_0$ as the linear functions represented by the tangent lines at the points $(x_0, f(x_0 +))$ and $(x_0, g(x_0 +))$ with slopes $f'_R(x_0)$ and $g'_R(x_0)$, respectively. Those extensions of f and g are differentiable at x_0, with derivatives equal to the right-hand derivatives. Thus the derivative of their product exists there; its value is the right-hand derivative of $f(x)g(x)$ at x_0.

Likewise, if $f'_L(x_0)$ and $g'_L(x_0)$ exist, the left-hand derivative of the product $f(x)g(x)$ exists at x_0.

A further property of one-sided derivatives will be useful in our theory of Fourier series and integrals.

Suppose that both f and its derivative f' are piecewise continuous functions on some closed interval, and let the interval $a \leq x \leq b$ be any one of the subintervals interior to which both f and f' are continuous and have one-sided limits from the interior at the end points. If we define $f(a)$ as $f(a+)$ and $f(b)$ as $f(b-)$, then f is continuous on the closed interval $a \leq x \leq b$. Also, f' exists on the open interval $a < x < b$; so the law of the mean applies to any interval $a \leq x \leq a + h$ where $0 < h < b - a$. That is, there exists a number θ_h $(0 < \theta_h < 1)$ such that

(4)
$$\frac{f(a + h) - f(a+)}{h} = f'(a + \theta_h h).$$

Since $f'(a+)$ exists, the limit of $f'(a + \theta_h h)$ as $h \to 0$ exists and has that value. The difference quotient on the left in equation (4) therefore has the same limit; that is, $f'_R(a) = f'(a+)$. Similarly, $f'_L(b) = f'(b-)$.

Thus *at each point x_0 of a closed interval on which both f and f' are piecewise continuous the one-sided derivatives of f, from the interior of the interval, exist and are the same as the corresponding one-sided limits of f':*

(5)
$$f'_R(x_0) = f'(x_0 +), \qquad f'_L(x_0) = f'(x_0 -).$$

The continuous function

(6)
$$f(x) = \begin{cases} x^2 \sin \dfrac{1}{x} & \text{when } x \neq 0, \\ 0 & \text{when } x = 0 \end{cases}$$

illustrates the distinction between one-sided derivatives and one-sided limits of derivatives. Here $f'_R(0) = f'_L(0) = 0$, while the one-sided limits $f'(0+)$ and $f'(0-)$ do not exist. The verification of this is left as a problem.

40. Preliminary Theory

We begin our discussion of the convergence of Fourier series with two preliminary theorems, or lemmas. The first is often referred to as the *Riemann-Lebesgue lemma*, and we present it in somewhat greater generality than we actually need in order that it can be used as well in Chap. 7, where the convergence of Fourier integrals is treated, and also in Chap. 8.

Lemma 1. *If a function $G(u)$ is piecewise continuous on an interval $(0,c)$, then*

(1)
$$\lim_{r \to \infty} \int_0^c G(u) \sin ru \, du = \lim_{r \to \infty} \int_0^c G(u) \cos ru \, du = 0.$$

We shall verify only the first limit here and leave verification of the second, which is similar, to the problems.

To verify the first limit, it is sufficient to show that if $G(u)$ is continuous at each point of an interval $a \leq u \leq b$, then

(2)
$$\lim_{r \to \infty} \int_a^b G(u) \sin ru \, du = 0.$$

For, in view of the discussion of integrals of piecewise continuous functions in Sec. 24, the integral in the first of limits (1) can be expressed as the sum of a finite number of integrals of the type appearing in equation (2).

Assuming, then, that $G(u)$ is continuous on the closed bounded interval $a \leq u \leq b$, we note that it must also be *uniformly* continuous there. That is, for each positive number ε there exists a positive number δ such that $|G(u) - G(v)| < \varepsilon$ whenever u and v lie in the interval and satisfy the inequality $|u - v| < \delta$.† Writing

$$\varepsilon = \frac{\varepsilon_0}{2(b-a)}$$

---

† See, for example, Taylor and Mann (1972, pp. 558–561) or Buck (1978, Sec. 2.3), listed in the Bibliography.

where ε_0 is an arbitrary positive number, we are thus assured that there is a positive number δ such that

(3) $$\left| G(u) - G(v) \right| < \frac{\varepsilon_0}{2(b-a)} \quad \text{whenever } |u - v| < \delta.$$

To obtain the limit (2), divide the interval $a \leq u \leq b$ into N subintervals of equal length $(b-a)/N$ by means of the points $a = u_0, u_1, u_2, \ldots, u_N = b$, where $u_0 < u_1 < u_2 < \cdots < u_N$, and let N be so large that the length of each subinterval is less than the number δ in condition (3). Then write

$$\int_a^b G(u) \sin ru \, du = \sum_{n=1}^N \int_{u_{n-1}}^{u_n} G(u) \sin ru \, du$$

$$= \sum_{n=1}^N \int_{u_{n-1}}^{u_n} [G(u) - G(u_n)] \sin ru \, du + \sum_{n=1}^N G(u_n) \int_{u_{n-1}}^{u_n} \sin ru \, du,$$

or

(4) $$\left| \int_a^b G(u) \sin ru \, du \right|$$

$$\leq \sum_{n=1}^N \int_{u_{n-1}}^{u_n} |G(u) - G(u_n)| \, |\sin ru| \, du + \sum_{n=1}^N |G(u_n)| \left| \int_{u_{n-1}}^{u_n} \sin ru \, du \right|.$$

In view of condition (3) and the fact that $|\sin ru| \leq 1$, it is easy to see that

$$\int_{u_{n-1}}^{u_n} |G(u) - G(u_n)| \, |\sin ru| \, du < \frac{\varepsilon_0}{2(b-a)} \frac{b-a}{N} = \frac{\varepsilon_0}{2N}$$
$$(n = 1, 2, \ldots, N).$$

Also, since $G(u)$ is continuous on the closed interval $a \leq u \leq b$, it is bounded there; that is, there is a positive number M such that $|G(u)| \leq M$ for all u between a and b inclusive. Furthermore,

$$\left| \int_{u_{n-1}}^{u_n} \sin ru \, du \right| \leq \frac{|\cos ru_n| + |\cos ru_{n-1}|}{r} \leq \frac{2}{r}$$
$$(n = 1, 2, \ldots, N),$$

where it is understood that $r > 0$. With these observations, we find that inequality (4) yields the statement

$$\left| \int_a^b G(u) \sin ru \, du \right| < \frac{\varepsilon_0}{2} + \frac{2MN}{r}.$$

Now write $R = 4MN/\varepsilon_0$ and observe that if $r > R$, then $2MN/r < \varepsilon_0/2$. Consequently,

$$\left| \int_a^b G(u) \sin ru \, du \right| < \frac{\varepsilon_0}{2} + \frac{\varepsilon_0}{2} = \varepsilon_0 \qquad \text{whenever } r > R;$$

and limit (2) is established.

Note that the lemma is, in particular, valid if r denotes only positive integers, of half integers, and tends to infinity through those values rather than continuously.

Our second lemma involves the *Dirichlet kernel*

$$(5) \qquad\qquad D_m(u) = \frac{1}{2} + \sum_{n=1}^{m} \cos nu,$$

which is a continuous function defined for all u and where m is allowed to be any positive integer. Note that $D_m(u)$ is even and periodic with period 2π. The Dirichlet kernel plays a central role in our theory, and the following two properties will be useful:

$$(6) \qquad\qquad \int_0^\pi D_m(u) \, du = \frac{\pi}{2},$$

$$(7) \qquad\qquad D_m(u) = \frac{\sin\left[(m + \frac{1}{2})u\right]}{2 \sin (u/2)} \qquad (u \neq 0, \pm 2\pi, \pm 4\pi, \ldots).$$

Property (6) is obvious upon integrating each side of equation (5), and property (7) is identity (9), Sec. 29, in only slightly different notation.

Lemma 2. *Suppose that a function $F(u)$ is piecewise continuous on the interval $(0,\pi)$ and that the right-hand derivative $F'_R(0)$ exists. Then*

$$(8) \qquad\qquad \lim_{m \to \infty} \int_0^\pi F(u)D_m(u) \, du = \frac{\pi}{2} F(0+),$$

where $D_m(u)$ is defined by equation (5).

Our proof is based on the fact that, for each positive integer m, we may write

$$(9) \qquad\qquad \int_0^\pi F(u)D_m(u) \, du = I_m + J_m$$

where

$$I_m = \int_0^\pi [F(u) - F(0+)]D_m(u) \, du, \qquad J_m = \int_0^\pi F(0+)D_m(u) \, du.$$

In view of expression (7), the first of these integrals can be put into the form

(10)
$$I_m = \int_0^\pi \frac{F(u) - F(0+)}{2 \sin (u/2)} \sin \left[\left(m + \frac{1}{2} \right) u \right] du.$$

Observe that the function

$$G(u) = \frac{F(u) - F(0+)}{2 \sin (u/2)}$$

is the quotient of two functions which are piecewise continuous on the interval $(0,\pi)$. Although the denominator vanishes at the point $u = 0$, the existence of $F'_R(0)$ ensures the existence of $G(0+)$:

$$\lim_{\substack{h \to 0 \\ h > 0}} G(0 + h) = \lim_{\substack{h \to 0 \\ h > 0}} \frac{F(0 + h) - F(0+)}{h} \lim_{\substack{h \to 0 \\ h > 0}} \frac{(h/2)}{\sin (h/2)} = F'_R(0).$$

Hence $G(u)$ is itself piecewise continuous on the interval $(0,\pi)$. Applying Lemma 1 to integral (10), we therefore conclude that

(11)
$$\lim_{m \to \infty} I_m = 0.$$

With property (6) of the Dirichlet kernel, we know that $J_m = \frac{\pi}{2} F(0+)$, or

(12)
$$\lim_{m \to \infty} J_m = \frac{\pi}{2} F(0+).$$

The desired result (8) now follows from equation (9) and limits (11) and (12).

41. A Fourier Theorem

A theorem that gives conditions under which a Fourier series converges to its function is called a *Fourier theorem*. One such theorem will now be established. Although it is stated for periodic functions of period 2π, it applies as well to functions defined only on the fundamental interval $(-\pi,\pi)$; for we need only consider the periodic extensions of those functions, with period 2π.

Theorem 1. *Let $f(x)$ be a function which is piecewise continuous on the interval $(-\pi,\pi)$ and periodic with period 2π. Its Fourier series*

(1)
$$\frac{1}{2\pi} \int_{-\pi}^\pi f(s) \, ds + \frac{1}{\pi} \sum_{n=1}^\infty \int_{-\pi}^\pi f(s) \cos [n(s - x)] \, ds$$

converges to the value

$$(2) \qquad\qquad \frac{1}{2}[f(x+) + f(x-)]$$

at each point x $(-\infty < x < \infty)$ *where the one-sided derivatives* $f'_R(x)$ *and* $f'_L(x)$ *both exist.*

As already pointed out in Sec. 35, series (1) is an alternative form of the basic Fourier series

$$\frac{1}{2}a_0 + \sum_{n=1}^{\infty} (a_n \cos nx + b_n \sin nx).$$

Note that the piecewise continuity of f ensures the existence of integrals (2), Sec. 35, which define the coefficients a_n and b_n. Also note that the quantity (2) is simply the mean value of the one-sided limits of f at x and is actually $f(x)$ if f is continuous at the point x.

It should be emphasized that the conditions given in the theorem are only sufficient, and there is no claim that they are necessary conditions. Indeed, there are well-known examples of functions which even become unbounded at certain points but which nevertheless have valid Fourier series representations.† The conditions we give are, however, broad enough for use in many physical applications and are suitable for solving the boundary value problems treated in this book.

Turning now to the proof of the theorem, we let $S_m(x)$ denote the sum of the first $m + 1$ terms, where $m \geq 1$, in series (1):

$$S_m(x) = \frac{1}{2\pi} \int_{-\pi}^{\pi} f(s)\, ds + \frac{1}{\pi} \sum_{n=1}^{m} \int_{-\pi}^{\pi} f(s) \cos [n(s - x)]\, ds.$$

Using the Dirichlet kernel, defined in equation (5) of the previous section, we can express $S_m(x)$ more compactly in the form

$$S_m(x) = \frac{1}{\pi} \int_{-\pi}^{\pi} f(s) D_m(s - x)\, ds.$$

Observe that the periodicity of the integrand in this last integral allows us to change the interval of integration to any interval of length 2π without altering the value of the integral (see Problem 15, Sec. 42). Thus

$$(3) \qquad\qquad S_m(x) = \frac{1}{\pi} \int_{x-\pi}^{x+\pi} f(s) D_m(s - x)\, ds,$$

---

† See, for instance, Tolstov (1976, pp. 91–94), listed in the Bibliography.

where the point x is at the center of the interval we have chosen; and, according to equation (3),

$$(4) \qquad\qquad S_m(x) = \frac{1}{\pi}[I_m(x) + J_m(x)]$$

where

$$(5) \qquad\qquad I_m(x) = \int_x^{x+\pi} f(s)D_m(s - x)\, ds$$

and

$$(6) \qquad\qquad J_m(x) = \int_{x-\pi}^x f(s)D_m(s - x)\, ds.$$

If we replace the variable of integration s in integral (5) by the new variable $u = s - x$, that integral becomes

$$I_m(x) = \int_0^\pi f(x + u)D_m(u)\, du.$$

Assuming that $f'_R(x)$ exists, we can now write $F(u) = f(x + u)$ and apply Lemma 2 (see Problem 13, Sec. 42). The result is

$$(7) \qquad\qquad \lim_{m\to\infty} I_m(x) = \frac{\pi}{2} f(x+).$$

If, on the other hand, we substitute the variable $u = x - s$ into integral (6) and recall that $D_m(u)$ is an even function of u, we find that

$$J_m(x) = \int_0^\pi f(x - u)D_m(u)\, du.$$

This time we write $F(u) = f(x - u)$ and note that $F'_R(0) = -f'_L(x)$ when $f'_L(x)$ exists. Lemma 2 then reveals that

$$(8) \qquad\qquad \lim_{m\to\infty} J_m(x) = \frac{\pi}{2} f(x-).$$

We now conclude from equations (4), (7), and (8) that when $f'_R(x)$ and $f'_L(x)$ exist,

$$\lim_{m\to\infty} S_m(x) = \frac{1}{2}[f(x+) + f(x-)];$$

and the proof of the theorem is complete.

42. Discussion of the Theorem

Suppose that a function f is defined only on the interval $(-\pi,\pi)$ and that it is piecewise continuous there. As pointed out just prior to the statement of Theorem 1, that theorem then applies to the periodic extension of f. Thus at each *interior* point x of the interval $-\pi \leq x \leq \pi$ where both one-sided derivatives exist the Fourier series for f converges to the mean value

$$\frac{1}{2}[f(x+) + f(x-)] \qquad (-\pi < x < \pi).$$

But *at both end points* $x = \pm\pi$ *it converges to the value*

(1)
$$\frac{1}{2}[f(-\pi+) + f(\pi-)],$$

provided $f'_R(-\pi)$ and $f'_L(\pi)$ exist, because that is the mean value of the ons-sided limits of the periodic extension at each of thos points. The mean value (1) reduces to $f(-\pi+)$, or to $f(\pi-)$, if and only if $f(-\pi+) = f(\pi-)$.

Note that the theorem assumes the existence of both one-sided derivatives of the function only at those points where it ensures convergence of the series to the mean value of the function. The function $f(x) = x^{2/3}$, for instance, is continuous on the interval $-\pi \leq x \leq \pi$, and it has one-sided derivatives there except at the point $x = 0$; also $f(-\pi) = f(\pi)$. Hence Theorem 1 shows that the Fourier series for f converges to $f(x)$ when $-\pi \leq x < 0$ and when $0 < x \leq \pi$; it does *not* ensure convergence when $x = 0$.

If in Theorem 1 both f and f' are piecewise continuous on the interval $-\pi \leq x \leq \pi$, then the one-sided derivatives of the periodic extension of f exist at every point (see Sec. 39); so *the series converges everywhere* to the mean value of the limits from the right and left for the periodic extension.

In Sec. 36 we wrote the Fourier series for the function

(2)
$$f(x) = \begin{cases} 0 & \text{when } -\pi < x \leq 0, \\ x & \text{when } 0 < x < \pi. \end{cases}$$

Both f and f' are piecewise continuous on the interval $-\pi \leq x \leq \pi$; so the one-sided derivatives of the periodic extension of f, with period 2π, exist everywhere. Also, $f(-\pi+) = 0$ and $f(\pi-) = \pi$. Theorem 1 therefore shows that the Fourier series found in Sec. 36 converges to $f(x)$ when $-\pi < x < \pi$ and to $(0 + \pi)/2$, or $\pi/2$, when $x = \pm\pi$. The periodic graph in Fig. 10, with the points $(\pm\pi,\pi/2)$, $(\pm 3\pi,\pi/2)$, ... included, represents the sum of the series for all x.

Representations of piecewise continuous functions on the interval $(0,\pi)$ by their Fourier cosine series and Fourier sine series are often ensured by

Theorem 1 because it applies to the even and odd extensions of those functions to functions defined on the interval $(-\pi, \pi)$. The series for such extensions reduce to the cosine series and the sine series, respectively (Sec. 37). We state the special Fourier theorems as a corollary.

Corollary 1. *Let a function* $f(x)$ *be piecewise continuous on the interval* $0 < x < \pi$ *and, for convenience, let the value of f be defined at each point where it is discontinuous as the mean value of its right-hand and left-hand limits there. Then at each point x of that interval where the one-sided derivatives* $f'_R(x)$ *and* $f'_L(x)$ *exist, f is represented by its Fourier cosine series*

$$(3) \qquad f(x) = \frac{1}{2} a_0 + \sum_{n=1}^{\infty} a_n \cos nx \qquad (0 < x < \pi),$$

where

$$(4) \qquad a_n = \frac{2}{\pi} \int_0^{\pi} f(x) \cos nx \, dx \qquad (n = 0, 1, 2, \ldots);$$

it is also represented by its Fourier sine series

$$(5) \qquad f(x) = \sum_{n=1}^{\infty} b_n \sin nx \qquad (0 < x < \pi),$$

where

$$(6) \qquad b_n = \frac{2}{\pi} \int_0^{\pi} f(x) \sin nx \, dx \qquad (n = 1, 2, \ldots).$$

In view of the even periodic function represented by the cosine series (3), that series converges at the point $x = 0$ to $f(0+)$ if $f'_R(0)$ exists; at $x = \pi$ it converges to $f(\pi-)$ if $f'_L(\pi)$ exists. The sum of the sine series (5) is clearly zero when $x = 0$ and when $x = \pi$.

Broader conditions than those given in Theorem 1, under which a Fourier series converges to its function, are stated in Chap. 5 and in a number of the references listed in the Bibliography.

PROBLEMS

1. Use Theorem 1 and series (1), Sec. 38, to show that, for every value of x,

$$|\sin x| = \frac{2}{\pi} - \frac{4}{\pi} \sum_{n=1}^{\infty} \frac{\cos 2nx}{4n^2 - 1}.$$

By writing $x = 0$ and $x = \pi/2$ in this expansion, obtain the following summations:

$$\sum_{n=1}^{\infty} \frac{1}{4n^2 - 1} = \frac{1}{2}, \qquad \sum_{n=1}^{\infty} \frac{(-1)^n}{4n^2 - 1} = \frac{1}{2} - \frac{\pi}{4}.$$

2. With the aid of Theorem 1, state why the series found in Problem 4, Sec. 38, for the function described by the equations $f(x) = 0$ when $-\pi \le x \le 0$ and $f(x) = \sin x$ when $0 < x \le \pi$ must converge to $f(x)$ everywhere in the interval $-\pi \le x \le \pi$.

3. Show that the functions described in Problems 2 and 5 of Sec. 38 satisfy conditions under which the series found there actually converge on the interval $-\pi \le x \le \pi$. Give the sum of the first series when $x = 0$, $\pm\pi$ and the sum of the second when $x = \pm\pi$.

Ans. Prob. 2: $x = 0$, $\pm\pi$, sum $= -\pi/2$. Prob. 5: $x = \pm\pi$, sum $= \cosh a\pi$.

4. State why the function $|x|$ is represented by its Fourier series everywhere in the interval $-\pi \le x \le \pi$. From that series, found in Problem 7, Sec. 38, show that

$$\sum_{n=1}^{\infty} \frac{1}{(2n-1)^2} = \frac{\pi^2}{8}.$$

5. Describe graphically the function to which the sine series found in Problem 11, Sec. 38, must converge for all values of x $(-\infty < x < \infty)$.

6. Show that the odd function $f(x) = x^{1/3}$ $(-\pi < x < \pi)$ does not have one-sided derivatives at the point $x = 0$. Without finding its Fourier series, state why that series converges to $f(x)$ on the interval $-\pi < x < \pi$, including the point $x = 0$. Thus illustrate the fact that the existence of one-sided derivatives is not a necessary condition for convergence.

7. Use the series in Problem 12, Sec. 38, to show that

$$\sum_{n=1}^{\infty} \frac{(-1)^{n+1}}{n^2} = \frac{\pi^2}{12}, \qquad \sum_{n=1}^{\infty} \frac{1}{n^2} = \frac{\pi^2}{6}.$$

8. Write $x = \pi$ in the series in Problem 13, Sec. 38, and then, with the aid of the second result in Problem 7 above, show that

$$\sum_{n=1}^{\infty} \frac{1}{n^4} = \frac{\pi^4}{90}.$$

9. With the aid of the series in Problem 14, Sec. 38, obtain the expansion

$$\frac{a\pi}{\sin a\pi} = 1 + 2a^2 \sum_{n=1}^{\infty} \frac{(-1)^{n+1}}{n^2 - a^2} \qquad (a \ne 0, \pm 1, \pm 2, \ldots).$$

10. Show that the function

$$f(x) = \begin{cases} x \sin \dfrac{1}{x} & \text{when } x \ne 0, \\ 0 & \text{when } x = 0 \end{cases}$$

is continuous at $x = 0$ but that neither of the one-sided derivatives exists at that point.

11. Let f be the function defined by equations (6), Sec. 39; namely, $f(x) = x^2 \sin(1/x)$ when $x \ne 0$ and $f(0) = 0$. Show that $f'_R(0) = 0$ but that $f'(0+)$ fails to exist. Likewise, show that $f'_L(0) = 0$ but that $f'(0-)$ does not exist.

12. With the aid of l'Hospital's rule, find $f(0+)$ and $f'_R(0)$ for the function

$$f(x) = \frac{e^x - 1}{x} \qquad (x \ne 0).$$

Ans. $1, \frac{1}{2}$.

13. Suppose that a function $f(x)$ has a right-hand derivative $f'_R(x_0)$ at a point x_0. Show that the function $F(x) = f(x_0 + x)$ has a right-hand derivative at $x = 0$ and that $F'_R(0) = f'_R(x_0)$. Likewise, suppose that $f'_L(x_0)$ exists and show that if $F(x) = f(x_0 - x)$, then $F'_R(0) = -f'_L(x_0)$.

14. Lemma 1 in Sec. 40 was needed in order to prove Lemma 2 in that section; but only the first limit in Lemma 1 occurring when $c = \pi$ and $r = m + 1/2$, where m denotes positive integers, was used. Obtain that special case of Lemma 1 from limits (7) and (8), Sec. 27; that is, use those limits to show that if $G(u)$ is piecewise continuous on the interval $(0,\pi)$, then

$$\lim_{m\to\infty} \int_0^\pi G(u) \sin\left[\left(m + \frac{1}{2}\right)u\right] du = 0.$$

Suggestion: Start by rewriting the integrand in the above integral according to the trigonometric identity

$$\sin (A + B) = \sin A \cos B + \cos A \sin B.$$

15. Let f denote a function which is piecewise continuous on an interval $(-c,c)$ and periodic with period $2c$. Show that, for any number a,

$$\int_{-c}^c f(x) \, dx = \int_{a-c}^{a+c} f(x) \, dx.$$

Suggestion: Write

$$\int_{-c}^c f(x) \, dx = \int_{-c}^{a+c} f(x) \, dx + \int_{a+c}^c f(x) \, dx$$

and then substitute the variable $s = x - 2c$ into the second integral on the right-hand side of this equation.

16. Prove property (7), Sec. 40, of the Dirichlet kernel by writing $A = nu$ and $B = u/2$ in the trigonometric identity

$$2 \cos A \sin B = \sin (A + B) - \sin (A - B)$$

and then summing each side of the result from $n = 1$ to $n = m$.

17. Given that the right-hand derivatives of two functions f and g exist at a point x_0, prove that the product fg of those functions has a right-hand derivative there by inserting the term $f(x_0 + h)g(x_0 +)$ and its negative into the numerator of the difference quotient

$$\frac{f(x_0 + h)g(x_0 + h) - f(x_0+)g(x_0+)}{h}.$$

18. Verify the second limit in Lemma 1, Sec. 40.

43. Other Forms of Fourier Series

Let c denote any positive number and f a periodic function of period $2c$ that is piecewise continuous on the interval $(-c,c)$. We define $f(x)$ at each point of discontinuity x as the mean value of $f(x+)$ and $f(x-)$. A Fourier theorem for f can be obtained from Theorem 1 by changing the unit of length on the x axis.

To do this, we introduce a new independent variable

(1)
$$s = \frac{\pi x}{c}.$$

Then $-\pi < s < \pi$ when $-c < x < c$, and $f(x) = f(cs/\pi)$; we write $F(s) = f(cs/\pi)$. The function $F(s)$ is periodic with period 2π, and it is continuous at a point s if $f(x)$ is continuous at the corresponding point x. From the piecewise continuity of $f(x)$ on the interval $(-c,c)$ we can conclude that $F(s)$ is piecewise continuous on $(-\pi,\pi)$; and, from the definition of $f(x)$ at points of discontinuity, we find that the value $F(s)$ is the mean of the values $F(s+)$ and $F(s-)$ at points s of discontinuity. Moreover, the existence of one-sided derivatives of $f(x)$ at a point x ensures the existence of those derivatives of $F(s)$ at the corresponding point s.

According to Theorem 1, $F(s)$ is represented by its Fourier series at each point s where $F'_R(s)$ and $F'_L(s)$ both exist. That is,

(2) $$f\left(\frac{cs}{\pi}\right) = \frac{1}{2} a_0 + \sum_{n=1}^{\infty} (a_n \cos ns + b_n \sin ns)$$

where

(3) $$a_n = \frac{1}{\pi} \int_{-\pi}^{\pi} f\left(\frac{cs}{\pi}\right) \cos ns \, ds \qquad (n = 0, 1, 2, \ldots),$$

$$b_n = \frac{1}{\pi} \int_{-\pi}^{\pi} f\left(\frac{cs}{\pi}\right) \sin ns \, ds \qquad (n = 1, 2, \ldots).$$

With the substitution (1), equation (2) becomes

(4) $$f(x) = \frac{1}{2} a_0 + \sum_{n=1}^{\infty} \left(a_n \cos \frac{n\pi x}{c} + b_n \sin \frac{n\pi x}{c} \right).$$

Formulas (3) for the coefficients can be written as follows, by making the substitution (1) for the variable of integration:

(5) $$a_n = \frac{1}{c} \int_{-c}^{c} f(x) \cos \frac{n\pi x}{c} \, dx \qquad (n = 0, 1, 2, \ldots),$$

$$b_n = \frac{1}{c} \int_{-c}^{c} f(x) \sin \frac{n\pi x}{c} \, dx \qquad (n = 1, 2, \ldots).$$

Note that the right-hand side of equation (4) can then be written

(6) $$\frac{1}{2c} \int_{-c}^{c} f(s) \, ds + \frac{1}{c} \sum_{n=1}^{\infty} \int_{-c}^{c} f(s) \cos \left[\frac{n\pi}{c} (s - x) \right] ds,$$

where here s is a new variable of integration and is not to be confused with the variable s defined in equation (1). [Compare series (6), Sec. 35.]

The series in equation (4) with coefficients (5), or the series (6), *is the Fourier series for periodic functions of period 2c.* The set of functions

(7) $$\left\{ 1, \cos \frac{m\pi x}{c}, \sin \frac{n\pi x}{c} \right\} \qquad (m, n = 1, 2, \ldots)$$

used in expansion (4) is the orthogonal set of eigenfunctions of the eigenvalue problem (Problem 10, Sec. 34)

$$(8) \qquad X'' + \lambda X = 0, \qquad X(-c) = X(c), \qquad X'(-c) = X'(c).$$

The representation (4) is used for functions defined only on the interval $(-c,c)$, as well as for periodic functions. But, as in the special case $c = \pi$, we state the Fourier theorem as follows in terms of periodic extensions of the functions.

Corollary 2. *If a function $f(x)$ is periodic with period $2c$, piecewise continuous on the interval $(-c,c)$, and defined as its mean value from the right and left at its points of discontinuity, then at each point x where $f'_R(x)$ and $f'_L(x)$ exist $f(x)$ has the Fourier series representation (4) with the coefficients (5).*

If f is an even function, expansion (4) reduces to the *Fourier cosine series* representation

$$(9) \qquad f(x) = \frac{1}{2} a_0 + \sum_{n=1}^{\infty} a_n \cos \frac{n\pi x}{c}$$

where

$$(10) \qquad a_n = \frac{2}{c} \int_0^c f(x) \cos \frac{n\pi x}{c} \, dx \qquad (n = 0, 1, 2, \ldots).$$

This is the representation of a function f on the interval $(0,c)$ in terms of the eigenfunctions of the Sturm-Liouville problem (Problem 2, Sec. 34)

$$(11) \qquad X'' + \lambda X = 0, \qquad X'(0) = 0, \qquad X'(c) = 0.$$

When f is odd, the series in representation (4) becomes the *Fourier sine series* for the interval $(0,c)$, introduced in Sec. 16.

If we express $\cos(n\pi x/c)$ and $\sin(n\pi x/c)$ in terms of complex exponential functions (Sec. 29) and write

$$(12) \qquad \gamma_0 = \frac{1}{2} a_0, \qquad \gamma_n = \frac{1}{2}(a_n - ib_n), \qquad \gamma_{-n} = \frac{1}{2}(a_n + ib_n)$$

$$(n = 1, 2, \ldots),$$

representation (4) takes the form

$$(13) \qquad f(x) = \lim_{m \to \infty} \sum_{n=-m}^{m} \gamma_n \exp \frac{in\pi x}{c}.$$

This is the *exponential form of the Fourier series* expansion of a periodic function with period $2c$. Equations (12) can be written

(14)
$$\gamma_n = \frac{1}{2c} \int_{-c}^{c} f(x) \exp\left(\frac{-in\pi x}{c}\right) dx$$

$$(n = 0, \pm 1, \pm 2, \ldots).$$

Note that the sum in equation (13) includes the term γ_0, corresponding to $n = 0$. The limit there is the *principal value* of the series

(15)
$$\sum_{n=-\infty}^{\infty} \gamma_n \exp\frac{in\pi x}{c} = \gamma_0 + \sum_{n=1}^{\infty} \gamma_n \exp\frac{in\pi x}{c} + \sum_{n=1}^{\infty} \gamma_{-n} \exp\frac{-in\pi x}{c},$$

obtained by grouping γ_0 with the first m terms of each of the last two series and then taking the limit, as $m \to \infty$, of the sum of terms in that group. The principal value of a series summed from $n = -\infty$ to $n = \infty$ sometimes exists when the series diverges; but if the series converges, its sum is the same as the principal value. Details are left to the problems.

44. The Orthonormal Trigonometric Functions

We use the symbol $C'_p(a,b)$ to denote the set, or function space, of all functions $f(x)$ such that f and f' are both piecewise continuous on the interval (a,b). At every point x in the interval $a \leq x \leq b$ the one-sided derivatives of f, from the interior of that interval, therefore exist; also, the number of discontinuities of f in the interval is finite.

Corollary 2 applies to the periodic extension, with period $2c$, of each function f in the space $C'_p(-c,c)$ to show that $f(x)$ has the Fourier series representation (4), Sec. 43, at all points x $(-c < x < c)$ where f is continuous. That series is the generalized Fourier series with respect to the set

(1)
$$\left\{ \frac{1}{\sqrt{2c}}, \frac{1}{\sqrt{c}} \cos\frac{m\pi x}{c}, \frac{1}{\sqrt{c}} \sin\frac{n\pi x}{c} \right\} \qquad (m, n = 1, 2, \ldots),$$

which is orthonormal on the interval $(-c,c)$ [see Problem 2(c), Sec. 25] and is contained in the space $C'_p(-c,c)$. Since the representation is ensured except for at most a finite number of points in the interval $(-c,c)$, we can state the following result in the terminology introduced in Chap. 3.

Corollary 3. *In the function space $C'_p(-c,c)$ the orthonormal set of functions* (1) *is closed in the sense of pointwise convergence. It is also complete.*

It was pointed out in Sec. 28 that such a set must be complete if it is closed. Note that the corollary is stated for functions whose one-sided derivatives exist everywhere in the interval.

From the representations of functions on the interval $(0,c)$ by Fourier cosine and sine series, corresponding statements follow for the orthonormal sets

$$(2) \qquad \left| \frac{1}{\sqrt{c}}, \sqrt{\frac{2}{c}} \cos \frac{n\pi x}{c} \right| \qquad\qquad (n = 1, 2, \ldots)$$

(Problem 1, Sec. 25) and

$$(3) \qquad \left| \sqrt{\frac{2}{c}} \sin \frac{n\pi x}{c} \right| \qquad\qquad (n = 1, 2, \ldots),$$

(see Sec. 25) on that interval.

Corollary 4. *In the function space $C'_p(0,c)$ each of the orthonormal sets* (2) *and* (3) *is closed in the sense of pointwise convergence, and each is complete.*

In particular, the set (2) is complete in the space of continuous functions with continuous derivatives on the interval $0 \le x \le c$ (see Problem 1, Sec. 28).

PROBLEMS

1. Show that if $f(x + 2c) = f(x)$ for all x and $f(x) = -1$ when $-c < x < 0, f(x) = 1$ when $0 < x < c$, and $f(0) = f(c) = 0$, then

$$f(x) = \frac{4}{\pi} \sum_{n=1}^{\infty} \frac{1}{2n-1} \sin \frac{(2n-1)\pi x}{c} \qquad (-\infty < x < \infty).$$

2. Suppose that $f(x) = 0$ when $-2 < x < 1$ and $f(x) = 1$ when $1 < x < 2$; also, $f(-2) = f(1) = f(2) = \frac{1}{2}$. Show that, for all x in the interval $-2 \le x \le 2$,

$$f(x) = \frac{1}{4} - \frac{1}{\pi} \sum_{n=1}^{\infty} \frac{1}{n} \left[\sin \frac{n\pi}{2} \cos \frac{n\pi x}{2} + \left(\cos n\pi - \cos \frac{n\pi}{2} \right) \sin \frac{n\pi x}{2} \right].$$

3. Prove that when $0 \le x \le c$,

$$x^2 = \frac{c^2}{3} + \frac{4c^2}{\pi^2} \sum_{n=1}^{\infty} \frac{(-1)^n}{n^2} \cos \frac{n\pi x}{c}.$$

Suggestion: The expansion in Problem 12, Sec. 38, can be used here, and no new integrals need be evaluated.

4. Show that

$$x^2 = 2c^2 \sum_{n=1}^{\infty} \left[\frac{(-1)^{n+1}}{n\pi} - 2 \frac{1 - (-1)^n}{n^3 \pi^3} \right] \sin \frac{n\pi x}{c} \qquad (0 \le x < c).$$

5. Show that if $f(x) = 0$ when $-3 < x < 0$, $f(x) = 1$ when $0 < x < 3$, and $f(0) = \frac{1}{2}$, then

$$f(x) = \frac{1}{2} + \frac{2}{\pi} \sum_{n=1}^{\infty} \frac{1}{2n-1} \sin \frac{(2n-1)\pi x}{3} \qquad (-3 < x < 3).$$

Suggestion: The expansion in Problem 1 can be used here.

6. Show that when $-1 < x < 1$,

$$x + x^2 = \frac{1}{3} + \frac{2}{\pi} \sum_{n=1}^{\infty} (-1)^n \left(\frac{2}{n^2 \pi} \cos n\pi x - \frac{1}{n} \sin n\pi x \right).$$

7. Suppose that $f(x) = \cos \pi x$ when $0 < x < 1$, $f(x) = 0$ when $1 < x < 2$, $f(0) = \frac{1}{2}$, $f(1) = -\frac{1}{2}$, and $f(x + 2) = f(x)$ for all x. Write the Fourier series for f and state why it must converge everywhere to $f(x)$.

$$\text{Ans.} \quad \frac{1}{2} \cos \pi x + \frac{4}{\pi} \sum_{n=1}^{\infty} \frac{n}{4n^2 - 1} \sin 2n\pi x.$$

8. Obtain the Fourier sine series representation

$$\cos \pi x = \frac{8}{\pi} \sum_{n=1}^{\infty} \frac{n}{4n^2 - 1} \sin 2n\pi x \qquad (0 < x < 1).$$

9. Suppose that

$$f(x) = \begin{cases} \dfrac{c}{4} - x & \text{when } 0 \le x \le \dfrac{c}{2}, \\[2ex] x - \dfrac{3c}{4} & \text{when } \dfrac{c}{2} < x \le c. \end{cases}$$

Establish the Fourier cosine series representation

$$f(x) = \frac{2c}{\pi^2} \sum_{n=1}^{\infty} \frac{1}{(2n-1)^2} \cos \frac{(4n-2)\pi x}{c} \qquad (0 \le x \le c).$$

10. State why the set of cosine functions $\{\cos (n\pi x/c)\}$ $(n = 1, 2, \ldots)$, excluding a constant function, is closed in the sense of pointwise convergence in the subspace of the function space $C'_p(0,c)$ consisting of all functions f of that space for which

$$\int_0^c f(x)\, dx = 0.$$

11. Let f be piecewise continuous on an interval $(0,c)$ and let b_n denote the coefficients in the corresponding Fourier sine series for f (Sec. 16). Use Bessel's inequality (Sec. 27) to prove that the series of terms b_n^2 converges and that

$$\sum_{n=1}^{\infty} b_n^2 \le \frac{2}{c} \int_0^c [f(x)]^2\, dx.$$

Thus deduce that $b_n \to 0$ as $n \to \infty$, and show how this conclusion also follows from Lemma 1.

12. Let f be piecewise continuous on an interval $(0,c)$ and prove that the coefficients (10), Sec. 43, satisfy the condition

$$\frac{1}{2} a_0^2 + \sum_{n=1}^{\infty} a_n^2 \le \frac{2}{c} \int_0^c [f(x)]^2\, dx.$$

(Compare Problem 11.) Deduce that $a_n \to 0$ as $n \to \infty$.

13. Let f be piecewise continuous on an interval $(-c,c)$ and show that the coefficients (5), Sec. 43, satisfy the condition

$$\frac{1}{2} a_0^2 + \sum_{n=1}^{\infty} (a_n^2 + b_n^2) \le \frac{1}{c} \int_{-c}^c [f(x)]^2\, dx.$$

(Compare Problems 11 and 12.)

14. (a) The coefficients b_n in the Fourier sine series for a piecewise continuous function f on an interval $(0,c)$ are those for which a finite linear combination of the sine functions becomes the best approximation in the mean to $f(x)$ over the interval $(0,c)$. Show how this follows from Theorem 1 in Sec. 27.

(b) Find the values of C_1, C_2, and C_3 such that the function

$$y = C_1 \sin \frac{\pi x}{2} + C_2 \sin \frac{2\pi x}{2} + C_3 \sin \frac{3\pi x}{2}$$

is the best approximation in the mean to the function $f(x) = 1$ over the interval $(0,2)$. Also, draw the graph of y versus x using the coefficients found, and compare it to the graph of $y = f(x)$.

$$Ans. \quad C_1 = \frac{4}{\pi}, C_2 = 0, C_3 = \frac{4}{3\pi}.$$

15. Theorem 1 ensures the convergence of Fourier series with corresponding cosine and sine terms grouped as a single term

$$(a_n \cos nx + b_n \sin nx).$$

Show that when f satisfies the conditions stated in that theorem and when x is a point such that $f'(x)$ and $f'(-x)$ both exist, then

$$\frac{1}{2}[f(x) + f(-x)] = \frac{1}{2}a_0 + \sum_{n=1}^{\infty} a_n \cos nx.$$

Thus the cosine terms themselves form a convergent series at the point. Why must the series of sine terms converge there also?

16. In Sec. 43, write the Fourier representation (4) in the form

$$f(x) = \frac{1}{2}a_0 + \lim_{m \to \infty} \sum_{n=1}^{m} \left(a_n \cos \frac{n\pi x}{c} + b_n \sin \frac{n\pi x}{c} \right);$$

then obtain the exponential form (13) of that representation and expression (14) for the coefficients γ_n. Also, show how those same values of γ_n can be found formally from equation (13) by using the hermitian orthogonality (see Problem 7, Sec. 30) of the functions $\exp(in\pi x/c)$ on the interval $(-c,c)$.

17. (a) The series on the left-hand side of equation (15), Sec. 43, is said to converge if each of the individual series on the right-hand side converges. Prove that if the left-hand side converges, then its principal value exists.

(b) Use the exponential form (13), Sec. 43, of Fourier series representations, when $c = \pi$, to show that

$$\frac{\pi}{\sinh \pi} e^x = \sum_{n=-\infty}^{\infty} \frac{(-1)^n}{1 - in} e^{inx} \qquad (-\pi < x < \pi),$$

where the principal value of the series is taken.

(c) Show how it follows that the principal value of the series in part (b) exists, with value $\pi \coth \pi$, when $x = \pi$. Show that when $x = \pi$, the series does not, however, converge in the sense described in part (a).[†] The converse of the statement proved in part (a) is therefore not true.

[†] For details regarding the convergence of series with complex terms, see, for example, Churchill, Brown, and Verhey (1974, Chap. 6), listed in the Bibliography.

FIVE

FURTHER PROPERTIES OF FOURIER SERIES

45. Uniform Convergence

In establishing the uniqueness of solutions of boundary value problems in partial differential equations (Chap. 10) it is sometimes helpful to know conditions on a function under which its Fourier series will converge uniformly over the fundamental interval. We present such conditions in this section. The development in Secs. 46 through 49, dealing mostly with differentiation and integration of Fourier series, will be used only occasionally later on. The reader can skip those sections at this time without disruption and refer back to results that will be specifically cited as needed.

Our treatment of the uniform convergence of Fourier series will depend on an important inequality, and we now present a simple but indirect derivation of it.

Let A_n and B_n $(n = 1, 2, \ldots, m)$ denote real numbers where at least one of the A_n, say A_N, is nonzero. If the quadratic equation

$$x^2 \sum_{n=1}^{m} A_n^2 + 2x \sum_{n=1}^{m} A_n B_n + \sum_{n=1}^{m} B_n^2 = 0$$

in the variable x has a real root x_0, we need only write that equation in the form

$$\sum_{n=1}^{m} (A_n x + B_n)^2 = 0$$

to see that $A_n x_0 + B_n = 0$ for each n, in particular when $n = N$. Consequently, the only possible value of x_0 is $x_0 = -B_N/A_N$, and we find that our quadratic equation cannot have two *distinct* real roots. Its discriminant is therefore negative or zero; that is,

(1)
$$\left(\sum_{n=1}^{m} A_n B_n \right)^2 \le \left(\sum_{n=1}^{m} A_n^2 \right) \left(\sum_{n=1}^{m} B_n^2 \right).$$

This result is obviously valid even when all of the numbers A_n are zero.

Condition (1) is known as *Cauchy's inequality.* When $m = 3$, it simply states that the square of the inner product of two vectors in three-dimensional space does not exceed the product of the squares of their norms. The corresponding property for inner products of functions is the Schwarz inequality (Problem 9, Sec. 25). We shall use Cauchy's inequality in proving the following theorem.

Theorem 1. *Let f be a continuous function on the interval $-\pi \le x \le \pi$ such that $f(-\pi) = f(\pi)$, and let its derivative f' be piecewise continuous on that interval. If a_n and b_n are the Fourier coefficients*

$$a_n = \frac{1}{\pi} \int_{-\pi}^{\pi} f(x) \cos nx \, dx, \qquad b_n = \frac{1}{\pi} \int_{-\pi}^{\pi} f(x) \sin nx \, dx,$$

the series

(2)
$$\sum_{n=1}^{\infty} \sqrt{a_n^2 + b_n^2}$$

converges.

From the comparison test we note that each of the series

(3)
$$\sum_{n=1}^{\infty} |a_n|, \qquad \sum_{n=1}^{\infty} |b_n|$$

converges as a consequence of the convergence of series (2).

We begin the proof of the theorem with the observation that the Fourier coefficients of f',

(4)
$$\alpha_n = \frac{1}{\pi} \int_{-\pi}^{\pi} f'(x) \cos nx \, dx, \qquad \beta_n = \frac{1}{\pi} \int_{-\pi}^{\pi} f'(x) \sin nx \, dx,$$

exist because of the piecewise continuity of f'. Since f is continuous and $f(-\pi) = f(\pi)$,

(5)
$$\alpha_0 = \frac{1}{\pi} \int_{-\pi}^{\pi} f'(x) \, dx = \frac{1}{\pi} [f(\pi) - f(-\pi)] = 0.$$

Also, when $n = 1, 2, \ldots$, integration by parts reveals that

(6)
$$\alpha_n = \frac{n}{\pi} \int_{-\pi}^{\pi} f(x) \sin nx \, dx + \frac{1}{\pi} [f(x) \cos nx]_{-\pi}^{\pi}$$

$$= nb_n + \frac{\cos n\pi}{\pi} [f(\pi) - f(-\pi)] = nb_n,$$

(7)
$$\beta_n = -\frac{n}{\pi} \int_{-\pi}^{\pi} f(x) \cos nx \, dx + \frac{1}{\pi} [f(x) \sin nx]_{-\pi}^{\pi} = -na_n.$$

Note that the condition $f(-\pi) = f(\pi)$, under which the periodic extension of f is also continuous, is necessary if expression (6) is to reduce to the
form $\alpha_n = nb_n$.

Now let S_m denote the sum of the first m terms of the infinite series (2). In
view of relations (6) and (7),

$$S_m = \sum_{n=1}^{m} \sqrt{a_n^2 + b_n^2} = \sum_{n=1}^{m} \frac{1}{n}\sqrt{\alpha_n^2 + \beta_n^2}.$$

Since $S_m \geq 0$, it follows from Cauchy's inequality (1) that

$$(8) \qquad S_m \leq \left[\sum_{n=1}^{m} \frac{1}{n^2} \sum_{n=1}^{m} (\alpha_n^2 + \beta_n^2) \right]^{1/2}.$$

The first sum on the right here is bounded for all m because the infinite series
of positive terms $1/n^2$ converges.

Using the piecewise continuous function f' in Bessel's inequality (see
Sec. 27 and Problem 13, Sec. 44) with respect to the orthonormal set

$$\left\{ \frac{1}{\sqrt{2\pi}}, \frac{\cos x}{\sqrt{\pi}}, \frac{\cos 2x}{\sqrt{\pi}}, \ldots, \frac{\sin x}{\sqrt{\pi}}, \frac{\sin 2x}{\sqrt{\pi}}, \ldots \right\}$$

on the interval $(-\pi, \pi)$, we find that for every m it is true that

$$\sum_{n=1}^{m} (\alpha_n^2 + \beta_n^2) \leq \frac{1}{\pi} \int_{-\pi}^{\pi} [f'(x)]^2 \, dx,$$

the coefficient α_0 being zero. The right-hand side of inequality (8) is therefore
bounded for all m, and so is S_m.

Since S_m is a bounded nondecreasing sequence, its limit as m tends to
infinity exists; that is, series (2) converges, as stated in Theorem 1.

The next theorem is closely related to the first.

Theorem 2. *Under the conditions stated in Theorem 1 the convergence of
the Fourier series*

$$(9) \qquad \frac{1}{2} a_0 + \sum_{n=1}^{\infty} (a_n \cos nx + b_n \sin nx)$$

*to $f(x)$ on the interval $-\pi \leq x \leq \pi$ is absolute and uniform with respect to x on
that interval.*

The conditions on f and f' ensure the continuity, and also the existence
of one-sided derivatives, of the periodic extension of f for all x. It follows
from the Fourier theorem in Sec. 41 that series (9) converges to $f(x)$ everywhere in the interval $-\pi \leq x \leq \pi$. Now, to prove Theorem 2, we observe
that

$$|a_n \cos nx + b_n \sin nx| \leq |a_n| + |b_n|$$

and also that the series of constants $|a_n| + |b_n|$ converges because each of series (3) converges. The comparison test and the Weierstrass M-test (Sec. 14) therefore apply to show that the convergence of series (9) is indeed absolute and uniform.

The tests apply as well to show the convergence of the series of cosine or sine terms only; in fact, the convergence is absolute and uniform. Therefore series (9) is the sum of those series; that is,

$$(10) \qquad f(x) = \frac{1}{2} a_0 + \sum_{n=1}^{\infty} a_n \cos nx + \sum_{n=1}^{\infty} b_n \sin nx \quad (-\pi \leq x \leq \pi),$$

and both series here converge absolutely and uniformly.

46. Observations

A Fourier series cannot converge uniformly on an interval that contains a discontinuity of its sum because a uniformly convergent series of continuous functions always converges to a continuous function. Hence some continuity requirement on f, such as the continuity assumed in Theorem 2, is necessary in order to ensure uniform convergence of the Fourier series to $f(x)$.

Modifications of Theorems 1 and 2 for cosine series or sine series, or for Fourier series on an interval $(-c,c)$, are apparent. For instance, it follows from Theorem 2 that the Fourier cosine series for a continuous function f, on the interval $0 \leq x \leq \pi$, converges uniformly to $f(x)$ on that interval if f' is piecewise continuous on the interval. For the sine series, however, the further conditions $f(0) = f(\pi) = 0$ are needed.

Consider the space of functions satisfying the conditions stated in Theorem 1. Parseval's equation (Sec. 28) with respect to the orthonormal trigonometric functions on the interval $(-\pi,\pi)$ is satisfied by each function f of that space. To see this, we multiply the Fourier series expansion of f by $f(x)$, thus leaving the series still uniformly convergent, and then integrate:

$$\int_{-\pi}^{\pi} [f(x)]^2 \, dx$$

$$= \frac{1}{2} a_0 \int_{-\pi}^{\pi} f(x) \, dx + \sum_{n=1}^{\infty} \left[a_n \int_{-\pi}^{\pi} f(x) \cos nx \, dx + b_n \int_{-\pi}^{\pi} f(x) \sin nx \, dx \right].$$

This is Parseval's equation

$$(1) \qquad \frac{1}{\pi} \int_{-\pi}^{\pi} [f(x)]^2 \, dx = \frac{1}{2} a_0^2 + \sum_{n=1}^{\infty} (a_n^2 + b_n^2).$$

From it we conclude (Sec. 28) that the orthonormal set of trigonometric functions is closed in the sense of mean convergence in the function space.

47. Differentiation of Fourier Series

We have seen why the Fourier series for the function $f(x) = x$ on the interval $-\pi < x < \pi$ converges to $f(x)$ at each point of the interval; that is, (Problem 1, Sec. 38)

$$x = 2 \sum_{n=1}^{\infty} (-1)^{n+1} \frac{\sin nx}{n} \qquad (-\pi < x < \pi).$$

But the series here is not differentiable. The differentiated series

$$2 \sum_{n=1}^{\infty} (-1)^{n+1} \cos nx$$

does not converge since its nth term fails to approach zero as n tends to infinity. The periodic extension of f with period 2π, represented by the first series for all x, has discontinuities at the points $x = \pm \pi, \pm 3\pi, \ldots$.

Continuity of the periodic functions is an important condition for differentiability of Fourier series. Sufficient conditions can be stated as follows.

Theorem 3. *Let f be a continuous function on the interval $-\pi \leq x \leq \pi$ such that $f(-\pi) = f(\pi)$, and let f' be piecewise continuous on that interval. Then the Fourier series in the representation*

(1) $$f(x) = \frac{1}{2} a_0 + \sum_{n=1}^{\infty} (a_n \cos nx + b_n \sin nx) \qquad (-\pi \leq x \leq \pi),$$

where

$$a_n = \frac{1}{\pi} \int_{-\pi}^{\pi} f(x) \cos nx \, dx, \qquad b_n = \frac{1}{\pi} \int_{-\pi}^{\pi} f(x) \sin nx \, dx,$$

is differentiable at each point x in the interval $-\pi < x < \pi$ where $f''(x)$ exists:

(2) $$f'(x) = \sum_{n=1}^{\infty} n(-a_n \sin nx + b_n \cos nx).$$

To prove this, we note first that since the Fourier theorem in Sec. 41 can be applied to the function f', that function is represented by its Fourier series at each point $\dot{x}$ $(-\pi < x < \pi)$ where its derivative $f''(x)$ exists. At such a point f' is continuous, and so

(3) $$f'(x) = \frac{1}{2} \alpha_0 + \sum_{n=1}^{\infty} (\alpha_n \cos nx + \beta_n \sin nx)$$

where α_n and β_n are the coefficients (4), Sec. 45. But when f satisfies the conditions stated in Theorem 3, we know from Sec. 45 that

(4) $$\alpha_0 = 0, \qquad \alpha_n = nb_n, \qquad \beta_n = -na_n \qquad (n = 1, 2, \ldots).$$

When these substitutions are made, equation (3) takes the form (2). This completes the proof of the theorem.

At a point x where $f''(x)$ does not exist, but where f' has derivatives from the right and left, the differentiation is still valid in the sense that the series in equation (2) converges to the mean of the values $f'(x+)$ and $f'(x-)$. This is also true for the periodic extension of f.

Theorem 3 applies with natural changes to other forms of Fourier series. For instance, if f is continuous and f' is piecewise continuous on an interval $0 \le x \le c$, then the Fourier cosine series for f on that interval is differentiable at each point where $f''(x)$ exists.

48. Integration of Fourier Series

Integration of a Fourier series is possible under much more general conditions than those for differentiation. This is to be expected because an integration introduces a factor n in the denominator of the general term. In the following theorem it is not even essential that the original series converge to its function in order that the integrated series converge to the integral of the function. The integrated series is not, however, a Fourier series if $a_0 \ne 0$; for it contains a term $a_0 x/2$.

Theorem 4. *Let f be piecewise continuous on the interval $(-\pi, \pi)$. Then, whether the Fourier series corresponding to f,*

(1) $$f(x) \sim \frac{1}{2} a_0 + \sum_{n=1}^{\infty} (a_n \cos nx + b_n \sin nx),$$

converges or not, the following equality is true when $-\pi \le x \le \pi$: → closed interval.

(2) $$\int_{-\pi}^{x} f(s)\, ds = \frac{1}{2} a_0(x + \pi) + \sum_{n=1}^{\infty} \frac{1}{n} [a_n \sin nx - b_n(\cos nx - \cos n\pi)].$$

The latter series is obtained by integrating the former term by term.

Our proof starts with the fact that since f is piecewise continuous, the function

(3) $$F(x) = \int_{-\pi}^{x} f(s)\, ds - \frac{1}{2} a_0 x$$

is continuous; moreover,

$$F'(x) = f(x) - \frac{1}{2} a_0$$

except at points where f is discontinuous. Therefore F' is piecewise continuous on the interval $(-\pi,\pi)$. Also,

$$F(\pi) = \int_{-\pi}^{\pi} f(s)\, ds - \frac{1}{2} a_0 \pi = a_0 \pi - \frac{1}{2} a_0 \pi = \frac{1}{2} a_0 \pi$$

and $F(-\pi) = a_0 \pi/2$; hence $F(\pi) = F(-\pi)$. According to our Fourier
theorem (Sec. 41), then, for all x in the interval $-\pi \le x \le \pi$ it is true that

$$F(x) = \frac{1}{2} A_0 + \sum_{n=1}^{\infty} (A_n \cos nx + B_n \sin nx)$$

where

$$A_n = \frac{1}{\pi} \int_{-\pi}^{\pi} F(x) \cos nx\, dx, \qquad B_n = \frac{1}{\pi} \int_{-\pi}^{\pi} F(x) \sin nx\, dx.$$

When $n \neq 0$, we integrate the last two integrals by parts, using the fact
that F is continuous and F' is piecewise continuous. Thus

$$A_n = \frac{1}{n\pi} [F(x) \sin nx]_{-\pi}^{\pi} - \frac{1}{n\pi} \int_{-\pi}^{\pi} F'(x) \sin nx\, dx$$

$$= -\frac{1}{n\pi} \int_{-\pi}^{\pi} \left[f(x) - \frac{1}{2} a_0 \right] \sin nx\, dx = -\frac{1}{n} b_n.$$

Similarly, $B_n = a_n/n$; hence

(4) $$F(x) = \frac{1}{2} A_0 + \sum_{n=1}^{\infty} \frac{1}{n} (a_n \sin nx - b_n \cos nx)$$

when $-\pi \le x \le \pi$. But, since $F(\pi) = a_0 \pi/2$,

$$\frac{1}{2} a_0 \pi = \frac{1}{2} A_0 - \sum_{n=1}^{\infty} \frac{1}{n} b_n \cos n\pi.$$

With the value of A_0 given here, equation (4) becomes

$$F(x) = \frac{1}{2} a_0 \pi + \sum_{n=1}^{\infty} \frac{1}{n} [a_n \sin nx - b_n(\cos nx - \cos n\pi)].$$

In view of equation (3), equation (2) follows at once.

The theorem can be written for the integral from x_0 to x, where
$-\pi \le x_0 \le \pi$ and $-\pi \le x \le \pi$, by noting that

$$\int_{x_0}^{x} f(s)\, ds = \int_{-\pi}^{x} f(s)\, ds - \int_{-\pi}^{x_0} f(s)\, ds.$$

49. More General Conditions

A few of the many more general results in the theory of Fourier series will be noted here. They are stated without proof; our purpose is only to inform the reader of the existence of such theorems. We first introduce some concepts involved in the theorems.

A function g defined on a closed interval is *monotone nondecreasing* there if its value $g(x)$ never decreases when x is increased. One example, on the interval $-\pi \le x \le \pi$, is the step function

$$(1) \quad G(x) = \begin{cases} 0 & \text{when } -\pi \le x < \dfrac{\pi}{2}, \\[2ex] \dfrac{n}{n+1}\pi & \text{when } \dfrac{n}{n+1}\pi \le x < \dfrac{n+1}{n+2}\pi \quad (n = 1, 2, \ldots), \\[2ex] \pi & \text{when } x = \pi. \end{cases}$$

This function is not piecewise continuous on the interval; graphically, it has steps up to the line $y = x$ at the infinite set of points $x = n\pi/(n+1)$. The function $-G(x)$ is monotone nonincreasing.

A function f of *bounded variation* on an interval $a \le x \le b$ can be defined as one that is the sum of two monotone functions g and h,

$$f(x) = g(x) + h(x) \qquad (a \le x \le b),$$

where g is nondecreasing and h is nonincreasing. Each such function f has the following properties.† The one-sided limits $f(x+)$ and $f(x-)$ from the interior of the interval exist at each point; f has at most a countable infinity of discontinuities in the interval, and f is bounded and integrable over the interval.

(a) *Another Fourier Theorem.* Let f denote a periodic function of period 2π whose integral from $-\pi$ to π exists. If that integral is improper, let it be absolutely convergent. There is a theorem which states that at each point x which is interior to an interval on which f is of bounded variation, the Fourier series for the function converges to the value

$$\frac{1}{2}[f(x+) + f(x-)].$$

The periodic extension of the function G defined on the interval $-\pi \le x \le \pi$ by equations (1), for example, satisfies the conditions in this theorem, but not those in our Fourier theorem in Sec. 41. For that periodic

---

† See, for instance, Franklin (1964, pp. 255 ff) or Apostol (1974, pp. 127 ff), listed in the Bibliography.

function the series converges at every point to the mean value of the limits from the right and left.

The cosine series for the function $\sqrt{x}$ on the interval $0 \leq x \leq \pi$ converges to $\sqrt{x}$ over that interval, including the point $x = 0$ where the right-hand derivative fails to exist. Also, the series for the unbounded function $x^{-1/3}$ on the interval $(-\pi, \pi)$ represents the function over the interval except at the point $x = 0$.

(b) *Uniform Convergence.* Let the periodic integrable function f described in (a) above satisfy the additional condition that on some interval $a \leq x \leq b$ it is continuous and of bounded variation. Then its Fourier series converges uniformly to $f(x)$ on each closed interval interior to the interval (a,b). The series for the function $f(x) = x$ on the interval $(-\pi, \pi)$, for example, converges uniformly to x on the interval $-3 \leq x \leq 3$.

We noted earlier (Sec. 46) that a Fourier series for a periodic function f cannot converge *uniformly* to $f(x)$ on an interval that contains a point where f is discontinuous. The nature of the deviation of the partial sums $S_m(x)$ from $f(x)$ on such intervals is known as the *Gibbs phenomenon*.† Suppose, for example, that $f(x) = -\pi$ when $-\pi < x < 0$ and $f(x) = 0$ when $0 < x < \pi$. The sequence of sums $S_m(x)$ of the first $m + 1$ nonzero terms in the Fourier series for f on the interval $(-\pi, \pi)$ converges, of course, to zero for all x in the subinterval $0 < x < \pi$. It can be shown, however, that there exists a fixed positive number σ such that for large m there are always positive numbers x in that subinterval for which the number $S_m(x)$ is close to σ. See Fig. 13, which shows how "spikes" in the graphs of the partial sums form near the point $x = 0$, where f is discontinuous, as m increases. The tips of these spikes tend toward the point $(0,\sigma)$ on the vertical axis. The behavior of the partial sums is similar on the subinterval $-\pi < x < 0$. This illustrates how special

---

† See Carslaw (1930, Chap. IX) for a detailed analysis of this phenomenon.

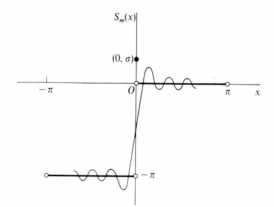

Figure 13

care must be taken when a function is approximated by the partial sums of its Fourier series near a discontinuity.

(c) *Integration.* Parseval's equation

$$(2) \qquad \frac{1}{2} a_0^2 + \sum_{n=1}^{\infty} (a_n^2 + b_n^2) = \frac{1}{\pi} \int_{-\pi}^{\pi} [f(x)]^2 \, dx$$

is satisfied whenever f is bounded and integrable over the interval $(-\pi,\pi)$. That is, this series of squares of the Fourier coefficients of f converges to $\|f\|^2/\pi$.

Proofs of the results stated in (a), (b), and (c) above are given in some of the books listed in the Bibliography.†

Let a function F also satisfy the conditions on f stated under (c), and let A_n and B_n denote its Fourier coefficients. Then $a_n + A_n$ and $b_n + B_n$ are the coefficients for $f + F$ and, according to Parseval's equation,

$$\frac{1}{\pi} \int_{-\pi}^{\pi} [f(x) + F(x)]^2 \, dx = \frac{1}{2} (a_0 + A_0)^2 + \sum_{n=1}^{\infty} [(a_n + A_n)^2 + (b_n + B_n)^2].$$

When we subtract the corresponding equation for $f - F$, we obtain *Parseval's equation for the inner product:*

$$(3) \qquad \frac{1}{\pi} \int_{-\pi}^{\pi} f(x)F(x) \, dx = \frac{1}{2} a_0 A_0 + \sum_{n=1}^{\infty} (a_n A_n + b_n B_n).$$

In equation (3) let us write

$$F(x) = \begin{cases} g(x) & \text{when } -\pi < x < t, \\ 0 & \text{when } t < x < \pi \end{cases} \qquad (-\pi \le t \le \pi),$$

where g is bounded and integrable on the interval $(-\pi,\pi)$. Then equation (3) takes the form

$$(4) \quad \int_{-\pi}^{t} f(x)g(x) \, dx$$

$$= \frac{1}{2} a_0 \int_{-\pi}^{t} g(x) \, dx + \sum_{n=1}^{\infty} \left[a_n \int_{-\pi}^{t} g(x) \cos nx \, dx + b_n \int_{-\pi}^{t} g(x) \sin nx \, dx \right].$$

So it follows from statement (c) that if the Fourier series corresponding to any bounded integrable function f is multiplied by any other function g of the same class and then integrated term by term, the resulting series converges to the integral of the product $f(x)g(x)$. When $g(x) = 1$, we have a general theorem for integration of a Fourier series.

---

† See, in particular, Chap. 9 of the book by Whittaker and Watson (1963) that is listed there.

According to Sec. 28, statement (c) implies that, in the space of bounded integrable functions on the interval $(-\pi, \pi)$, the orthonormal set of trigonometric functions (4), Sec. 35, is closed in the sense of mean convergence.

PROBLEMS

1. Show that the function

$$f(x) = \begin{cases} 0 & \text{when } -\pi \le x \le 0, \\ \sin x & \text{when } 0 < x \le \pi \end{cases}$$

satisfies all the conditions in Theorems 1 and 2. Verify directly from the Fourier series for f, found in Problem 4, Sec. 38, that the series converges uniformly for all x. Also, state why that series is differentiable on the interval $-\pi \le x \le \pi$ except at the points $x = 0, \pm\pi$, and describe the function represented by the differentiated series for all x.

2. Differentiate the Fourier cosine series for the function $f(x) = x$ on the interval $0 < x < \pi$ (Problem 7, Sec. 38) to obtain the Fourier sine series expansion of the function $f'(x) = 1$ on that interval. State why the procedure is reliable in this case.

3. State Theorem 3 as it applies to Fourier sine series on the interval $(0,\pi)$. Point out, in particular, why the conditions $f(0) = f(\pi) = 0$ are present in this case.

4. Prove that the Fourier coefficients a_n and b_n for the function f in Theorem 1 satisfy the conditions

$$\lim_{n \to \infty} na_n = 0, \qquad \lim_{n \to \infty} nb_n = 0.$$

5. Integrate from $s = 0$ to $s = x$ $(-\pi \le x \le \pi)$ the Fourier series

$$2 \sum_{n=1}^{\infty} \frac{(-1)^{n+1}}{n} \sin ns$$

and

$$-\frac{\pi}{2} + 2 \sum_{n=1}^{\infty} \frac{\sin (2n-1)s}{2n-1},$$

obtained in Problems 1 and 2, respectively, of Sec. 38. In each case describe the function represented by the new series.

6. Let f be the function described by the equations $f(x) = -\pi$ when $-\pi < x < 0$ and $f(x) = 0$ when $0 < x < \pi$ (Fig. 13).

(a) Using the results in Problem 2, Sec. 38, and Problem 4, Sec. 30, show that the sum $S_m(x)$ of the first $m + 1$ $(m \ge 1)$ nonzero terms of the Fourier series for f on the interval $(-\pi, \pi)$ can be written

$$S_m(x) = -\frac{\pi}{2} + \int_0^x \frac{\sin 2mu}{\sin u} \, du.$$

(b) Write

$$\sigma = \int_0^\pi \frac{\sin s}{s} \, ds - \frac{\pi}{2},$$

the value of σ being approximately 0.28.† Then, using the integral representation for $S_m(x)$ found in part (a), show that

$$S_m\left(\frac{\pi}{2m}\right) - \sigma = \int_0^{\pi/(2m)} \left(\frac{u - \sin u}{u \sin u}\right) \sin 2mu \; du.$$

(c) Show that the function $(u - \sin u)/(u \sin u)$ is piecewise continuous on the interval $0 \le u \le \pi/2$ and is therefore bounded on that interval (see Problem 5, Sec. 25). Then show how it follows from the result in part (b) that

$$\lim_{m \to \infty} S_m\left(\frac{\pi}{2m}\right) = \sigma.$$

(d) Suppose that the sequence $S_m(x)$ converges *uniformly* to 0 on the interval $0 < x < \pi$. There is then a positive integer m_1 such that, for all x in the interval, $S_m(x) < \sigma/2$ whenever $m > m_1$, σ being the number defined in part (b). By using the main result in part (c), show that there is a positive integer m_2 such that

$$S_m\left(\frac{\pi}{2m}\right) > \frac{\sigma}{2} \qquad\qquad \text{whenever } m > m_2,$$

and reach a contradiction. The convergence is therefore *not* uniform.

---

† The integral here occurs as a particular value of the so-called sine integral function $Si(x)$, which is tabulated, for instance, on p. 244 of the handbook edited by Abramowitz and Stegun (1965) that is listed in the Bibliography. Approximation methods for evaluating definite integrals can also be used to find σ.

BOUNDARY VALUE PROBLEMS

50. Formal and Rigorous Solutions

The foregoing theory of representing prescribed functions by Fourier series enables us to use the method of separation of variables to solve important types of boundary value problems in partial differential equations. The method is illustrated in this chapter for a variety of boundary value problems which are mathematical formulations of problems in physics.

We illustrate ways of proving that the function we find truly satisfies the partial differential equation and all boundary conditions and continuity requirements. When that is done, our function is rigorously established as *a* solution of the problem. The physical problem may indicate that there should be *only one* solution. In Chap. 10 we shall give some attention to that question of uniqueness of solutions.

Even for some of the simpler problems, the full treatment, consisting of verifying a solution and proving that it is the only one, may be lengthy or difficult. We shall give the full treatment in only a few cases. Most of the problems will be solved *formally* in the sense that we may not verify our solution fully or that we may not prove that our solution is unique.

51. The Vibrating String Initially Displaced

In Sec. 15 we found, but did not verify, an expression for the transverse displacements $y(x,t)$ of a string, stretched between the origin and the fixed point $(c,0)$, after the string is released at rest from a position $y = f(x)$ in the xy plane (Fig. 14). The function y was therefore required to satisfy all conditions of the boundary value problem

$$(1) \qquad\qquad y_{tt}(x,t) = a^2 y_{xx}(x,t) \qquad\qquad (0 < x < c, \, t > 0),$$

$$(2) \qquad\qquad y(0,t) = 0, \qquad y(c,t) = 0 \qquad\qquad (t \geq 0),$$

$$(3) \qquad\qquad y(x,0) = f(x), \qquad y_t(x,0) = 0 \qquad\qquad (0 \leq x \leq c).$$

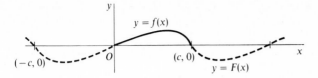

Figure 14

We used the method of separation of variables, and also superposition and the orthogonality of the functions sin $(n\pi x/c)$, to arrive at the formal solution

$$(4) \qquad y(x,t) = \sum_{n=1}^{\infty} b_n \sin \frac{n\pi x}{c} \cos \frac{n\pi at}{c}$$

where

$$(5) \qquad b_n = \frac{2}{c} \int_0^c f(x) \sin \frac{n\pi x}{c} \, dx \qquad\qquad (n = 1, 2, \ldots).$$

The given function f was assumed to be continuous on the interval $0 \le x \le c$; also, $f(0) = f(c) = 0$. Assuming further that f' is at least piece-wise continuous on the interval, we now know that f is represented by its Fourier sine series there. The coefficients in that series are the numbers b_n given by equation (5). Hence when $t = 0$, the series in expression (4) converges to $f(x)$; that is, $y(x,0) = f(x)$ when $0 \le x \le c$.

The nature of the problem calls for a solution $y(x,t)$ that is continuous in x and t when $0 \le x \le c$ and $t \ge 0$ and is such that $y_t(x,t)$ is continuous in t at $t = 0$. Thus the prescribed boundary values are limiting values at the boundaries of the domain $0 < x < c$, $t > 0$: $y(0,t) = y(0+,t)$, $y(c,t) = y(c-,t)$, and so on.

To verify that expression (4) represents a solution, we must prove that the series there converges to a continuous function $y(x,t)$ which satisfies the wave equation (1) and all the boundary conditions. But the series may not be twice differentiable with respect to x and t even though it has a sum $y(x,t)$ that may satisfy the wave equation. In the case of the plucked string in Sec. 17, for instance, the coefficients b_n are proportional to $(1/n^2) \sin (n\pi/2)$; so after the series for $y(x,t)$ is differentiated twice with respect to either x or t, the resulting series fails to converge.

It is possible to sum the series in expression (4); that is, we can write its sum without using infinite series. This will simplify the verification of the solution.

Since

$$2 \sin \frac{n\pi x}{c} \cos \frac{n\pi at}{c} = \sin \frac{n\pi(x + at)}{c} + \sin \frac{n\pi(x - at)}{c},$$

equation (4) can be written

(6) $$y(x,t) = \frac{1}{2} \sum_{n=1}^{\infty} b_n \sin \frac{n\pi(x + at)}{c} + \frac{1}{2} \sum_{n=1}^{\infty} b_n \sin \frac{n\pi(x - at)}{c}.$$

Let $F(x)$ be defined for *all* real x by the sine series for f:

(7) $$F(x) = \sum_{n=1}^{\infty} b_n \sin \frac{n\pi x}{c} \qquad (-\infty < x < \infty).$$

Then $F(x)$ is the odd periodic extension, with period $2c$, of $f(x)$; that is,

(8)
$$F(x) = f(x) \qquad \text{when } 0 \le x \le c,$$
$$F(-x) = -F(x), \qquad F(x + 2c) = F(x) \qquad \text{for all } x.$$

In view of equation (7), expression (6) can be written

(9) $$y(x,t) = \frac{1}{2}[F(x + at) + F(x - at)].$$

Thus series (4) is summed with the aid of the function F defined by equations (8). The convergence of series (6) and (4) follows from the convergence of series (7).

Solution Verified. Our conditions on f are such that its extension F is continuous for all x (Fig. 14). Hence $F(x + at)$ and $F(x - at)$, and therefore the function $y(x,t)$ given by equation (9), are continuous functions of x and t for all x and t. While it is evident from expression (4) that our function $y(x,t)$ satisfies the conditions $y(0,t) = 0$, $y(c,t) = 0$, and $y(x,0) = f(x)$, expression (9) can also be used to verify this. Note, for example, that when $x = c$ in expression (9), we can write $F(c - at) = -F(at - c) = -F(at + c)$; therefore $y(c,t) = 0$.

Since $F(-x) = -F(x)$, then $-F'(-x) = -F'(x)$ whenever $F'(x)$ exists, where the prime denotes the derivative with respect to the argument of F. That is, F' is an even function. Likewise, F'' is an odd function.

If f' and f'' *are continuous when* $0 \le x \le c$ *and*

$$f''(0) = f''(c) = 0,$$

then $F'(x)$ and $F''(x)$ are continuous for all x, as indicated in Fig. 15. Thus

$$y_t(x,t) = \frac{a}{2}[F'(x + at) - F'(x - at)],$$

y_t is continuous for all x and t, and $y_t(x,0) = 0$. According to Sec. 19, $F(x + at)$ and $F(x - at)$ satisfy the wave equation (1); therefore $y(x,t)$ satisfies that equation, as well as all boundary conditions. The function $y(x,t)$ given by equation (9) is then fully verified as a solution of our boundary

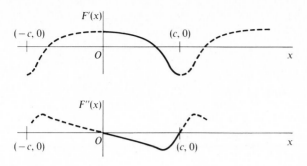

Figure 15

value problem. In Chap. 10 we shall show why it is the only possible solution which, together with its derivatives of the first and second order, is everywhere continuous.

If the conditions on f' and f'' are relaxed by merely requiring those two functions to be *piecewise continuous*, we find that at each instant t there may be a finite number of points x $(0 \le x \le c)$ where the partial derivatives of y fail to exist. Except at those points, our function satisfies the wave equation and the condition $y_t(x,0) = 0$. The other boundary conditions are satisfied as before. In this case we have a solution of our boundary value problem in a broader sense.

52. Discussion of the Solution

From either equation (4) or its alternate form (9) in the previous section we can see that for each fixed x the displacement $y(x,t)$ is a periodic function of time t, with period

$$
(1) \qquad\qquad T_0 = \frac{2c}{a}.
$$

The period is independent of the initial displacement $f(x)$. Since $a^2 = H/\delta$, where H is the magnitude of the horizontal component of the tensile force and δ is the mass of the string per unit length (Sec. 3), the period varies directly with c and $\sqrt{\delta}$ and inversely with $\sqrt{H}$.

From either equation (4) or (9) in Sec. 51 it is also evident that, for a given length c and initial displacement $f(x)$, the displacement y depends only on the value of x and the value of the product at. That is, $y = \phi(x,at)$ where the function ϕ is the same function regardless of the value of the constant a. Let a_1 and a_2 denote different values of that constant, and let $y_1(x,t)$ and $y_2(x,t)$ be the corresponding displacements. Then

$$
(2) \qquad\qquad y_1(x,t_1) = y_2(x,t_2) \qquad \text{if } a_1 t_1 = a_2 t_2 \qquad (0 \le x \le c).
$$

In particular, suppose that only the constant H differs, with values H_1 and H_2. The same set of instantaneous positions is then taken by the string when $H = H_1$ and when $H = H_2$, but the times t_1 and t_2 required to reach any one position have the ratio

$$(3) \qquad \frac{t_1}{t_2} = \sqrt{\frac{H_2}{H_1}}.$$

Approximate Solutions. Except for the nonhomogeneous condition

$$(4) \qquad y(x,0) = f(x),$$

our boundary value problem is satisfied by any partial sum

$$(5) \qquad y_m(x,t) = \sum_{n=1}^{m} b_n \sin \frac{n\pi x}{c} \cos \frac{n\pi a t}{c}$$

of the infinite series (4), Sec. 51. Instead of meeting requirement (4), however, it satisfies the condition

$$(6) \qquad y_m(x,0) = \sum_{n=1}^{m} b_n \sin \frac{n\pi x}{c}.$$

The sum on the right-hand side of equation (6) is, of course, the partial sum consisting of the first m terms of the Fourier sine series for $f(x)$ on the interval $(0,c)$. Since the odd periodic extension of f is continuous and f' is piecewise continuous, that series converges uniformly to $f(x)$ on the interval $0 \le x \le c$ (Chap. 5). Hence, if m is taken sufficiently large, the sum $y_m(x,0)$ can be made to approximate $f(x)$ arbitrarily closely for all values of x in that interval.

The function $y_m(x,t)$, which is everywhere continuous together with all its partial derivatives, is therefore established as a solution of the approximating problem obtained by replacing condition (4) in the original problem by condition (6).

Corresponding approximations can be made to other problems. But a remarkable feature in the present case is that $y_m(x,t)$ never deviates from the actual displacement $y(x,t)$ by more than the greatest deviation of $y_m(x,0)$ from $f(x)$. This is true because

$$y_m(x,t) = \frac{1}{2} \left[\sum_{n=1}^{m} b_n \sin \frac{n\pi(x + at)}{c} + \sum_{n=1}^{m} b_n \sin \frac{n\pi(x - at)}{c} \right]$$

and two sums here are those of the first m terms of the sine series for the odd periodic extension F of f with arguments $x + at$ and $x - at$. But the greatest deviation of the first sum from $F(x + at)$, or of the second from $F(x - at)$, is the same as the greatest deviation of $y_m(x,0)$ from $f(x)$.

PROBLEMS

1. A string is stretched between the fixed points $(0,0)$ and $(1,0)$ and released at rest from the position $y = A \sin \pi x$, where A is a constant. Find an expression for its subsequent displacements $y(x,t)$, and verify the result fully. Sketch the position of the string at several instants of time.

Ans. $y = A \sin \pi x \cos \pi a t.$

2. Solve Problem 1 when the initial displacement there is changed to $y = B \sin 2\pi x$, where B is a constant.

Ans. $y = B \sin 2\pi x \cos 2\pi a t.$

3. Show why the sum of the two functions $y(x,t)$ found in Problems 1 and 2 represents the displacements after the string is released at rest from the position $y = A \sin \pi x + B \sin 2\pi x$.

4. For the initially displaced string of length c considered in Secs. 51 and 52, show why the *frequency* v of vibration in cycles per unit time has the value

$$v = \frac{a}{2c} = \frac{1}{2c}\sqrt{\frac{H}{\delta}}.$$

Show that if $H = 200$ lb, the weight per foot is 0.01 lb ($g\delta = 0.01$, $g = 32$), and the length is 2 ft, then $v = 200$ cycles/sec.

5. In Sec. 51 the position of the string at each instant can be shown graphically by moving the graph of the periodic function $\frac{1}{2}F(x)$ to the right with velocity a and an identical curve to the left at the same rate and adding ordinates, on the interval $0 \le x \le c$, of the two curves so obtained at the instant t. Show how this follows from expression (9), Sec. 51.

6. Plot some positions of the plucked string considered in Sec. 17 by the method described in Problem 5 to verify that the string assumes such positions as indicated by the bold line segments in Fig. 16.

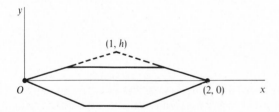

Figure 16

7. Write the boundary value problem (1) through (3), Sec. 51, in terms of the two independent variables x and $\tau = at$ to show that the problem in y as a function of x and τ does not involve the constant a. Thus, without solving the problem, deduce that the solution has the form $y = \phi(x,\tau) = \phi(x,at)$ and hence that the relation (2), Sec. 52, follows.

53. Prescribed Initial Velocity

When, initially, the string has some prescribed distribution of velocities $g(x)$ parallel to the y axis in its position of equilibrium $y = 0$, the boundary value problem for the displacements $y(x,t)$ becomes

$$(1) \qquad y_{tt}(x,t) = a^2 y_{xx}(x,t) \qquad\qquad (0 < x < c, \, t > 0),$$

$$(2) \qquad y(0,t) = 0, \qquad y(c,t) = 0 \qquad\qquad (t \geq 0),$$

$$(3) \qquad y(x,0) = 0, \qquad y_t(x,0) = g(x) \qquad\qquad (0 \leq x \leq c).$$

If the xy plane, with the string lying on the x axis, is moving parallel to the y axis and is brought to rest at the instant $t = 0$, the function $g(x)$ is a constant. The hammer action in a piano may produce approximately a uniform initial velocity over a short span of a piano wire, in which case $g(x)$ may be considered to be a step function.

As in Sec. 15, we find the functions of type $X(x)T(t)$ that satisfy all *homogeneous* equations in the boundary value problem. The Sturm-Liouville problem in X is the same as the one found in Sec. 15; thus $\lambda = n^2\pi^2/c^2$ and $X = \sin(n\pi x/c)$ where $n = 1, 2, \ldots$. The conditions on T become

$$T''(t) + \left(\frac{n\pi a}{c}\right)^2 T(t) = 0, \qquad T(0) = 0;$$

hence $T = \sin(n\pi at/c)$. The homogeneous equations are then formally satisfied by the function

$$y(x,t) = \sum_{n=1}^{\infty} B_n \sin\frac{n\pi x}{c} \sin\frac{n\pi at}{c},$$

where the constants B_n are to be determined from the second of conditions (3):

$$(4) \qquad \sum_{n=1}^{\infty} B_n \frac{n\pi a}{c} \sin\frac{n\pi x}{c} = g(x) \qquad\qquad (0 \leq x \leq c).$$

Under the assumption that g and g' are piecewise continuous, the series in equation (4) is the Fourier sine series for $g(x)$ on the interval $(0,c)$ if $B_n(n\pi a/c) = b_n$ where

$$(5) \qquad b_n = \frac{2}{c} \int_0^c g(x) \sin\frac{n\pi x}{c} \, dx.$$

Thus $B_n = b_n(c/n\pi a)$ and

$$(6) \qquad y(x,t) = \frac{c}{\pi a} \sum_{n=1}^{\infty} \frac{b_n}{n} \sin\frac{n\pi x}{c} \sin\frac{n\pi at}{c}.$$

We can sum the series here by first writing

$$y_t(x,t) = \sum_{n=1}^{\infty} b_n \sin\frac{n\pi x}{c} \cos\frac{n\pi at}{c} = \frac{1}{2}G(x + at) + \frac{1}{2}G(x - at)$$

where G is the odd periodic extension, with period $2c$, of the given function g (compare Sec. 51). Then, since $y(x,0) = 0$,

$$y(x,t) = \frac{1}{2} \int_0^t G(x + a\tau) \, d\tau + \frac{1}{2} \int_0^t G(x - a\tau) \, d\tau$$

$$= \frac{1}{2a} \int_x^{x+at} G(s) \, ds - \frac{1}{2a} \int_x^{x-at} G(s) \, ds;$$

and, in terms of the function

(7) $$H(x) = \int_0^x G(s) \, ds \qquad (-\infty < x < \infty),$$

(8) $$y(x,t) = \frac{1}{2a}[H(x + at) - H(x - at)].$$

The verification of the solution in the form (8) is left to the problems.

Superposition of Solutions. If the string is given both a nonzero initial displacement and a nonzero initial velocity,

(9) $$y(x,0) = f(x), \qquad y_t(x,0) = g(x),$$

the displacements $y(x,t)$ can be written as a superposition of the solution (9), Sec. 51, and the solution (8) above:

(10) $$y(x,t) = \frac{1}{2}[F(x + at) + F(x - at)] + \frac{1}{2a}[H(x + at) - H(x - at)].$$

Note that both terms satisfy the homogeneous equations (1) and (2), while their sum clearly satisfies the nonhomogeneous conditions (9). (Compare Problem 5, Sec. 20.)

In general the solution of a linear problem containing more than one nonhomogeneous condition can be written as a sum of solutions of problems each of which contains only one nonhomogeneous condition. The resolution of the original problem in this way, though not an essential step, often simplifies the process of solving the problem.

54. Nonhomogeneous Differential Equations

The substitution of a new unknown function sometimes reduces a nonhomogeneous partial differential equation to one that is homogeneous. Then, if variables can be separated and if the new two-point boundary conditions can be made homogeneous by properly selecting the new function so that a Sturm-Liouville problem arises, our method involving superposition may be effective.

To illustrate, consider the displacements in a stretched string upon which an external force per unit length acts, parallel to the y axis. Let that force be proportional to the distance from one end, and let the initial displacement and velocity be zero. Units for x and t can be chosen so that the problem becomes

$$(1) \qquad\qquad y_{tt}(x,t) = y_{xx}(x,t) + Ax \qquad (0 < x < 1, t > 0),$$

$$(2) \qquad\qquad y(0,t) = y(1,t) = 0, y(x,0) = y_t(x,0) = 0,$$

where A is a constant.

In terms of a new function $Y(x,t)$ where

$$y(x,t) = Y(x,t) + \phi(x)$$

and $\phi(x)$ is yet to be determined, equation (1) becomes

$$Y_{tt}(x,t) = Y_{xx}(x,t) + \phi''(x) + Ax.$$

This will be homogeneous if

$$(3) \qquad\qquad \phi''(x) = -Ax.$$

The two-point boundary conditions on Y are

$$Y(0,t) + \phi(0) = 0, \qquad Y(1,t) + \phi(1) = 0;$$

they are homogeneous if

$$(4) \qquad\qquad \phi(0) = 0, \qquad \phi(1) = 0.$$

From equations (3) and (4) we find that

$$(5) \qquad\qquad \phi(x) = \frac{A}{6} x(1 - x^2) \qquad\qquad (0 \le x \le 1).$$

The boundary value problem in Y now consists of the equations $Y_{tt} = Y_{xx}$, $Y(0,t) = Y(1,t) = 0$, and

$$Y(x,0) = -\phi(x), \qquad Y_t(x,0) = 0.$$

This is a special case of the problem solved in Sec. 51. The solution of our problem can therefore be written

$$(6) \qquad\qquad y(x,t) = \phi(x) - \frac{1}{2}[\Phi(x + t) + \Phi(x - t)]$$

where $\Phi(x)$ is defined for all finite x as the odd periodic extension of $\phi(x)$, with period 2.

Analogous substitutions can sometimes be made in order to transform nonhomogeneous two-point boundary conditions into homogeneous ones, so that a Sturm-Liouville problem arises. (See Problem 11, Sec. 55, and also Sec. 57.)

55. Elastic Bar

A cylindrical bar of natural length c is initially stretched by an amount bc (Fig. 17) and at rest. The initial longitudinal displacements of its sections are then proportional to the distance from the fixed end $x = 0$. At the instant $t = 0$ both ends are released and *kept free*. The boundary value problem in the *longitudinal displacements* $y(x,t)$ is (Sec. 5)

$$(1) \qquad\qquad y_{tt}(x,t) = a^2 y_{xx}(x,t) \qquad (0 < x < c,\, t > 0;\, a^2 = E/\delta),$$

$$(2) \qquad\qquad y_x(0,t) = 0, \qquad y_x(c,t) = 0,$$

$$(3) \qquad\qquad y(x,0) = bx, \qquad y_t(x,0) = 0.$$

The homogeneous two-point boundary conditions (2) state that the force per unit area $E\,\partial y/\partial x$ on the end sections is zero.

Functions $X(x)T(t)$ satisfy all the homogeneous equations above when X is an eigenfunction of the problem

$$(4) \qquad X''(x) + \lambda X(x) = 0, \qquad X'(0) = 0, \qquad X'(c) = 0$$

and when, for the same eigenvalue λ,

$$(5) \qquad\qquad T''(t) + \lambda a^2 T(t) = 0, \qquad T'(0) = 0.$$

The Sturm-Liouville problem (4) generates the eigenfunctions used in Fourier cosine series on the interval $(0,c)$. The eigenvalues, all real, are $\lambda_0 = 0$, $\lambda_n = n^2\pi^2/c^2$ $(n = 1, 2, \ldots)$; also $X_0(x) = 1$, $T_0(t) = 1$ and

$$X_n(x) = \cos\frac{n\pi x}{c}, \qquad T_n(t) = \cos\frac{n\pi a t}{c}. \qquad (n = 1, 2, \ldots).$$

Formally, then, the generalized linear combination

$$y(x,t) = \frac{1}{2}a_0 + \sum_{n=1}^{\infty} a_n \cos\frac{n\pi x}{c}\cos\frac{n\pi a t}{c}$$

satisfies all the equations (1) through (3) provided that

$$(6) \qquad\qquad bx = \frac{1}{2}a_0 + \sum_{n=1}^{\infty} a_n \cos\frac{n\pi x}{c} \qquad (0 < x < c).$$

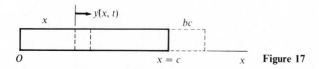

0 $x = c$ x **Figure 17**

The function bx is such that it is represented by the Fourier cosine series (6) on the interval $0 \le x \le c$, where

$$a_n = \frac{2b}{c} \int_0^c x \cos \frac{n\pi x}{c} \, dx \qquad (n = 0, 1, 2, \ldots).$$

We find that $a_0 = bc$, $a_n = -2bc[1 - (-1)^n]/(n\pi)^2$ $(n = 1, 2, \ldots)$; hence

$$(7) \quad y(x,t) = \frac{1}{2} bc - \frac{4bc}{\pi^2} \sum_{n=1}^{\infty} \frac{1}{(2n-1)^2} \cos \frac{(2n-1)\pi x}{c} \cos \frac{(2n-1)\pi at}{c}.$$

If $P(x)$ denotes the *even* periodic extension, with period $2c$, of the function bx $(0 \le x \le c)$, equation (7) can be written

$$(8) \qquad y(x,t) = \frac{1}{2} [P(x + at) + P(x - at)].$$

This reduction and its verification are left as a problem.

PROBLEMS

1. Show that, for each fixed x, the displacements y given by equation (10), Sec. 53, are periodic functions of t with period $2c/a$.

2. Show that the motion of each cross section of the elastic bar treated in Sec. 55 is periodic in t with period $2c/a$.

3. A string stretched between the points $(0,0)$ and $(\pi,0)$ is initially straight with velocity $y_t(x,0) = b \sin x$, where b is a constant. Write the boundary value problem in $y(x,t)$, solve it, and verify the solution.

$$\textit{Ans.} \quad y = \frac{b}{a} \sin x \sin at.$$

4. A wire stretched between the points $(0,0)$ and $(1,0)$ of a horizontal x axis is initially at rest along that axis, its subsequent motion being caused by the force of gravity. If the y axis is positive upwards, the equation of motion is $y_{tt} = a^2 y_{xx} - g$, where g is the acceleration due to gravity. Write the boundary value problem in the vertical displacements $y(x,t)$ and solve it. Show that

$$y(x,t) = \phi(x) - \frac{1}{2} [\Phi(x + at) + \Phi(x - at)]$$

where the function Φ is the odd periodic extension, with period 2, of the function

$$\phi(x) = \frac{g}{2a^2} x(x - 1) \qquad (0 \le x \le 1).$$

5. Display graphically the periodic functions $G(x)$ and $H(x)$ in Sec. 53 when all points of the string there have the same initial velocity $g(x) = v_0$. Then, using expression (8), Sec. 53, indicate some instantaneous positions of the string by means of bold line segments, which will be similar to those in Fig. 16.

6. Verify the solution (8), Sec. 53. Show that the function H defined by equation (7) in that section is continuous, even, and periodic and that $H'(x) = G(x)$ wherever G is continuous.

7. In Sec. 53 we noted that

$$G(s) = \sum_{n=1}^{\infty} b_n \sin \frac{n\pi s}{c} \qquad (-\infty < s < \infty).$$

Integrate that series (Sec. 48) to show that

$$H(x) = \frac{c}{\pi} \sum_{n=1}^{\infty} \frac{1}{n} b_n \left(1 - \cos \frac{n\pi x}{c}\right) \qquad (-\infty < x < \infty)$$

and hence that the function (8) in Sec. 53 is represented by the series (6) there.

8. Reduce the representation (7), Sec. 55, of the function $y(x,t)$ to the form (8) in that section and verify the solution in that form.

9. From expression (8), Sec. 55, show that $y(0,t) = P(at)$ and hence that the end $x = 0$ of the bar moves with constant velocity ab during the half period $0 < t < c/a$ and with velocity $-ab$ during the next half period.

10. Let the function $y(x,t)$ satisfying conditions (1) through (3), Sec. 55, be interpreted as representing transverse displacements in a stretched string, where $a^2 = H/\delta$. Describe the initial and end conditions for the string. [Note that conditions (2), Sec. 55, imply that no vertical force acts at the ends, as if the ends were looped about perfectly smooth rods along the lines $x = 0$ and $x = c$.]

11. The end $x = 0$ of an elastic bar is free, and a constant longitudinal force F_0 per unit area is applied at the end $x = c$. The bar is initially unstrained and at rest. Set up the boundary value problem for the longitudinal displacements $y(x,t)$, the condition at the end $x = c$ being $y_x(c,t) = F_0/E$ (Sec. 5).

(a) After noting that the method of separation of variables cannot be applied directly to find $y(x,t)$, make the substitution

$$y(x,t) = Y(x,t) + \phi(x) + \psi(t)$$

in the above boundary value problem and determine $\phi(x)$ and $\psi(t)$ so that Y satisfies the wave equation and the boundary conditions

$$Y_x(0,t) = 0, \qquad Y_x(c,t) = 0,$$

$$Y(x,0) = -\phi(x), \qquad Y_t(x,0) = 0.$$

Then show that

$$y(x,t) = \frac{F_0}{2cE}(x^2 + a^2t^2) - \frac{1}{2}[\Phi(x + at) + \Phi(x - at)]$$

where $\Phi(x)$ is the even periodic extension, with period $2c$, of the function

$$\phi(x) = \frac{F_0}{2cE} x^2.$$

(b) Use the result in part (a) to show that the end $x = 0$ of the bar remains at rest until time $t = c/a$ and then moves with velocity $v_0 = 2aF_0/E$ when $c/a < t < 3c/a$, with velocity $2v_0$ when $3c/a < t < 5c/a$, and so on.

56. Temperatures in a Bar

The lateral surface of a solid right cylinder of length π is insulated. The initial temperature distribution within the bar is a prescribed function f of the distance x from the end $x = 0$. At the instant $t = 0$ the temperature of both

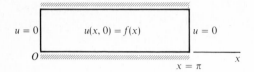

Figure 18

ends $x = 0$ and $x = \pi$ is brought to zero and kept at that value (Fig. 18). If no heat is generated in the solid, the temperatures should be the values of a function $u(x,t)$ that satisfies the heat equation

(1) $$u_t(x,t) = ku_{xx}(x,t) \qquad (0 < x < \pi, \; t > 0)$$

and the boundary conditions

(2) $$u(0+,t) = 0, \qquad u(\pi-,t) = 0 \qquad (t > 0),$$

(3) $$u(x,0+) = f(x) \qquad (0 < x < \pi).$$

The constant k is the thermal diffusivity of the material (Sec. 6). We have written the boundary conditions as limits in order to indicate continuity properties that should be satisfied by u. The problem is also that of determining temperatures $u(x,t)$ in a slab bounded by the planes $x = 0$ and $x = \pi$ and initially at temperature $f(x)$, with its faces kept at temperature zero.

By separation of variables we find that $X(x)T(t)$ satisfies the homogeneous equations (1) and (2) if

(4) $$X''(x) + \lambda X(x) = 0, \qquad X(0) = 0, \qquad X(\pi) = 0,$$

(5) $$T'(t) + \lambda k T(t) = 0.$$

The Sturm-Liouville problem (4) has eigenvalues $\lambda = n^2$ and eigenfunctions $X = \sin nx \; (n = 1, 2, \ldots)$. The corresponding functions from equation (5) are $T = \exp(-n^2 kt)$. Formally, then, the function

(6) $$u(x,t) = \sum_{n=1}^{\infty} b_n \exp(-n^2 kt) \sin nx$$

satisfies all equations in the problem, including (3), if

(7) $$f(x) = \sum_{n=1}^{\infty} b_n \sin nx \qquad (0 < x < \pi).$$

We assume that f and f' are piecewise continuous. Then f is represented by its Fourier sine series (7), where

(8) $$b_n = \frac{2}{\pi} \int_0^{\pi} f(x) \sin nx \; dx \qquad (n = 1, 2, \ldots).$$

Expression (6), with coefficients b_n defined by equation (8), is our formal solution of the boundary value problem.

(a) *Verification.* According to Sec. 27, $b_n \to 0$ as $n \to \infty$. Hence those coefficients are bounded for all n; that is, $|b_n| < B$ where B is some positive constant. So whenever $t \geq t_0$, where t_0 is a positive constant,

$$|b_n \exp(-n^2 kt) \sin nx| < B \exp(-n^2 kt_0).$$

Since the infinite series with constant terms $\exp(-n^2 kt_0)$ converges, according to the ratio test, the Weierstrass M-test ensures the uniform convergence of series (6) with respect to x and t when $0 \leq x \leq \pi, t \geq t_0 > 0$. The terms of that series are continuous functions; hence series (6) converges to a continuous function $u(x,t)$ when $t \geq t_0$, or whenever $t > 0$ since t_0 is an arbitrary positive number. In particular, $u(0+,t) = u(0,t)$ when $t > 0$; and since $u(0,t) = 0$, the first of conditions (2) is satisfied by the function u defined by equation (6). Similarly, the second of those conditions is satisfied.

The series with terms $n \exp(-n^2 kt_0)$, or $n^2 \exp(-n^2 kt_0)$, also converges. Hence series (6) can be differentiated twice with respect to x and once with respect to t, when $t > 0$, because the series of derivatives converge uniformly when $t \geq t_0$. But the terms of series (6) satisfy the heat equation (1), and so the sum $u(x,t)$ of the series satisfies that linear homogeneous differential equation.

It remains to show that u satisfies the initial condition (3). This can be done as follows, with the aid of a convergence test due to Abel† and to be derived in Chap. 10. For each fixed x $(0 < x < \pi)$ the series with terms $b_n \sin nx$ converges to $f(x)$; at a point of discontinuity we define $f(x)$ as the mean of the values $f(x+)$ and $f(x-)$. According to Abel's test, the new series formed by multiplying the terms of a convergent series by corresponding terms of a bounded sequence of functions of t whose values never increase with n, such as $\exp(-n^2 kt)$, is uniformly convergent with respect to t. Our series (6) therefore converges uniformly with respect to t when $t \geq 0$.

It follows that our function u is continuous in t $(t \geq 0)$; thus $u(x,0+) = u(x,0)$. Since $u(x,0) = f(x)$, condition (3) is satisfied. The function u defined by equation (6) is now verified as a solution.

(b) *Uniqueness.* Consider the somewhat special case in which f is continuous $(0 \leq x \leq \pi), f(0) = f(\pi) = 0$, and f' is piecewise continuous, so that the Fourier sine series (7) converges uniformly to $f(x)$. Then Abel's test shows uniform convergence of series (6) with respect to x and t together in the region $0 \leq x \leq \pi, t \geq 0$. Consequently, our function u is continuous in that region. As before, u_t, u_x, and u_{xx} are continuous when $t > 0$ and $0 \leq x \leq \pi$.

A simple uniqueness theorem for solutions of such boundary value problems is proved in Chap. 10. It shows that not more than one function $U(x,y,z,t)$ can satisfy the heat equation throughout the interior of a closed bounded region R in space, such as our cylindrical region, and meet the

† Niels H. Abel (pronounced Aȟ-bel), Norwegian, 1802–1829.

following requirements: U is to be a continuous function of x, y, z, and t together when $t \geq 0$ and when (x,y,z) is in R, where R includes its boundary surface; the derivatives of U that appear in the heat equation are continuous in R when $t > 0$; U is prescribed on part of the boundary and the normal derivative of U is prescribed on the rest when $t > 0$.

Our function $u(x,t)$, which is independent of y and z, satisfies all those requirements, including conditions (1) through (3). For each fixed t the gradient of u is parallel to the x axis; so the normal derivative of u at the lateral surface of the cylinder, or the flux of heat across that surface, is zero. Hence expression (6) represents *the* solution of our problem that satisfies the continuity conditions stated. We have now justified the assumption that u is a function of x and t only.

57. Other Boundary Conditions

We point out modifications of the procedure that are useful when the slab $0 \leq x \leq \pi$, or the bar with insulated lateral surface, is subjected to other simple thermal conditions at its boundary surfaces $x = 0$ and $x = \pi$. In each case the temperature function is to satisfy the heat equation in x and t within the slab:

$$(1) \qquad\qquad u_t(x,t) = ku_{xx}(x,t) \qquad\qquad (0 < x < \pi, t > 0).$$

(a) *One Face at Temperature u_0.* If the slab is initially at temperature zero throughout and the face $x = 0$ is kept at that temperature while the face $x = \pi$ is kept at a constant temperature u_0 when $t > 0$, then

$$(2) \qquad\qquad u(0,t) = 0, \qquad u(\pi,t) = u_0, \qquad u(x,0) = 0.$$

The boundary value problem consisting of equations (1) and (2) is not in proper form for the method of separation of variables because one of the two-point boundary conditions is not homogeneous. If we write (compare Sec. 54)

$$u(x,t) = U(x,t) + \phi(x),$$

however, those equations become

$$U_t(x,t) = kU_{xx}(x,t) + k\phi''(x)$$

and

$$U(0,t) + \phi(0) = 0, \qquad U(\pi,t) + \phi(\pi) = u_0, \qquad U(x,0) + \phi(x) = 0.$$

We thus find that if $\phi(x) = u_0 x/\pi$, the function U satisfies the conditions

$$(3) \qquad U_t = kU_{xx}, \qquad U(0,t) = U(\pi,t) = 0, \qquad U(x,0) = -\phi(x).$$

This is a special case of the problem solved in Sec. 56, where $f(x) = -u_0 x/\pi$.

Hence we obtain the result

(4) $$u(x,t) = \frac{u_0}{\pi}\left[x + 2\sum_{n=1}^{\infty}\frac{(-1)^n}{n}\exp\left(-n^2kt\right)\sin nx\right].$$

Note that the function $\phi(x)$ represents the *steady-state* temperatures in the slab.

   (*b*) *Insulated Faces.* The flux of heat across the faces $x = 0$ and $x = \pi$ is the value of $-Ku_x$ at those faces, where K is the thermal conductivity (Sec. 6). If both faces of the slab are insulated and the initial temperature is $f(x)$, the boundary value problem consists of equation (1) and the conditions

(5) $$u_x(0,t) = 0, \qquad u_x(\pi,t) = 0, \qquad u(x,0) = f(x) \qquad (0 < x < \pi).$$

   Separation of variables leads to the Sturm-Liouville problem associated with the Fourier cosine series,

$$X''(x) + \lambda X(x) = 0, \qquad X'(0) = 0, \qquad X'(\pi) = 0,$$

and to the equation $T'(t) + \lambda kT(t) = 0$. The temperature function is then found to be

(6) $$u(x,t) = \frac{1}{2}a_0 + \sum_{n=1}^{\infty}a_n\exp\left(-n^2kt\right)\cos nx$$

where $$a_n = \frac{2}{\pi}\int_0^{\pi}f(x)\cos nx\,dx \qquad (n = 0, 1, 2, \ldots).$$

   (*c*) *One Face Insulated.* If the face $x = \pi$ is insulated while the face $x = 0$ is kept at temperature zero (Fig. 19), then

(7) $$u(0,t) = 0, \qquad u_x(\pi,t) = 0, \qquad u(x,0) = f(x) \qquad (0 < x < \pi),$$

where $f(x)$ is the initial temperature. Upon separating variables, we obtain the Sturm-Liouville problem

$$X''(x) + \lambda X(x) = 0, \qquad X(0) = 0, \qquad X'(\pi) = 0$$

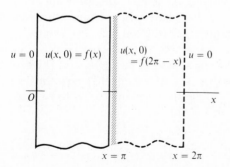

$u = 0$ $\quad u(x, 0) = f(x)$ $\qquad u(x, 0) = f(2\pi - x)$ $\quad u = 0$

$x = \pi$ $\qquad x = 2\pi$ $\qquad$ **Figure 19**

and the equation $T'(t) + \lambda k T(t) = 0$. The eigenvalues and normalized eigen-functions of that problem are found to be

$$(8) \qquad \lambda_n = \frac{(2n-1)^2}{4}, \qquad \phi_n(x) = \sqrt{\frac{2}{\pi}} \sin \frac{(2n-1)x}{2} \qquad (n = 1, 2, \ldots).$$

Note that the orthogonal sine functions here are *not* those used in the Fourier sine series for the interval $(0,\pi)$.

Our generalized linear combination of functions of type XT, written in terms of the functions ϕ_n, is

$$(9) \qquad u(x,t) = \sum_{n=1}^{\infty} c_n \exp\left[-\frac{(2n-1)^2 kt}{4}\right] \sqrt{\frac{2}{\pi}} \sin \frac{(2n-1)x}{2}.$$

It satisfies the condition $u(x,0) = f(x)$ if

$$(10) \qquad f(x) = \sum_{n=1}^{\infty} c_n \sqrt{\frac{2}{\pi}} \sin \frac{(2n-1)x}{2} \qquad (0 < x < \pi).$$

Series (10) is the generalized Fourier series corresponding to f with respect to the eigenfunctions ϕ_n if

$$(11) \qquad c_n = \int_0^{\pi} f(x)\phi_n(x)\, dx = \sqrt{\frac{2}{\pi}} \int_0^{\pi} f(x) \sin \frac{(2n-1)x}{2}\, dx.$$

If we assume that the representation (10) is valid, as proved in references to Sturm-Liouville theory cited in Sec. 31, expression (9) with coefficients (11) represents a formal solution of our problem.

But we can actually establish the special representation (10) with the aid of Fourier sine series on the larger interval $(0,2\pi)$. Let the given function f and its derivative f' be piecewise continuous on the interval $(0,\pi)$. Then let f be defined on $(\pi,2\pi)$ so that its graph is symmetric with respect to the line $x = \pi$; that is, write

$$(12) \qquad f(x) = f(2\pi - x) \qquad \text{when } \pi < x < 2\pi.$$

This procedure is suggested by considering the temperature problem in the slab extended to the plane $x = 2\pi$ with that plane kept at temperature zero (Fig. 19). When the initial temperature distribution has the symmetry property (12), no heat should flow across the midsection $x = \pi$. So the temperature function u should be the one sought when $0 \le x \le \pi$. The new problem is essentially the one solved in Sec. 56 except that now the sine series on the interval $(0,2\pi)$, for the symmetric extension of f, is to be used.

That sine series representation on $(0,2\pi)$ is

$$(13) \qquad f(x) = \sum_{n=1}^{\infty} b_n \sin \frac{nx}{2} \qquad (0 < x < 2\pi)$$

where $b_n = \dfrac{1}{\pi} \displaystyle\int_0^{2\pi} f(x) \sin \dfrac{nx}{2}\, dx$

$$= \frac{1}{\pi} \int_0^\pi f(x) \sin \frac{nx}{2}\, dx + \frac{1}{\pi} \int_\pi^{2\pi} f(2\pi - s) \sin \frac{ns}{2}\, ds.$$

In the last integral we substitute $x = 2\pi - s$ and note that $\sin (ns/2)$ then becomes $-(-1)^n \sin (nx/2)$. Thus

$$b_n = \frac{1}{\pi}[1 - (-1)^n] \int_0^\pi f(x) \sin \frac{nx}{2}\, dx;$$

that is,

(14) $\qquad b_{2n-1} = \dfrac{2}{\pi} \displaystyle\int_0^\pi f(x) \sin \dfrac{(2n-1)x}{2}\, dx, \quad b_{2n} = 0 \qquad (n = 1, 2, \ldots).$

The representation (13) with coefficients (14) is therefore valid for the given function f on the interval $(0,\pi)$. It is the same as the representation (10) with coefficients (11).

PROBLEMS†

1. The faces $x = 0$ and $x = c$ of a slab which is initially at temperatures $f(x)$ are kept at temperature zero (compare Sec. 56). Derive the temperature formula

$$u(x,t) = \frac{2}{c} \sum_{n=1}^\infty \exp\left(-\frac{n^2\pi^2 kt}{c^2}\right) \sin \frac{n\pi x}{c} \int_0^c f(s) \sin \frac{n\pi s}{c}\, ds.$$

2. Let the initial temperature distribution be uniform over the slab in Problem 1, so that $f(x) = u_0$. Write the formula for $u(x,t)$ and find the flux $-Ku_x(x_0,t)$ across a plane $x = x_0$ $(0 \le x_0 \le c, t > 0)$. Show that no heat flows across the center plane $x = c/2$.

3. Suppose that $f(x) = \sin (\pi x/c)$ in Problem 1. Find $u(x,t)$ and verify the result fully.

$$\text{Ans.} \quad u(x,t) = \exp\left(-\frac{\pi^2 kt}{c^2}\right) \sin \frac{\pi x}{c}.$$

4. (a) Show that if A is a constant and $f(x) = A$ when $0 < x < c/2$ and $f(x) = 0$ when $c/2 < x < c$, the temperature formula in Problem 1 becomes

$$u(x,t) = \frac{4A}{\pi} \sum_{n=1}^\infty \frac{\sin^2 (n\pi/4)}{n} \exp\left(-\frac{n^2\pi^2 kt}{c^2}\right) \sin \frac{n\pi x}{c}.$$

(b) Two slabs of iron ($k = 0.15$ centimeter-gram-second unit), each 20 cm thick, are such that one is at temperature 100°C and the other at 0°C throughout. They are placed face to face in perfect contact, and their outer faces are kept at 0°C. Compute the temperature at their common face 10 min after contact has been made to show that it is then approximately 36°C.

---

† Only formal solutions of the boundary value problems here and in the sets of problems to follow are expected, unless the problem states that the solution found is to be fully verified. Partial verification is often easy and helpful.

(c) Show that if the slabs in part (b) are made of concrete ($k = 0.005$ cgs unit), it takes 5 hr for the common face to reach that temperature of 36°C. [Note that $u(x,t)$ depends on the product kt.]

5. The faces $x = 0$ and $x = c$ of a slab which is initially at temperatures $f(x)$ are insulated [compare Sec. 57(b)]. Derive the temperature formula

$$u(x,t) = \frac{1}{2} a_0 + \sum_{n=1}^{\infty} a_n \exp\left(-\frac{n^2 \pi^2 kt}{c^2}\right) \cos \frac{n\pi x}{c}$$

where

$$a_n = \frac{2}{c} \int_0^c f(x) \cos \frac{n\pi x}{c} dx \qquad (n = 0, 1, 2, \ldots).$$

Show how it follows that $u(x,t) = u_0$ when $f(x) = u_0$, where u_0 is a constant.

6. Verify the result in Problem 5 [compare Sec. 56(a)].

7. Let $v(x,t)$ denote temperatures in a slender wire lying along the x axis. Variations of the temperature over each cross section are to be neglected. At the lateral surface the linear law of surface heat transfer between the wire and its surroundings is assumed to apply (see Problem 7, Sec. 7). Let the surroundings be at temperature zero; then

$$v_t(x,t) = k v_{xx}(x,t) - h v(x,t) \qquad (h \text{ constant}, h > 0).$$

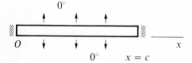

Figure 20

The ends $x = 0$ and $x = c$ of the wire are insulated (Fig. 20) and the initial temperature distribution is $f(x)$. Solve the boundary value problem for v and then show that

$$v(x,t) = e^{-ht} u(x,t)$$

where u is the temperature function found in Problem 5.

8. Use the substitution $v(x,t) = e^{-ht} u(x,t)$ to reduce the boundary value problem in Problem 7 to the one in Problem 5.

9. Assuming that the ends of the wire in Problem 7 are not insulated, but kept at temperature zero instead, find the temperature function.

10. Heat is generated at a constant rate uniformly throughout a slab which is initially at temperatures $f(x)$ and whose faces $x = 0$ and $x = \pi$ are kept at temperature zero. Then the heat equation has the form (Sec. 7)

$$u_t(x,t) = k u_{xx}(x,t) + C$$

where C is a positive constant. Solve the boundary value problem to obtain the temperature formula

$$u(x,t) = \frac{Cx(\pi - x)}{2k} + \sum_{n=1}^{\infty} b_n \exp\left(-n^2 kt\right) \sin nx$$

where

$$b_n = \frac{2}{\pi} \int_0^\pi \left[\frac{Cx(x - \pi)}{2k} + f(x)\right] \sin nx \, dx.$$

In particular, find $u(x,t)$ when $f(x) = Cx(\pi - x)/(2k)$.

11. Show that when $f(x) = 0$ $(0 < x < \pi)$ in Problem 10, the temperature formula there can be written

$$u(x,t) = \frac{4C}{\pi k} \sum_{n=1}^{\infty} \frac{1 - \exp\left[-(2n-1)^2 kt\right]}{(2n-1)^3} \sin\,(2n-1)x.†$$

Suggestion: Note that when $f(x) = 0$ $(0 < x < \pi)$, the constants b_n in Problem 10 are the coefficients in the Fourier sine series representation of $Cx(x - \pi)/(2k)$ on the interval $(0,\pi)$. That representation is readily obtained by combining the series found in Problem 1, Sec. 38, and Problem 4, Sec. 44.

12. Solve this boundary value problem for $u(x,t)$:

$$u_t(x,t) = u_{xx}(x,t) - hu(x,t) \quad (0 < x < \pi, \, t > 0; \, h \text{ constant}, \, h > 0),$$

$$u(0,t) = 0, \qquad u(\pi,t) = 1 \qquad\qquad\qquad (t > 0),$$

$$u(x,0) = 0 \qquad\qquad\qquad\qquad (0 < x < \pi).$$

Ans. $u(x,t) = \dfrac{\sinh\,(x\sqrt{h})}{\sinh\,(\pi\sqrt{h})} + \dfrac{2}{\pi} \exp\,(-ht) \displaystyle\sum_{n=1}^{\infty} \dfrac{(-1)^n n}{n^2 + h} \exp\,(-n^2 t)\, \sin\, nx.$

13. Use superposition of known solutions (Secs. 56 and 57) to write a formula for temperatures in a slab whose faces $x = 0$ and $x = \pi$ are kept at temperatures 0 and u_0, respectively, and whose initial temperature distribution is $f(x)$.

14. The face $x = 0$ of a slab is kept at temperature zero while heat is supplied or extracted at the other face $x = \pi$ at a constant rate per unit area so that $u_x(\pi,t) = A$. The slab is initially at temperature zero and the unit of time is chosen so that we can write $k = 1$. Derive the temperature formula

$$u(x,t) = Ax + \frac{8A}{\pi} \sum_{n=1}^{\infty} \frac{(-1)^n}{(2n-1)^2} \exp\left[-\frac{(2n-1)^2 t}{4}\right] \sin\frac{(2n-1)x}{2}$$

with the aid of the series representation (10), Sec. 57.

15. Let f and f' be piecewise continuous on an interval $(0,c)$. By the process used in Sec. 57(c), establish the representation

$$f(x) = \frac{2}{c} \sum_{n=1}^{\infty} \sin\frac{(2n-1)\pi x}{2c} \int_0^c f(s) \sin\frac{(2n-1)\pi s}{2c}\, ds \qquad (0 < x < c)$$

of f in terms of the eigenfunctions of the Sturm-Liouville problem

$$X''(x) + \lambda X(x) = 0, \qquad X(0) = 0, \qquad X'(c) = 0.$$

58. Dirichlet Problems

As already noted in Sec. 10, a boundary value problem in u is said to be a Dirichlet problem when it consists of Laplace's equation $\nabla^2 u = 0$, which states that u is harmonic in the interior of a given region, together with

---

† This result occurs, for example, in the theory of gluing of wood with the aid of radio-frequency heating. See G. H. Brown, *Proc. Inst. Radio Engrs.*, vol. 31, no. 10, pp. 537–548, 1943, where operational methods are used.

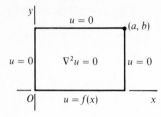

Figure 21

prescribed values of u on the boundaries. We now illustrate the use of the method of separation of variables in solving such problems for regions in the plane.

(a) *Cartesian coordinates.* Let u be harmonic in the interior of a rectangular region $0 \le x \le a$, $0 \le y \le b$, so that

$$(1) \qquad u_{xx}(x,y) + u_{yy}(x,y) = 0 \qquad (0 < x < a, 0 < y < b),$$

with these values prescribed on the boundary (Fig. 21):

$$(2) \qquad u(0,y) = 0, \qquad u(a,y) = 0 \qquad (0 < y < b),$$
$$(3) \qquad u(x,0) = f(x), \qquad u(x,b) = 0 \qquad (0 < x < a).$$

Separation of variables leads to the Sturm-Liouville problem

$$(4) \qquad X''(x) + \lambda X(x) = 0, \qquad X(0) = 0, \qquad X(a) = 0,$$

where $\lambda = n^2\pi^2/a^2$ and $X = \sin(n\pi x/a)$ $(n = 1, 2, \ldots)$, and to the conditions

$$(5) \qquad Y''(y) - \left(\frac{n\pi}{a}\right)^2 Y(y) = 0, \qquad Y(b) = 0.$$

Thus Y can be written $Y = \sinh[n\pi(b - y)/a]$, and the function

$$(6) \qquad u(x,y) = \sum_{n=1}^{\infty} B_n \sinh \frac{n\pi(b - y)}{a} \sin \frac{n\pi x}{a}$$

formally satisfies all the equations (1) through (3) if

$$(7) \qquad f(x) = \sum_{n=1}^{\infty} B_n \sinh \frac{n\pi b}{a} \sin \frac{n\pi x}{a} \qquad (0 < x < a).$$

We assume that f and f' are piecewise continuous. Then series (7) is the Fourier sine series representation of f on the interval $(0,a)$ provided $B_n \sinh(n\pi b/a) = b_n$ where

$$b_n = \frac{2}{a} \int_0^a f(x) \sin \frac{n\pi x}{a} \, dx.$$

The function defined by equation (6) with coefficients

$$(8) \qquad B_n = \frac{2}{a \sinh{(n\pi b/a)}} \int_0^a f(x) \sin\frac{n\pi x}{a}\, dx$$

is therefore our formal solution. The result can be verified by the method used in Sec. 56. We defer the verification, however, till Chap. 10, where uniqueness is also considered.

If y is replaced by the new variable $b - y$ in the above problem and solution, and if $f(x)$ is replaced by $g(x)$, the nonhomogeneous condition satisfied by u becomes $u(x,b) = g(x)$. An interchange of x and y places the nonhomogeneous conditions at the boundaries $x = 0$ or $x = a$. Superposition of the four solutions then gives the harmonic function whose values are prescribed as functions of position along the entire boundary of the rectangle.

From the equations (1) through (3) we note that $u(x,y)$ represents the *steady-state temperatures* in a rectangular plate with insulated faces when $u = f(x)$ on the edge $y = 0$ and $u = 0$ on the other three edges. The function u also represents the *electrostatic potential* in the space bounded by the planes $x = 0$, $x = a$, $y = 0$, and $y = b$ when the space is free of charges and the planar boundaries are kept at potentials given by conditions (2) and (3). If f is continuous and $f(0) = f(a) = 0$, $u(x,y)$ represents the *static transverse displacements in a membrane* stretched over a rectangular frame after the supporting side along $y = 0$ is displaced according to the equation $u(x,0) = f(x)$.

(*b*) *Cylindrical coordinates.* The same Sturm-Liouville problem can arise in seemingly unrelated boundary value problems. Problem (4) above, for example, first arose in finding transverse displacements of a string (Sec. 15); and a special case of it arose in finding time-dependent temperatures in a bar (Sec. 56). As we shall now see, the same Sturm-Liouville problem can arise even when a different coordinate system is used.

Let $u(\rho,\phi)$ denote a function of the cylindrical coordinates ρ and ϕ which is harmonic in the region $1 < \rho < b$, $0 < \phi < \pi$ of the plane $z = 0$ (Fig. 22). That is (Sec. 9),

$$(9) \qquad \rho^2 u_{\rho\rho}(\rho,\phi) + \rho u_\rho(\rho,\phi) + u_{\phi\phi}(\rho,\phi) = 0 \qquad (1 < \rho < b, 0 < \phi < \pi).$$

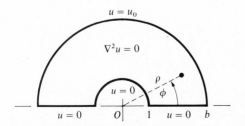

Figure 22

Suppose further that

(10)
$$u(\rho,0) = 0, \qquad u(\rho,\pi) = 0 \qquad (1 < \rho < b),$$

(11)
$$u(1,\phi) = 0, \qquad u(b,\phi) = u_0 \qquad (0 < \phi < \pi),$$

where u_0 is a constant.

Substituting $u = R(\rho)\Phi(\phi)$ into the homogeneous equations and separating variables, we find that

(12)
$$\rho^2 R''(\rho) + \rho R'(\rho) - \lambda R(\rho) = 0, \qquad R(1) = 0$$

and

(13)
$$\Phi''(\phi) + \lambda\Phi(\phi) = 0, \qquad \Phi(0) = 0, \qquad \Phi(\pi) = 0.$$

Evidently, then, $\lambda = n^2$ and $\Phi = \sin n\phi$ ($n = 1, 2, \ldots$). The differential equation in R is a Cauchy-Euler equation [see Problem 13(a), Sec. 34], and it follows from equations (12) that $R = \rho^n - \rho^{-n}$ when $\lambda = n^2$. Thus, formally,

$$u(\rho,\phi) = \sum_{n=1}^{\infty} B_n(\rho^n - \rho^{-n}) \sin n\phi$$

where, according to the second of conditions (11), the constants B_n are such that

$$u_0 = \sum_{n=1}^{\infty} B_n(b^n - b^{-n}) \sin n\phi \qquad (0 < \phi < \pi).$$

It readily follows from the Fourier sine series expansion [see Problem 9(b), Sec. 38]

$$u_0 = \frac{4u_0}{\pi} \sum_{n=1}^{\infty} \frac{\sin (2n-1)\phi}{2n-1} \qquad (0 < \phi < \pi)$$

that

$$B_{2n}(b^{2n} - b^{-2n}) = 0 \qquad (n = 1, 2, \ldots)$$

and

$$B_{2n-1}[b^{2n-1} - b^{-(2n-1)}] = \frac{4u_0}{\pi} \frac{1}{2n-1} \qquad (n = 1, 2, \ldots).$$

The solution of this Dirichlet problem is therefore

(14)
$$u(\rho,\phi) = \frac{4u_0}{\pi} \sum_{n=1}^{\infty} \left[\frac{\rho^{2n-1} - \rho^{-(2n-1)}}{b^{2n-1} - b^{-(2n-1)}} \right] \frac{\sin (2n-1)\phi}{2n-1}.$$

59. Fourier Series in Two Variables

Let $z(x,y,t)$ denote the transverse displacement at each point (x,y) at time t in a membrane stretched across a rigid square frame in the xy plane. To simplify the notation, we select the origin and the point (π,π) as ends of a diagonal of the frame. If the membrane is released at rest with a given initial displacement $f(x,y)$ which is continuous and vanishes on the boundary of the square, then

(1) $$z_{tt} = a^2(z_{xx} + z_{yy}) \quad (0 < x < \pi, \, 0 < y < \pi, \, t > 0),$$

(2) $$z(0,y,t) = z(\pi,y,t) = z(x,0,t) = z(x,\pi,t) = 0,$$

(3) $$z(x,y,0) = f(x,y), \quad z_t(x,y,0) = 0 \quad (0 \le x \le \pi, \, 0 \le y \le \pi).$$

Let us assume that the partial derivatives $f_x(x,y)$ and $f_y(x,y)$ are also continuous.

Functions of the type $X(x)Y(y)T(t)$ satisfy equation (1) if

$$\frac{X''(x)}{X(x)} = \frac{T''(t)}{a^2 T(t)} - \frac{Y''(y)}{Y(y)}.$$

Thus $X''/X = -\lambda$ and $Y''/Y = T''/(a^2T) + \lambda$, where λ is a constant; also $Y''/Y = -\mu$ and $T''/(a^2T) + \lambda = -\mu$, where μ is a constant independent of λ. Separation of variables in the homogeneous equations therefore leads to the two Sturm-Liouville problems

$$X''(x) + \lambda X(x) = 0, \qquad X(0) = 0, \qquad X(\pi) = 0,$$

$$Y''(y) + \mu Y(y) = 0, \qquad Y(0) = 0, \qquad Y(\pi) = 0$$

and to the conditions

$$T''(t) + a^2(\lambda + \mu)T(t) = 0, \qquad T'(0) = 0.$$

Consequently $\lambda = m^2$, $X = \sin mx$ $(m = 1, 2, \ldots)$ and $\mu = n^2$, $Y = \sin ny$ $(n = 1, 2, \ldots)$; also, $T = \cos(at\sqrt{m^2 + n^2})$.

The formal solution of our problem is therefore

(4) $$z(x,y,t) = \sum_{m=1}^{\infty} \sum_{n=1}^{\infty} B_{mn} \sin mx \sin ny \cos(at\sqrt{m^2 + n^2})$$

where the coefficients B_{mn} need to be determined so that

(5) $$f(x,y) = \sum_{m=1}^{\infty} \sum_{n=1}^{\infty} B_{mn} \sin mx \sin ny$$

when $0 \le x \le \pi$ and $0 \le y \le \pi$. By grouping terms in this iterated sine series so as to display the total coefficient of sin mx for each m, we can write, formally,

$$(6) \qquad f(x,y) = \sum_{m=1}^{\infty} \left(\sum_{n=1}^{\infty} B_{mn} \sin ny \right) \sin mx.$$

For each fixed y $(0 \leq y \leq \pi)$ equation (6) is the Fourier sine series representation of the function $f(x,y)$ of the variable x $(0 \leq x \leq \pi)$, provided that

$$(7) \qquad \sum_{n=1}^{\infty} B_{mn} \sin ny = \frac{2}{\pi} \int_0^{\pi} f(x,y) \sin mx \, dx \qquad (m = 1, 2, \ldots).$$

The right-hand side here is a sequence of functions $F_m(y)$ $(m = 1, 2, \ldots)$, each represented by its Fourier sine series (7) on the interval $(0,\pi)$ when

$$B_{mn} = \frac{2}{\pi} \int_0^{\pi} F_m(y) \sin ny \, dy \qquad (n = 1, 2, \ldots).$$

The coefficients B_{mn} therefore have the values

$$(8) \qquad B_{mn} = \frac{4}{\pi^2} \int_0^{\pi} \sin ny \int_0^{\pi} f(x,y) \sin mx \, dx \, dy.$$

A formal solution of our membrane problem is given by equation (4) with the coefficients defined by equation (8).

Since the numbers $\sqrt{m^2 + n^2}$ do not change by integral multiples of some fixed number as m and n vary through integral values, the cosine functions in equation (4) have no common period in the variable t; so the displacement z is not generally a periodic function of t. Consequently the vibrating membrane, in contrast to the vibrating string, generally does not produce a musical note. It can be made to do so, however, by giving it the proper initial displacement. If, for example,

$$z(x,y,0) = A \sin x \sin y$$

where A is a constant, the displacements (4) are given by a single term

$$z(x,y,t) = A \sin x \sin y \cos (a\sqrt{2}\,t);$$

then z is periodic in t with period $\pi\sqrt{2}/a$.

PROBLEMS

1. One edge of a square plate with insulated faces is kept at a uniform temperature u_0, and the other three edges are kept at temperature zero. Without solving a boundary value problem, but by superposition of solutions of like problems to obtain the trivial case in which all four edges are at temperature u_0, show why the steady temperature at the center of the given plate must be $u_0/4$.

2. A square plate has its faces and its edges $x = 0$ and $x = \pi$ $(0 < y < \pi)$ insulated. Its edges $y = 0$ and $y = \pi$ $(0 < x < \pi)$ are kept at temperatures zero and $f(x)$, respectively. Let $u(x,y)$ denote its steady temperatures and derive the expression

$$u(x,y) = \frac{1}{2\pi} a_0 y + \sum_{n=1}^{\infty} a_n \frac{\sinh ny}{\sinh n\pi} \cos nx$$

where

$$a_n = \frac{2}{\pi} \int_0^\pi f(x) \cos nx \, dx \qquad (n = 0, 1, 2, \ldots).$$

Find $u(x,y)$ when $f(x) = u_0$, where u_0 is a constant.

3. Find the harmonic function $u(x,y)$ in the semi-infinite strip $0 < x < \pi$, $y > 0$ such that

$$u(0,y) = u(\pi,y) = 0 \qquad\qquad\qquad (y > 0),$$

$$u(x,0) = 1 \qquad\qquad\qquad (0 < x < \pi),$$

and $|u(x,y)| < M$ where M is some constant. (This boundedness requirement serves as a condition at the missing upper boundary of the strip.)

$$Ans. \quad u(x,y) = \frac{4}{\pi} \sum_{n=1}^{\infty} \exp\left[-(2n-1)y\right] \frac{\sin(2n-1)x}{2n-1}.$$

4. Given that the imaginary part of $\log(a + ib)$ is $\tan^{-1}(b/a)$ when a and b are real numbers and that

$$\log\frac{1+z}{1-z} = 2\sum_{n=1}^{\infty} \frac{1}{2n-1} z^{2n-1} \qquad (|z| < 1),\dagger$$

write $z = e^{-y}e^{ix}(y > 0)$ and equate imaginary parts in the above expansion to show that the answer to Problem 3 can be written

$$u(x,y) = \frac{2}{\pi} \tan^{-1}\left(\frac{\sin x}{\sinh y}\right) \qquad (0 < x < \pi, y > 0).$$

Verify the answer in this form.

5. Derive an expression for the bounded steady temperatures $u(x,y)$ in a semi-infinite wall $0 < x < c$, $y > 0$ whose planar faces $x = 0$ and $x = c$ are insulated and where $u(x,0) = f(x)$. Assume that f and f' are piecewise continuous on $(0,c)$.

6. Let ρ, ϕ, z be cylindrical coordinates. Find the harmonic function $u(\rho,\phi)$ in the region $1 < \rho < b$, $0 < \phi < \pi/2$ of the plane $z = 0$ when $u = 0$ and $u = f(\phi)$ on the arcs $\rho = 1$, $0 < \phi < \pi/2$ and $\rho = b$, $0 < \phi < \pi/2$, respectively, and $u_\phi(\rho,0) = u_\phi(\rho,\pi/2) = 0$ $(1 < \rho < b)$. Give a physical interpretation of this problem.

$$Ans. \quad u(\rho,\phi) = \frac{1}{2}\frac{\log \rho}{\log b} a_0 + \sum_{n=1}^{\infty} a_n \frac{\rho^{2n} - \rho^{-2n}}{b^{2n} - b^{-2n}} \cos 2n\phi$$

where

$$a_0 = \frac{4}{\pi} \int_0^{\pi/2} f(\phi) \, d\phi$$

and

$$a_n = \frac{4}{\pi} \int_0^{\pi/2} f(\phi) \cos 2n\phi \, d\phi \qquad (n = 1, 2, \ldots).$$

7. Let the faces of a plate in the shape of a wedge $0 \le \rho \le a$, $0 \le \phi \le \alpha$ be insulated. Find the steady temperatures $u(\rho,\phi)$ in the plate when $u = 0$ on the two rays $\phi = 0$, $\phi = \alpha$ $(0 < \rho < a)$

---

† See, for example, Churchill, Brown, and Verhey (1974), listed in the Bibliography.

and $u = f(\phi)$ on the arc $\rho = a$, $0 < \phi < \alpha$. Assume that f and f' are piecewise continuous and that u is bounded.

$$Ans. \quad u(\rho,\phi) = \frac{2}{\alpha} \sum_{n=1}^{\infty} \left(\frac{\rho}{a}\right)^{n\pi/\alpha} \sin\frac{n\pi\phi}{\alpha} \int_0^\alpha f(\theta) \sin\frac{n\pi\theta}{\alpha} d\theta.$$

8. All four faces of an infinitely long rectangular prism bounded by the planes $x = 0$, $x = a$, $y = 0$, and $y = b$ are kept at temperature zero. Let the initial temperature distribution be $f(x,y)$ and derive this expression for the temperatures $u(x,y,t)$ in the prism:

$$u(x,y,t) = \sum_{m=1}^{\infty} \sum_{n=1}^{\infty} B_{mn} \exp\left[-\pi^2 kt\left(\frac{m^2}{a^2} + \frac{n^2}{b^2}\right)\right] \sin\frac{m\pi x}{a} \sin\frac{n\pi y}{b}$$

where
$$B_{mn} = \frac{4}{ab} \int_0^b \sin\frac{n\pi y}{b} \int_0^a f(x,y) \sin\frac{m\pi x}{a} dx\, dy.$$

9. Write $f(x,y) = g(x)h(y)$ in Problem 8 and show that the double series obtained there for u reduces to the product

$$u(x,y,t) = v(x,t)w(y,t)$$

of two series where v and w represent temperatures in the slabs $0 \le x \le a$ and $0 \le y \le b$ with their faces at temperature zero and with initial temperatures $g(x)$ and $h(y)$, respectively.

10. Let the functions $v(x,t)$ and $w(y,t)$ satisfy the heat equation for one-dimensional flow: $v_t = kv_{xx}$, $w_t = kw_{yy}$. Show by differentiation that their product $u = vw$ satisfies the heat equation $u_t = k(u_{xx} + u_{yy})$. Use this fact to arrive at the solution in Problem 9.

60. Observations and Further Examples

Not all linear homogeneous partial differential equations yield to the method of separation of variables. Conditions for the existence of solutions $V = X(x)Y(y)$ of the equation

$$V_{xx}(x,y) + (x + y)^2 V_{yy}(x,y) = 0,$$

for instance, do not lead to differential equations of the Sturm-Liouville type in either X or Y.

Although the separation process does apply to Laplace's equation in rectangular, cylindrical, or spherical coordinates, to certain wave equations and heat equations involving the laplacian operator in those coordinates, and to many other equations, the general test consists of trying out the process on each equation.

(a) *Periodic Boundary Conditions.* The following example will illustrate problems that involve expansions in the basic Fourier series containing both sine and cosine terms.

Let $u(\rho,\phi)$ denote the steady temperatures in a long solid cylinder $\rho \le 1$, $-\infty < z < \infty$ when the temperature of the surface $\rho = 1$ is a given function $f(\phi)$; the variables ρ, ϕ and z are, of course, cylindrical coordinates. Then

(1) $$\rho^2 u_{\rho\rho}(\rho,\phi) + \rho u_\rho(\rho,\phi) + u_{\phi\phi}(\rho,\phi) = 0 \quad (\rho < 1, \ -\pi < \phi \le \pi),$$

(2) $$u(1,\phi) = f(\phi) \qquad (-\pi < \phi < \pi).$$

Also, u and its partial derivatives of the first order are to be continuous interior to the cylinder.

If functions of the type $R(\rho)\Phi(\phi)$ are to satisfy equation (1) and the continuity requirements, we and that

(3) $$\rho^2 R''(\rho) + \rho R'(\rho) - \lambda R(\rho) = 0 \qquad (0 \le \rho < 1),$$

(4) $$\Phi''(\phi) + \lambda\Phi(\phi) = 0, \qquad \Phi(-\pi) = \Phi(\pi), \qquad \Phi'(-\pi) = \Phi'(\pi).$$

The eigenvalues of problem (4), whose boundary conditions are of periodic type, are the numbers $\lambda = 0$ and $\lambda = n^2$ ($n = 1, 2, \ldots$); and the set

$$\{1, \cos x, \cos 2x, \ldots, \sin x, \sin 2x, \ldots\}$$

is the set of corresponding eigenfunctions, orthogonal on the interval $(-\pi,\pi)$. (See Sec. 33.) Equation (3) is of the Cauchy-Euler type. When $\lambda = 0$, $R = A \log \rho + B$; and when $\lambda = n^2$ ($n = 1, 2, \ldots$), the general solution is $R = C_1 \rho^n + C_2 \rho^{-n}$. If R is to be continuous on the axis $\rho = 0$, then $A = C_2 = 0$.

The generalized linear combination of our continuous functions $R\Phi$ is therefore

(5) $$u(\rho,\phi) = \frac{1}{2} a_0 + \sum_{n=1}^{\infty} \rho^n(a_n \cos n\phi + b_n \sin n\phi).$$

This satisfies boundary condition (2) if a_n and b_n, including a_0, are the coefficients in the Fourier series for f on the interval $(-\pi,\pi)$:

(6) $$a_n = \frac{1}{\pi} \int_{-\pi}^{\pi} f(\phi) \cos n\phi \, d\phi, \qquad b_n = \frac{1}{\pi} \int_{-\pi}^{\pi} f(\phi) \sin n\phi \, d\phi.$$

We assume that f satisfies the conditions in our Fourier theorem (Sec. 41).

(b) *Sturm-Liouville Series.* Even when separation of variables applies, the superposition process may lead to a *generalized* Fourier series representation of a given function. Our formal procedure may then include an assumption that the representation is valid.

To illustrate, let $u(x,t)$ denote temperatures in a slab, bounded by two planes $x = 0, x = 1$ and initially at temperatures $f(x)$, when $u = 0$ on the face $x = 0$ and surface heat transfer takes place at the face $x = 1$ into a medium at temperature zero (Fig. 23). The rate of heat transfer per unit area into that medium is assumed to be proportional to $u(1,t) - 0$, so that $u_x(1,t) = -hu(1,t)$ where h is a positive constant (Sec. 7). If the thermal diffusivity k is unity, then

(7) $$u_t(x,t) = u_{xx}(x,t) \qquad (0 < x < 1, \, t > 0),$$

(8) $$u(0,t) = 0, \qquad u_x(1,t) = -hu(1,t), \qquad u(x,0) = f(x).$$

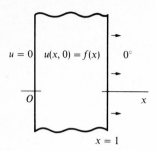

$$x = 1 \qquad \textbf{Figure 23}$$

If we let functions of the type $X(x)T(t)$ satisfy the homogeneous equations here, we have the Sturm-Liouville problem

$$X''(x) + \lambda X(x) = 0, \qquad X(0) = 0, \qquad hX(1) + X'(1) = 0,$$

along with the condition $T'(t) + \lambda T(t) = 0$. The full set of eigenvalues, obtained in Problem 5, Sec. 34, is $\lambda_n = \alpha_n^2$ $(n = 1, 2, \ldots)$ where α_n are the consecutive positive roots of the equation $\tan \alpha = -\alpha/h$; the normalized eigenfunctions are found to be

$$(9) \qquad \phi_n(x) = \left(\frac{2h}{h + \cos^2 \alpha_n} \right)^{1/2} \sin \alpha_n x \qquad (n = 1, 2, \ldots).$$

Hence the formal solution of our problem is

$$(10) \qquad u(x,t) = \sum_{n=1}^{\infty} c_n \exp\left(-\alpha_n^2 t\right)\phi_n(x),$$

where, in order that $u(x,0) = f(x)$ when $0 < x < 1$,

$$(11) \qquad c_n = \int_0^1 f(x)\phi_n(x)\, dx = \left(\frac{2h}{h + \cos^2 \alpha_n} \right)^{1/2} \int_0^1 f(x) \sin \alpha_n x\, dx.$$

We have *assumed* that our function f is represented by its generalized Fourier series with respect to the orthonormal eigenfunctions (9), so that

$$f(x) = \sum_{n=1}^{\infty} c_n \phi_n(x) \qquad (0 < x < 1)$$

where $c_n = (f, \phi_n)$. Representation theorems in Sturm-Liouville theory, proved in references cited in Sec. 31, show that the series does converge to $f(x)$ at each point x $(0 < x < 1)$ where f is continuous, provided that f and f' are piecewise continuous on the interval $(0,1)$.

PROBLEMS

1. Show that when $f(x) = 1$ $(0 < x < 1)$ in the boundary value problem consisting of equations (7) and (8) in Sec. 60, the solution obtained there reduces to

$$u(x,t) = 2h \sum_{n=1}^{\infty} \frac{1 - \cos \alpha_n}{\alpha_n(h + \cos^2 \alpha_n)} \exp\left(-\alpha_n^2 t\right) \sin \alpha_n x.$$

2. Solve the boundary value problem

$$u_t(x,t) = k u_{xx}(x,t) \qquad\qquad (-\pi < x < \pi, \, t > 0),$$

$$u(-\pi,t) = u(\pi,t), \qquad u_x(-\pi,t) = u_x(\pi,t), \qquad u(x,0) = f(x).$$

The solution $u(x,t)$ represents, for example, temperatures in an insulated wire of length 2π which is bent into a unit circle and which has a given temperature distribution along it. For convenience, the wire is thought of as being cut at one point and laid on the x axis between the points $x = -\pi$ and $x = \pi$. The variable x then measures the distance along the wire starting at the point $x = -\pi$, the points $x = -\pi$ and $x = \pi$ denote the same point on the circle, and the first two boundary conditions in the problem state that the temperatures and the flux must be the same for each of those values of x. This problem was of considerable interest to Fourier himself, and the wire has come to be known as *Fourier's ring*.

$$Ans. \quad u(x,t) = \frac{1}{2} a_0 + \sum_{n=1}^{\infty} a_n e^{-n^2 kt}(a_n \cos nx + b_n \sin nx)$$

where

$$a_n = \frac{1}{\pi} \int_{-\pi}^{\pi} f(x) \cos nx \, dx \qquad\qquad (n = 0, 1, 2, \ldots)$$

and

$$b_n = \frac{1}{\pi} \int_{-\pi}^{\pi} f(x) \sin nx \, dx \qquad\qquad (n = 1, 2, \ldots).$$

3. A string stretched between the points $(0,0)$ and $(\pi,0)$ is released at rest from an initial displacement $y = f(x)$. Its motion is opposed by air resistance proportional to the velocity at each point. Let the unit of time be chosen so that the equation of motion becomes

$$y_{tt}(x,t) = y_{xx}(x,t) - 2h y_t(x,t),$$

where h is a constant. Assuming that $0 < h < 1$, derive the expression

$$y(x,t) = e^{-ht} \sum_{n=1}^{\infty} b_n \left(\cos K_n t + \frac{h}{K_n} \sin K_n t\right) \sin nx,$$

where

$$K_n = \sqrt{n^2 - h^2}, \qquad b_n = \frac{2}{\pi} \int_0^{\pi} f(x) \sin nx \, dx,$$

for the transverse displacements.

4. Show formally that the expression obtained in Sec. 60 for the steady temperatures in the cylinder $\rho \le 1$, $-\infty < z < \infty$ whose surface $\rho = 1$ is kept at temperatures $f(\phi)$ can be written

$$u(\rho,\phi) = \frac{1}{2\pi} \int_{-\pi}^{\pi} f(\theta) \frac{1 - \rho^2}{1 - 2\rho \cos (\theta - \phi) + \rho^2} \, d\theta \qquad\qquad (\rho < 1).$$

This is known as *Poisson's integral formula*.

Suggestion: After replacing the variable of integration ϕ by θ in expressions (6), Sec. 60, substitute those expressions into equation (5) of that section. Then use the summation formula derived in Problem 5, Sec. 30, with a and θ there replaced by ρ and θ-ϕ, respectively.

5. Using cylindrical coordinates ρ, ϕ, and z, let $u(\rho,\phi)$ denote steady temperatures in a long hollow cylinder $a \le \rho \le b$, $-\infty < z < \infty$ when the temperatures on the inner surface $\rho = a$ are $f(\phi)$ and temperature of the outer surface $\rho = b$ is zero.

(a) Derive the temperature formula

$$u(\rho,\phi) = \frac{1}{2}\frac{\log(b/\rho)}{\log(b/a)}a_0 + \sum_{n=1}^{\infty}\frac{(b/\rho)^n - (\rho/b)^n}{(b/a)^n - (a/b)^n}(a_n\cos n\phi + b_n\sin n\phi)$$

where the coefficients a_n and b_n, including a_0, are given by equations (6), Sec. 60.

(b) Use the result in part (a) to show that if, in particular, $f(\phi) = A + B\sin\phi$ where A and B are constants, then

$$u(\rho,\phi) = A\frac{\log(b/\rho)}{\log(b/a)} + \frac{abB}{b^2 - a^2}\left(\frac{b}{\rho} - \frac{\rho}{b}\right)\sin\phi.$$

6. Find the harmonic function $u(\rho,\phi)$ in the region $1 < \rho < b$, $0 < \phi < \pi$ of the z plane such that $u = f(\rho)$ on the part of the boundary where $\phi = \pi$ and $u = 0$ on the rest of the boundary. [The Sturm-Liouville problem that arises is the one in Problem 13(b), Sec. 34.]

$$Ans. \quad u(\rho,\phi) = \sum_{n=1}^{\infty}C_n\left(\frac{\sinh\alpha_n\phi}{\sinh\alpha_n\pi}\right)\sin(\alpha_n\log\rho) \qquad \left(\alpha_n = \frac{n\pi}{\log b}\right)$$

where

$$C_n = \frac{2}{\log b}\int_1^b\frac{1}{\rho}f(\rho)\sin(\alpha_n\log\rho)\,d\rho \qquad (n = 1, 2, \ldots).$$

7. Show that (a) when $f(\rho) = \rho$ in Problem 6,

$$u(\rho,\phi) = 2\pi\sum_{n=1}^{\infty}\frac{n[1 - b(-1)^n]}{(\log b)^2 + (n\pi)^2}\left(\frac{\sinh\alpha_n\phi}{\sinh\alpha_n\pi}\right)\sin(\alpha_n\log\rho) \qquad \left(\alpha_n = \frac{n\pi}{\log b}\right);$$

(b) when u_0 is a constant and $f(\rho) = u_0$ in that problem,

$$u(\rho,\phi) = \frac{4u_0}{\pi}\sum_{n=1}^{\infty}\left(\frac{\sinh\beta_n\phi}{\sinh\beta_n\pi}\right)\frac{\sin(\beta_n\log\rho)}{2n - 1} \qquad \text{where } \beta_n = \frac{(2n-1)\pi}{\log b}.$$

Suggestion: Make the substitution $s = \log\rho$ in the integrals needed to find the coefficients C_n.

8. A bounded harmonic function $u(x,y)$ in the semi-infinite strip $x > 0$, $0 < y < 1$ is to satisfy the boundary conditions

$$u(0,y) = u_0 \qquad\qquad\qquad (0 < y < 1),$$

where u_0 is a constant, and

$$u_y(x,0) = 0, \qquad u_y(x,1) = -hu(x,1) \qquad\qquad (x > 0),$$

where h is a positive constant. Assuming the validity of the Sturm-Liouville expansion that arises, derive the expression

$$u(x,y) = 2hu_0\sum_{n=1}^{\infty}\frac{\sin\alpha_n}{\alpha_n(h + \sin^2\alpha_n)}\exp(-\alpha_n x)\cos\alpha_n y$$

where α_n are the consecutive positive roots of the equation $\tan\alpha = h/\alpha$ (see Sec. 34). Give a physical interpretation of the function $u(x,y)$ here.

9. Solve the boundary value problem

$$u_{xx}(x,y) + u_{yy}(x,y) = 0 \qquad\qquad (0 < x < a, \, 0 < y < b),$$

$$u_x(0,y) = 0, \qquad u_x(a,y) = -hu(a,y) \qquad\qquad (0 < y < b),$$

$$u(x,0) = 0, \qquad u(x,b) = f(x) \qquad\qquad (0 < x < a),$$

where h is a positive constant, and interpret $u(x,y)$ physically. (The Sturm-Liouville problem that arises is treated in Problem 12, Sec. 34.)

$$\textit{Ans.} \quad u(x,y) = 2h \sum_{n=1}^{\infty} \left(\frac{\cos \alpha_n x}{ah + \sin^2 \alpha_n a} \right) \left(\frac{\sinh \alpha_n y}{\sinh \alpha_n b} \right) \int_0^a f(s) \cos \alpha_n s \, ds$$

where α_n are the consecutive positive roots of the equation $\tan \alpha a = h/\alpha$.

10. Let $u(r,t)$ denote *temperatures in a solid sphere* bounded by the surface $r = c$, where r is the spherical coordinate, when the solid is initially at a uniform temperature u_0 and the surface is then kept at temperature zero. The function $u(r,t)$ satisfies the conditions

$$\frac{\partial u}{\partial t} = \frac{k}{r} \frac{\partial^2}{\partial r^2}(ru), \qquad u(c,t) = 0, \qquad u(r,0) = u_0 \qquad\qquad (0 \le r < c).$$

Introduce the new unknown function $v(r,t) = ru(r,t)$ and note that $v(0,t) = 0$ because u is continuous at the center $r = 0$. Thus derive the expression

$$u(r,t) = \frac{2u_0 c}{\pi} \sum_{n=1}^{\infty} (-1)^{n+1} \exp\left(-\frac{n^2\pi^2 kt}{c^2}\right) \frac{1}{nr} \sin \frac{n\pi r}{c}.$$

11. A spherical body 40 cm in diameter, initially at 100°C throughout, is cooled by keeping its surface at 0°C (see Problem 10). Find the approximate temperature of its center 10 min after cooling begins when the material is (*a*) iron, for which $k = 0.15$ cgs unit; (*b*) concrete, for which $k = 0.005$ cgs unit.

Ans. (*a*) 22°C; (*b*) 100°C.

12. The boundary $r = 1$ of a *solid sphere* is kept insulated. Initially the solid has temperatures $f(r)$. Assuming the validity of the Sturm-Liouville expansion that arises (see Problem 6, Sec. 34), derive this expression for the temperatures in the sphere:

$$u(r,t) = 3 \int_0^1 s^2 f(s) \, ds + \sum_{n=1}^{\infty} C_n \exp\left(-\alpha_n^2 kt\right) \frac{\sin \alpha_n r}{r}$$

where α_n are the consecutive positive roots of the equation $\tan \alpha = \alpha$ and

$$C_n = \frac{2}{\sin^2 \alpha_n} \int_0^1 sf(s) \sin \alpha_n s \, ds \, \alpha_n s \, ds. \qquad\qquad (n = 1, 2, \ldots).$$

13. Let ρ, ϕ, z denote cylindrical coordinates and solve the following boundary value problem in the region $1 \le \rho \le b, \, 0 \le \phi \le \pi$ of the plane $z = 0$:

$$\rho^2 u_{\rho\rho}(\rho,\phi) + \rho u_\rho(\rho,\phi) + u_{\phi\phi}(\rho,\phi) = 0 \qquad (1 < \rho < b, \, 0 < \phi < \pi),$$

$$u_\rho(1,\phi) = 0, \qquad u_\rho(b,\phi) = -hu(b,\phi) \qquad (0 < \phi < \pi),$$

$$u(\rho,0) = 0, \qquad u(\rho,\pi) = f(\rho) \qquad (1 < \rho < b),$$

where h is a positive constant. Interpret the function $u(\rho,\phi)$ physically. (The Sturm-Liouville problem that arises is the one solved in Problem 16, Sec. 34.)

$$Ans. \quad u(\rho,\phi) = \sum_{n=1}^{\infty} C_n \left(\frac{\sinh \alpha_n \phi}{\sinh \alpha_n \pi}\right) \cos (\alpha_n \log \rho)$$

where α_n are the consecutive positive roots of the equation $\tan (\alpha \log b) = bh/\alpha$ and

$$C_n = \frac{2bh}{bh \log b + \sin^2 (\alpha_n \log b)} \int_1^b \frac{1}{\rho} f(\rho) \cos (\alpha_n \log \rho) \, d\rho.$$

FOURIER INTEGRALS AND APPLICATIONS

61. The Fourier Integral Formula

From Sec. 43 we know conditions under which the Fourier series

(1) $$\frac{1}{2c}\int_{-c}^{c} f(s)\, ds + \frac{1}{c}\sum_{n=1}^{\infty}\int_{-c}^{c} f(s)\cos\left[\frac{n\pi}{c}(s-x)\right] ds$$

converges to $f(x)$ when $-c < x < c$. To be precise, it is sufficient that f be piecewise continuous on the interval $(-c,c)$, that f have one-sided derivatives at each point x in that interval, and that the value of f at x be the mean value of the limits $f(x+)$ and $f(x-)$.

Suppose that f satisfies such conditions on *every* bounded interval $(-c,c)$. Then c may be given any fixed positive value, arbitrarily large but finite, and series (1) will represent $f(x)$ over the large segment $-c < x < c$ of the x axis. But that series representation cannot apply over the rest of the x axis unless f is periodic with period $2c$ because the sum of the series has that periodicity.

In seeking a representation that is valid for all real x when f is not periodic, it is natural to try to extend the above representation by letting c tend to infinity. The first term in series (1) would then vanish, provided that f is such that the improper integral

$$\int_{-\infty}^{\infty} f(s)\, ds$$

exists. We write $\Delta\alpha = \pi/c$, and the remaining terms take the form

(2) $$\frac{1}{\pi}\sum_{n=1}^{\infty}\Delta\alpha\int_{-c}^{c} f(s)\cos\left[n\,\Delta\alpha(s-x)\right] ds \qquad (c = \pi/\Delta\alpha),$$

or

(3) $$\frac{1}{\pi}\sum_{n=1}^{\infty} g_c(n\,\Delta\alpha,\, x)\,\Delta\alpha \qquad (\Delta\alpha = \pi/c)$$

where

(4)
$$g_c(\alpha,x) = \int_{-c}^{c} f(s) \cos [\alpha(s - x)]\, ds.$$

Let the value of x be fixed and c be large, so that $\Delta\alpha$ is a small positive number. The points $n\,\Delta\alpha$ $(n = 1, 2, \ldots)$ are equally spaced along the entire positive α axis; and because of the resemblance of the series in expression (3) to a sum of the type used in defining definite integrals, one might expect that the sum of that series tends to

(5)
$$\int_{0}^{\infty} g_c(\alpha,x)\, d\alpha,$$

or possibly

(6)
$$\int_{0}^{\infty} g_\infty(\alpha,x)\, d\alpha,$$

as $\Delta\alpha$ tends to zero. As integral (6) indicates, however, the function $g_c(\alpha,x)$ changes with $\Delta\alpha$ because $c = \pi/\Delta\alpha$. Also, the limit of the series in expression (3) as $\Delta\alpha$ tends to zero is *not*, in fact, the definition of the improper integral (5) even if c could be kept fixed.

The above manipulations merely *suggest* that under appropriate conditions on f the function may have the representation

(7)
$$f(x) = \frac{1}{\pi} \int_{0}^{\infty} \int_{-\infty}^{\infty} f(s) \cos [\alpha(s - x)]\, ds\, d\alpha \qquad (-\infty < x < \infty).$$

This is the *Fourier integral formula* for the function f, to be established in the following two sections.

The formula can be written in terms of separate cosine and sine functions as follows:

(8)
$$f(x) = \int_{0}^{\infty} [A(\alpha) \cos \alpha x + B(\alpha) \sin \alpha x]\, d\alpha \qquad (-\infty < x < \infty)$$

where

(9)
$$A(\alpha) = \frac{1}{\pi} \int_{-\infty}^{\infty} f(x) \cos \alpha x\, dx,$$
$$B(\alpha) = \frac{1}{\pi} \int_{-\infty}^{\infty} f(x) \sin \alpha x\, dx.$$

Formulas (8) and (9) bear a resemblance to Fourier series representations and formulas for the coefficients a_n and b_n, stated at the beginning of Sec. 35.

62. Preliminary Lemmas

Just as we prefaced our Fourier theorem in Sec. 41 with some preliminary theory, we digress here with two lemmas that are essential to our proof of a *Fourier integral theorem* which gives conditions under which representation (7) in the previous section is valid. The statement and the proof of the theorem appear in the next section.

Our first lemma is an evaluation of an improper integral that occurs often in applied mathematics.

Lemma 1.
$$\int_0^\infty \frac{\sin x}{x}\, dx = \frac{\pi}{2}.†$$

Before evaluating the integral, we show that it actually converges. Note that the integrand $(\sin x)/x$ is piecewise continuous on any bounded interval; and since

$$(1) \qquad \int_0^c \frac{\sin x}{x}\, dx = \int_0^1 \frac{\sin x}{x}\, dx + \int_1^c \frac{\sin x}{x}\, dx,$$

where c is any positive number, it suffices to show that the integral

$$(2) \qquad \int_1^\infty \frac{\sin x}{x}\, dx$$

converges. To accomplish this, we apply the method of integration by parts to the function $(\sin x)/x = x^{-1}(d/dx)(-\cos x)$ and write

$$(3) \qquad \int_1^c \frac{\sin x}{x}\, dx = \cos 1 - \frac{\cos c}{c} - \int_1^c \frac{\cos x}{x^2}\, dx.$$

Since $|(\cos c)/c| \leq 1/c$, the second term on the right-hand side here tends to zero as c tends to infinity. Also, since $|(\cos x)/x^2| \leq 1/x^2$, the integral

$$(4) \qquad \int_1^\infty \frac{\cos x}{x^2}\, dx$$

is (absolutely) convergent. The limit of the left-hand side of equation (3) as c tends to infinity therefore exists; that is, integral (2) converges.

Now that we have established that the given integral converges to some number L, or that

$$\lim_{c \to \infty} \int_0^c \frac{\sin x}{x} = L,$$

---

† For another proof of Lemma 1 that is fairly standard, see, for example, Buck (1978, Sec. 6.4). A method of proof involving complex variables is pointed out in Churchill, Brown, and Verhey (1974, p. 193). Both books are listed in the Bibliography.

we note that, in particular,

(5)
$$\lim_{m \to \infty} \int_0^{(m+1/2)\pi} \frac{\sin x}{x} \, dx = L$$

as m passes through positive integers. That is,

(6)
$$\lim_{m \to \infty} \int_0^{\pi} \frac{\sin \left[(m + \frac{1}{2})u\right]}{u} \, du = L,$$

where the substitution $x = (m + \frac{1}{2})u$ has been made for the variable of integration. Observe that equation (6) can be written

(7)
$$\lim_{m \to \infty} \int_0^{\pi} F(u) D_m(u) \, du = L$$

where

(8)
$$F(u) = \frac{2 \sin (u/2)}{u}$$

and $D_m(u)$ is the Dirichlet kernel (Sec. 40)

$$D_m(u) = \frac{\sin \left[(m + \frac{1}{2})u\right]}{2 \sin (u/2)}.$$

The function $F(u)$, moreover, satisfies the conditions in Lemma 2, Sec. 40, and $F(0+) = 1$ (see Problem 1, Sec. 63). So, by that lemma, limit (7) has the value $\pi/2$; and, by uniqueness of limits, $L = \pi/2$. The proof of Lemma 1 in this section is now complete.

Our second lemma makes direct use of the first one.

Lemma 2. *Suppose that a function $F(u)$ is piecewise continuous on every bounded interval of the positive u axis and that the right-hand derivative $F'_R(0)$ exists. If the improper integral*

(9)
$$\int_0^{\infty} |F(u)| \, du$$

converges, then

(10)
$$\lim_{r \to \infty} \int_0^{\infty} F(u) \frac{\sin ru}{u} \, du = \frac{\pi}{2} F(0+).$$

Observe that the integrand appearing in equation (10) is piecewise continuous on the same intervals as $F(u)$ and that when $u \geq 1$,

$$\left| F(u) \frac{\sin ru}{u} \right| \leq |F(u)|.$$

Thus the convergence of integral (9) ensures the existence of the integral in equation (10).

We begin the proof of the lemma by demonstrating its validity when the range of integration is replaced by any bounded interval $(0,c)$. That is, we first show that if a function $F(u)$ is piecewise continuous on a bounded interval $(0,c)$ and $F_R'(0)$ exists, then

$$(11) \qquad \lim_{r \to \infty} \int_0^c F(u) \frac{\sin ru}{u} \, du = \frac{\pi}{2} F(0+).$$

To prove this, we write

$$\int_0^c F(u) \frac{\sin ru}{u} \, du = I(r) + J(r)$$

where

$$I(r) = \int_0^c \left[\frac{F(u) - F(0+)}{u} \right] \sin ru \, du$$

and

$$J(r) = \int_0^c F(0+) \frac{\sin ru}{u} \, du.$$

Since the function $G(u) = [F(u) - F(0+)]/u$ is piecewise continuous on the interval $(0,c)$, where $G(0+) = F_R'(0)$, we need only refer to Lemma 1 in Sec. 40 to see that

$$(12) \qquad \lim_{r \to \infty} I(r) = 0.$$

On the other hand, if we substitute $x = ru$ into the integral representing $J(r)$ and apply Lemma 1 of the present section, we find that

$$(13) \qquad \lim_{r \to \infty} J(r) = F(0+) \lim_{r \to \infty} \int_0^{cr} \frac{\sin x}{x} \, dx = F(0+) \frac{\pi}{2}.$$

Limit (11) is evidently now a consequence of limits (12) and (13).

To actually obtain limit (10), we note that

$$\left| \int_c^\infty F(u) \frac{\sin ru}{u} \, du \right| \leq \int_c^\infty |F(u)| \, du,$$

where we assume that $c \geq 1$. We then write

$$(14) \qquad \left| \int_0^\infty F(u) \frac{\sin ru}{u} \, du - \frac{\pi}{2} F(0+) \right|$$

$$\leq \left| \int_0^c F(u) \frac{\sin ru}{u} \, du - \frac{\pi}{2} F(0+) \right| + \int_c^\infty |F(u)| \, du,$$

choosing c to be so large that the second absolute value on the right, or the absolute value of the remainder of integral (9), is less than $\varepsilon/2$ where ε is an arbitrary positive number independent of the value of r. In view of limit (11), there exists a positive number R such that whenever $r > R$, the first absolute value on the right-hand side of inequality (14) is also less than $\varepsilon/2$. It then follows that

$$\left| \int_0^\infty F(u) \frac{\sin ru}{u} \, du - \frac{\pi}{2} F(0+) \right| < \frac{\varepsilon}{2} + \frac{\varepsilon}{2} = \varepsilon$$

whenever $r > R$, and this is the same as statement (10).

63. A Fourier Integral Theorem

The following theorem gives conditions under which the Fourier integral representation (7), Sec. 61, is valid.†

Theorem 1. *Let f denote a function which is piecewise continuous on every bounded interval of the x axis and suppose that it is absolutely integrable over that axis; that is, the improper integral*

$$\int_{-\infty}^\infty |f(x)| \, dx$$

converges. The Fourier integral

(1)
$$\frac{1}{\pi} \int_0^\infty \int_{-\infty}^\infty f(s) \cos[\alpha(s-x)] \, ds \, d\alpha$$

then converges to the value

(2)
$$\frac{1}{2}[f(x+) + f(x-)]$$

at each point $x(-\infty < x < \infty)$ where the one-sided derivatives $f'_R(x)$ and $f'_L(x)$ both exist.

We begin our proof with the observation that integral (1) represents the limit as r tends to infinity of the integral

(3)
$$\frac{1}{\pi} \int_0^r \int_{-\infty}^\infty f(s) \cos[\alpha(s-x)] \, ds \, d\alpha = \frac{1}{\pi}[I(r,x) + J(r,x)]$$

---

† For other conditions see, for instance, Carslaw (1930, pp. 315 ff) and Titchmarsh (1967, pp. 13 ff), both listed in the Bibliography.

where

$$I(r,x) = \int_0^r \int_x^\infty f(s) \cos\left[\alpha(s-x)\right] ds\, d\alpha$$

and

$$J(r,x) = \int_0^r \int_{-\infty}^x f(s) \cos\left[\alpha(s-x)\right] ds\, d\alpha.$$

We now show that the individual integrals $I(r,x)$ and $J(r,x)$ exist; and, assuming that $f'_R(x)$ and $f'_L(x)$ exist, we examine the behavior of these integrals as r tends to infinity.

Turning to $I(r,x)$ first, we introduce the new variable of integration $u = s - x$ and write that integral in the form

$$(4) \qquad I(r,x) = \int_0^r \int_0^\infty f(x+u) \cos \alpha u\, du\, d\alpha.$$

Since

$$\left| f(x+u) \cos \alpha u \right| \le \left| f(x+u) \right|$$

and because

$$\int_0^\infty \left| f(x+u) \right| du = \int_x^\infty \left| f(s) \right| ds \le \int_{-\infty}^\infty \left| f(s) \right| ds,$$

the Weierstrass M-test for improper integrals applies to show that the integral

$$\int_0^\infty f(x+u) \cos \alpha u\, du$$

converges uniformly with respect to the variable α. Consequently, not only does the iterated integral (4) exist, but the order of integration there can be reversed:

$$I(r,x) = \int_0^\infty \int_0^r f(x+u) \cos \alpha u\, d\alpha\, du = \int_0^\infty f(x+u) \frac{\sin ru}{u}\, du.\dagger$$

Observe now that the function $F(u) = f(x+u)$ satisfies all of the conditions in Lemma 2, Sec. 62. Hence, if we apply that lemma to this last integral, we find that

$$(5) \qquad \lim_{r\to\infty} I(r,x) = \frac{\pi}{2} f(x+).$$

---

† Theorems on improper integrals used here are developed in Buck's "Advanced Calculus" (1978, Sec. 6.4), listed in the Bibliography, as well as in most other texts with such titles. The theorems are usually given for integrals with continuous integrands, but they are also valid when the integrands are piecewise continuous.

The limit of $J(r,x)$ as r tends to infinity is treated similarly. Here we make the substitution $u = x - s$ and write

$$J(r,x) = \int_0^r \int_0^\infty f(x - u) \cos \alpha u \, du \, d\alpha = \int_0^\infty f(x - u) \frac{\sin ru}{u} \, du.$$

When $F(u) = f(x - u)$, the limit

(6) $$\lim_{r \to \infty} J(r,x) = \frac{\pi}{2} f(x-)$$

also follows from Lemma 2 in Sec. 62.

Finally, in view of limits (5) and (6), we see that the limit of the left-hand side of equation (3) as $r \to \infty$ has the value (2), which is then the value of integral (1) at any point where the one-sided derivatives of f exist.

Note that since the integrals in formulas (9), Sec. 61, for the coefficients $A(\alpha)$ and $B(\alpha)$ exist when f satisfies the conditions stated in the theorem, form (8), Sec. 61, of the Fourier integral formula is also justified.

PROBLEMS

1. Show that the function

$$F(u) = \frac{2 \sin (u/2)}{u},$$

used in equation (7), Sec. 62, satisfies the conditions in Lemma 2, Sec. 40. To be precise, show that it is piecewise continuous on the interval $(0,\pi)$ and that $F'_R(0) = 0$.

Suggestion: To obtain $F'_R(0)$, show that

$$F'_R(0) = \lim_{\substack{h \to 0 \\ h > 0}} \frac{2 \sin (h/2) - h}{h^2}$$

and apply l'Hospital's rule twice.

2. Verify that all the conditions in Theorem 1 are satisfied by the function

$$f(x) = \begin{cases} 1 & \text{when } |x| < 1, \\ \tfrac{1}{2} & \text{when } |x| = 1, \\ 0 & \text{when } |x| > 1. \end{cases}$$

Thus show that for every x $(-\infty < x < \infty)$

$$f(x) = \frac{1}{\pi} \int_0^\infty \frac{\sin [\alpha(1 + x)] + \sin [\alpha(1 - x)]}{\alpha} \, d\alpha = \frac{2}{\pi} \int_0^\infty \frac{\sin \alpha \cos \alpha x}{\alpha} \, d\alpha.$$

3. Show how it follows from Lemma 1, Sec. 62, that

$$\int_0^\infty \frac{\sin k\alpha}{\alpha}\, d\alpha = \begin{cases} \frac{\pi}{2} & \text{when } k > 0, \\[2mm] 0 & \text{when } k = 0, \\[2mm] -\frac{\pi}{2} & \text{when } k < 0. \end{cases}$$

Then use this result to evaluate the first integral written in Problem 2 when $|x| < 1$, when $|x| = 1$, and when $|x| > 1$, thereby making a direct verification of the integral representation found for the function f there.

4. Show that the function

$$f(x) = \begin{cases} 0 & \text{when } x < 0, \\[2mm] \frac{1}{2} & \text{when } x = 0, \\[2mm] e^{-x} & \text{when } x > 0 \end{cases}$$

satisfies the conditions in Theorem 1 and hence that

$$f(x) = \frac{1}{\pi} \int_0^\infty \frac{\cos \alpha x + \alpha \sin \alpha x}{1 + \alpha^2}\, d\alpha \qquad (-\infty < x < \infty).$$

Verify this representation directly at the point $x = 0$.

5. Prove that

$$\exp(-|x|) = \frac{2}{\pi} \int_0^\infty \frac{\cos \alpha x}{1 + \alpha^2}\, d\alpha \qquad (-\infty < x < \infty).$$

6. Prove that if $f(x) = \sin x$ when $0 \le x \le \pi$ and $f(x) = 0$ when $x < 0$ and when $x > \pi$, then

$$f(x) = \frac{1}{\pi} \int_0^\infty \frac{\cos \alpha x + \cos [\alpha(\pi - x)]}{1 - \alpha^2}\, d\alpha \qquad (-\infty < x < \infty).$$

In particular, write $x = \pi/2$ to show that

$$\int_0^\infty \frac{\cos (\alpha \pi/2)}{1 - \alpha^2}\, d\alpha = \frac{\pi}{2}.$$

7. Show why the Fourier integral formula fails to represent the function $f(x) = 1\ (-\infty < x < \infty)$. Also, point out which condition in Theorem 1 is not satisfied by that function.

8. Give the details showing that the integral $J(r,x)$ in Sec. 63 actually exists and that limit (6) in that section holds.

9. Let f be a nonzero function which is periodic, with period $2c$, for example. Point out why the integrals

$$\int_{-\infty}^\infty f(x)\, dx \qquad \text{and} \qquad \int_{-\infty}^\infty |f(x)|\, dx$$

fail to exist.

64. The Cosine and Sine Integrals

Let $f(x)$ be an *even* function that satisfies the conditions in Theorem 1. Since $f(x) \cos \alpha x$ and $f(x) \sin \alpha x$ are even and odd functions of x, respectively, the coefficients $A(\alpha)$ and $B(\alpha)$ given by formulas (9), Sec. 61, are

$$A(\alpha) = \frac{1}{\pi} \int_{-\infty}^{\infty} f(x) \cos \alpha x \, dx = \frac{2}{\pi} \int_{0}^{\infty} f(x) \cos \alpha x \, dx$$

and

$$B(\alpha) = \frac{1}{\pi} \int_{-\infty}^{\infty} f(x) \sin \alpha x \, dx = 0.$$

The Fourier integral formula

$$(1) \qquad f(x) = \int_{0}^{\infty} [A(\alpha) \cos \alpha x + B(\alpha) \sin \alpha x] \, d\alpha \qquad (-\infty < x < \infty)$$

thus reduces to the *Fourier cosine integral formula*

$$(2) \qquad f(x) = \frac{2}{\pi} \int_{0}^{\infty} \cos \alpha x \int_{0}^{\infty} f(s) \cos \alpha s \, ds \, d\alpha.$$

If, instead, f is an *odd* function, then

$$A(\alpha) = 0, \qquad B(\alpha) = \frac{2}{\pi} \int_{0}^{\infty} f(x) \sin \alpha x \, dx;$$

and formula (1) becomes the *Fourier sine integral formula*

$$(3) \qquad f(x) = \frac{2}{\pi} \int_{0}^{\infty} \sin \alpha x \int_{0}^{\infty} f(s) \sin \alpha s \, ds \, d\alpha.$$

If $f(x)$ is defined only on the half line $x > 0$, formulas (2) and (3) apply to the even and odd extensions of f, respectively, and represent $f(x)$ when $x > 0$ under the following conditions.

Corollary 1. *Let a function f be piecewise continuous on every bounded interval of the positive x axis; and, for convenience, let the value of f be defined at each point where it is discontinuous as the mean value of its one-sided limits there. Suppose further that the improper integral*

$$\int_{0}^{\infty} |f(x)| \, dx$$

converges. Then, at each point $x(x > 0)$ where the one-sided derivatives $f'_R(x)$ and $f'_L(x)$ exist, f is represented by its Fourier cosine integral (2) and its Fourier sine integral (3).

Under those conditions the cosine integral formula (2) represents $f(0+)$ when $x = 0$ if $f'_R(0)$ exists, as we see by considering the even extension of f. The sine integral clearly has the value zero when $x = 0$.

The eigenvalue problem

(4)
$$X''(x) + \lambda X(x) = 0 \qquad\qquad (x > 0),$$
$$X(0) = 0, \qquad |X(x)| < M \qquad\qquad (x > 0),$$

where M is some positive constant, is *singular* (Sec. 32) because its fundamental interval $x > 0$ is unbounded. If $\lambda = 0$, we have only the trivial solution. If λ is a real number such that $\lambda > 0$ and we write $\lambda = \alpha^2$ $(\alpha > 0)$, we readily find that, except for a constant factor, the eigenfunctions are $X = \sin \alpha x$ where α takes on all positive values. The eigenvalues $\lambda = \alpha^2$ are continuous rather than discrete. If $\lambda < 0$, or $\lambda = -\beta^2$ $(\beta > 0)$, the solution of the differential equation and the boundary condition at $x = 0$ is $X = D_1 \sinh \beta x$. This is, however, unbounded on the half line $x > 0$ unless $D_1 = 0$. So the case $\lambda < 0$ yields no new eigenfunctions. Cases other than those in which λ is real need not be considered since they yield unbounded solutions of the differential equation (see Problem 12, Sec. 66). Although the eigenfunctions $\sin \alpha x$ have no orthogonality property, the Fourier sine integral formula (3) gives representations of functions f on the interval $x > 0$ which are generalized linear combinations of those eigenfunctions.

Likewise, $\lambda = \alpha^2$ $(\alpha \geq 0)$ and $X = \cos \alpha x$ are the eigenvalues and eigenfunctions of the singular problem

(5)
$$X''(x) + \lambda X(x) = 0 \qquad\qquad (x > 0),$$
$$X'(0) = 0, \qquad |X(x)| < M \qquad\qquad (x > 0);$$

and formula (2) represents functions f in terms of $\cos \alpha x$.

65. The Exponential Form

Under the conditions stated in Theorem 1 the Fourier integral formula can be written

(1)
$$f(x) = \frac{1}{2\pi} \lim_{r \to \infty} \int_0^r \int_{-\infty}^{\infty} 2f(s) \cos [\alpha(s - x)] \, ds \, d\alpha.$$

Let the cosine function be expressed in terms of exponential functions. Since f is absolutely integrable, the integrals

$$\int_{-\infty}^{\infty} f(s) e^{i\alpha s} \, ds, \qquad \int_{-\infty}^{\infty} f(s) e^{-i\alpha s} \, ds$$

converge uniformly with respect to α for all positive α, according to the Weierstrass M-test for improper integrals. That uniform convergence together with the piecewise continuity of f implies that those integrals represent continuous functions of α (see Problem 9, Sec. 66). Thus the iterated integral in equation (1) can be written as the sum

$$(2) \qquad \int_0^r e^{-i\alpha x} \int_{-\infty}^\infty f(s)e^{i\alpha s} \, ds \, d\alpha + \int_0^r e^{i\alpha x} \int_{-\infty}^\infty f(s)e^{-i\alpha s} \, ds \, d\alpha$$

since the integrals with respect to α exist as integrals of continuous functions of α. (The improper integrals from $\alpha = 0$ to $\alpha = \infty$ may not exist, however.)

With the substitution $\beta = -\alpha$, the second term of the sum (2) takes the form

$$\int_{-r}^0 e^{-i\beta x} \int_{-\infty}^\infty f(s)e^{i\beta s} \, ds \, d\beta.$$

Evidently, then, the sum (2) can be written

$$\int_{-r}^r e^{-i\alpha x} \int_{-\infty}^\infty f(s)e^{i\alpha s} \, ds \, d\alpha.$$

Thus the *exponential form of the Fourier integral formula* (1) is

$$(3) \qquad f(x) = \frac{1}{2\pi} \lim_{r\to\infty} \int_{-r}^r e^{-i\alpha x} \int_{-\infty}^\infty f(s)e^{i\alpha s} \, ds \, d\alpha,$$

where $-\infty < x < \infty$. The minus sign in the exponent $-i\alpha x$ can be shifted to the exponent $i\alpha s$ by using $-\alpha$, instead of α, as a variable of integration.

The limit in formula (3) is called the *Cauchy principal value* of the improper integral from $-\infty$ to ∞ with respect to α. The principal value may exist when the improper integral does not. The integral

$$\int_{-\infty}^\infty \alpha \, d\alpha,$$

for example, fails to exist; but its principal value does exist, and it is equal to zero because

$$\int_{-r}^r \alpha \, d\alpha = 0$$

for each value of r. When the improper integral does exist, it is equal to its principal value.

Unless further conditions are imposed on f, the principal value used in formula (3) cannot be replaced by the improper integral itself, as the following example will show. The function

$$(4) \qquad f(x) = \begin{cases} 0 & \text{when } x < 0, \\ \frac{1}{2} & \text{when } x = 0, \\ e^{-x} & \text{when } x > 0 \end{cases}$$

satisfies all the conditions in our Fourier integral theorem. It is then represented by formula (3) at every point x, and at $x = 0$ in particular. In this case

$$(5) \qquad \int_{-\infty}^{\infty} f(s)e^{i\alpha s}\, ds = \int_{0}^{\infty} e^{-s(1-i\alpha)}\, ds = \frac{1}{1-i\alpha} = \frac{1+i\alpha}{1+\alpha^2}.$$

Thus when $x = 0$ in formula (3), the iterated integral becomes

$$\int_{-r}^{r} \frac{1+i\alpha}{1+\alpha^2}\, d\alpha = \left[\tan^{-1}\alpha + \frac{i}{2}\log\left(1+\alpha^2\right) \right]_{-r}^{r} = 2\tan^{-1} r,$$

which has the limit π as $r \to \infty$. The right-hand side of equation (3) therefore has the value $1/2$, which is $f(0)$. But the improper integral of the function (5) from $-\infty$ to ∞ does not exist since the function $\alpha/(1+\alpha^2)$ is not integrable from 0 to ∞.

The functions $X = e^{-i\alpha x}$ $(-\infty < \alpha < \infty)$ are eigenfunctions of the singular eigenvalue problem

$$(6) \qquad X''(x) + \lambda X(x) = 0, \qquad |X(x)| < M \quad (-\infty < x < \infty),$$

where M is some positive constant. The eigenvalues $\lambda = \alpha^2$ consist of all real nonnegative numbers, and representations of functions f in terms of the eigenfunctions are given by formula (3).

66. Fourier Transforms

The Fourier sine integral formula (3), Sec. 64, can be written in the form

$$(1) \qquad f(x) = \frac{2}{\pi}\int_{0}^{\infty} F_s(\alpha)\sin\alpha x\, d\alpha \qquad (x > 0)$$

where

$$(2) \qquad F_s(\alpha) = \int_{0}^{\infty} f(x)\sin\alpha x\, dx \qquad (\alpha > 0).$$

If f is a given function, then equation (1) is an *integral equation* in the unknown function F_s appearing in the integrand there. It is a *singular* integral equation because the integral is improper. Equation (2) gives a solution when f satisfies the conditions stated in Corollary 1, as we see by substituting the right-hand side of that equation, with the variable of integration x replaced by s, into equation (1).

The function F_s given by equation (2) is the *Fourier sine transform* of the function f. Transformation (2), which we may abbreviate as

$$(3) \qquad F_s(\alpha) = S_\alpha\{f\},$$

establishes a correspondence between functions f and F_s. Functions f satisfying the conditions in Corollary 1 have transforms F_s such that equation (1) gives f in terms of its transform; that is, equation (1) defines the inverse transformation.

Suppose that f, f', and f'' are continuous when $x \geq 0$, that $f(x)$ and $f'(x)$ tend to zero as x tends to infinity, and that

$$\int_0^\infty |f(x)| \, dx$$

exists. Then successive integrations by parts show that

$$(4) \qquad S_\alpha\{f''(x)\} = -\alpha^2 F_s(\alpha) + \alpha f(0)$$

(see Problem 7 of this section). This is the basic *operational property* of the sine transformation. It replaces the differential form $f''(x)$ by an algebraic form in the transform $F_s(\alpha)$ and involves the initial value $f(0)$.

Property (4), together with other operational properties of the transformation, may be used to reduce certain types of boundary value problems to simpler problems in the transforms of the unknown functions. The development of this operational method, using the various Fourier transforms and the related Laplace transform, is treated elsewhere.†

The *Fourier cosine transform* F_c of a function f is

$$(5) \qquad F_c(\alpha) = \int_0^\infty f(x) \cos \alpha x \, dx = C_\alpha\{f\} \qquad\qquad (\alpha > 0).$$

The inverse of the transformation $C_\alpha\{f\}$ is given by the Fourier cosine integral formula (2), Sec. 64; that is,

$$(6) \qquad f(x) = \frac{2}{\pi} \int_0^\infty F_c(\alpha) \cos \alpha x \, d\alpha \qquad\qquad (x > 0).$$

The basic operational property, valid under the conditions given above for property (4), involves $f'(0)$ rather than $f(0)$:

$$(7) \qquad C_\alpha\{f''(x)\} = -\alpha^2 F_c(\alpha) - f'(0).$$

Note that S_α and C_α are linear operators (Sec. 11).

The *exponential Fourier transform* is the function

$$(8) \qquad F(\alpha) = \int_{-\infty}^\infty f(x) e^{i\alpha x} \, dx = E_\alpha\{f\} \qquad\qquad (-\infty < \alpha < \infty)$$

whose inverse is obtained from formula (3), Sec. 65. There are also certain integral transformations over bounded intervals called *finite Fourier transforms*.

---

† See, for example, Churchill (1972), listed in the Bibliography.

PROBLEMS

1. By applying the Fourier sine integral formula to the function

$$f(x) = \begin{cases} 1 & \text{when } 0 < x < k, \\ \frac{1}{2} & \text{when } x = k, \\ 0 & \text{when } x > k, \end{cases}$$

obtain the representation

$$f(x) = \frac{2}{\pi} \int_0^\infty \frac{1 - \cos k\alpha}{\alpha} \sin \alpha x \, d\alpha \qquad (x > 0).$$

2. Use Corollary 1 and the Fourier sine integral formula to show that

$$e^{-x} \cos x = \frac{2}{\pi} \int_0^\infty \frac{\alpha^3 \sin \alpha x}{\alpha^4 + 4} \, d\alpha \qquad (x > 0).$$

3. Use the Fourier cosine integral formula to prove that

$$e^{-x} \cos x = \frac{2}{\pi} \int_0^\infty \frac{\alpha^2 + 2}{\alpha^4 + 4} \cos \alpha x \, d\alpha \qquad (x \geq 0).$$

4. Apply the operational property (4), Sec. 66, to the function e^{-kx}, where k is a positive constant, to show that the sine transform of that function is $\alpha/(\alpha^2 + k^2)$. Thus obtain the representation

$$e^{-kx} = \frac{2}{\pi} \int_0^\infty \frac{\alpha \sin \alpha x}{\alpha^2 + k^2} \, d\alpha \qquad (k > 0, \, x > 0).$$

5. Use the operational property (7), Sec. 66, to prove that the cosine transform of e^{-kx}, where k is a positive constant, is $k/(\alpha^2 + k^2)$. Thus show that

$$e^{-kx} = \frac{2k}{\pi} \int_0^\infty \frac{\cos \alpha x}{\alpha^2 + k^2} \, d\alpha \qquad (k > 0, \, x \geq 0).$$

6. By regarding the positive constant k in the equation obtained in **Problem 5** as a variable and then differentiating each side of that equation with respect to k, show *formally* that

$$(1 + x)e^{-x} = \frac{4}{\pi} \int_0^\infty \frac{\cos \alpha x}{(\alpha^2 + 1)^2} \, d\alpha \qquad (x \geq 0).$$

7. Assuming that f, f', and f'' are continuous $(x \geq 0)$, use integration by parts to prove that, for each positive number c,

$$\int_0^c f''(x) \sin \alpha x \, dx = f'(c) \sin \alpha c - \alpha f(c) \cos \alpha c + \alpha f(0) - \alpha^2 \int_0^c f(x) \sin \alpha x \, dx.$$

[The continuity of f'' here can be replaced by the condition that f'' be piecewise continuous on each interval $(0,c)$.] Assuming also that $f(x)$ and $f'(x)$ tend to zero as x tends to infinity and that the Fourier sine transform $F_s(\alpha)$ of f exists, show that the right-hand side of the above equation has the limit $\alpha f(0) - \alpha^2 F_s(\alpha)$ as c tends to infinity. Deduce that the sine transform of f'' exists and that it satisfies the operational property (4), Sec. 66.

8. Derive the operational property (7), Sec. 66, for the cosine transform of f''. (Compare **Problem 7.**)

9. Let $f(s)$ be piecewise continuous on each interval $(-c,c)$ of the s axis. State why the function

$$\phi(\alpha) = \int_{-c}^{c} f(s)e^{i\alpha s}\, ds$$

is continuous for all real α. If, in addition, f is absolutely integrable along the entire s axis, then the integral

$$\int_{-\infty}^{\infty} f(s)e^{i\alpha s}\, ds = F(\alpha)$$

is uniformly convergent with respect to α. Write

$$F(\alpha) = \phi(\alpha) + \int_{-\infty}^{-c} f(s)e^{i\alpha s}\, ds + \int_{c}^{\infty} f(s)e^{i\alpha s}\, ds$$

and prove that $F(\alpha + \Delta\alpha) - F(\alpha) \to 0$ as $\Delta\alpha \to 0$ by first making c large, independent of α and $\Delta\alpha$, and then taking $\Delta\alpha$ small to make $\phi(\alpha + \Delta\alpha) - \phi(\alpha)$ small. Thus establish the continuity of $F(\alpha)$ that was used at the beginning of Sec. 65.

10. Establish the following operational property of the exponential Fourier transform

$$F(\alpha) = \int_{-\infty}^{\infty} f(x)e^{i\alpha x}\, dx \qquad (-\infty < \alpha < \infty):$$

if f, f', and f'' are everywhere continuous, if the transform $F(\alpha)$ of f exists, and if $f(x)$ and $f'(x)$ both tend to zero as $|x|$ tends to ∞, then

$$E_\alpha\{f''(x)\} = -\alpha^2 F(\alpha) \qquad (-\infty < \alpha < \infty).$$

11. Verify the Fourier sine integral representation

$$\frac{x}{x^2 + k^2} = \frac{2}{\pi} \int_0^{\infty} \sin \alpha x \int_0^{\infty} \frac{s \sin \alpha s}{s^2 + k^2}\, ds\, d\alpha \qquad (k > 0)$$

by observing that according to Problem 4, the inner integral here has the value $(\pi/2)e^{-k\alpha}$ and the outer integral is then the sine transform of that exponential function of α. Point out why the function $x/(x^2 + k^2)$ is *not* absolutely integrable from $x = 0$ to $x = \infty$.

12. Using the fact that the function $|\sin(x + iy)|$ is unbounded as $|y| \to \infty$ (see Problem 2, Sec. 30), show that the eigenvalues of the singular Sturm-Liouville problem (4), Sec. 64, must be real.

67. More on Superposition of Solutions

In Sec. 11 we showed that linear combinations of solutions of linear homogeneous differential equations and boundary conditions are also solutions. In Sec. 13 we extended that result to include infinite series of solutions, thus providing the basis of the technique for solving boundary value problems in Chap. 6. Another useful extension is illustrated by the following example, where superposition consists of integration with respect to a parameter α instead of summation with respect to an index n. It will enable us to solve certain boundary value problems where Fourier integrals, rather than Fourier series, are required.

The functions of the set exp $(-\alpha y)$ sin αx, where the parameter α $(\alpha > 0)$ is independent of x and y, satisfy Laplace's equation

(1) $$u_{xx}(x,y) + u_{yy}(x,y) = 0 \qquad (x > 0,\ y > 0)$$

and the boundary condition

(2) $$u(0,y) = 0 \qquad (y > 0).$$

Those functions are bounded in the domain $x > 0$, $y > 0$; they are obtained from conditions (1) and (2) by the method of separation of variables when that boundedness condition is included (Problem 1, Sec. 69).

We now show that their combination of the type

(3) $$u(x,y) = \int_0^\infty g(\alpha)e^{-\alpha y} \sin \alpha x\, d\alpha \qquad (x > 0,\ y > 0)$$

also represents a solution of the homogeneous equations (1) and (2) which is bounded in the domain $x > 0$, $y > 0$ for each function $g(\alpha)$ that is bounded and continuous on the half line $\alpha > 0$ and absolutely integrable over it.

To accomplish this, we use tests for improper integrals that are analogous to those for infinite series.† The integral in equation (3) converges absolutely and uniformly with respect to x and y because

(4) $$|g(\alpha)e^{-\alpha y} \sin \alpha x| \le |g(\alpha)| \qquad (x \ge 0,\ y \ge 0)$$

and g is independent of x and y and absolutely integrable from zero to infinity with respect to α. Also

(5) $$|u(x,y)| \le \int_0^\infty |g(\alpha)e^{-\alpha y} \sin \alpha x|\, d\alpha \le \int_0^\infty |g(\alpha)|\, d\alpha,$$

so that u is bounded; it is also a continuous function of x and y $(x \ge 0,\ y \ge 0)$ because of the uniform convergence of the integral and the continuity of the integrand. Clearly $u = 0$ when $x = 0$.

When $y > 0$,

(6) $$\frac{\partial u}{\partial x} = \frac{\partial}{\partial x} \int_0^\infty g(\alpha)e^{-\alpha y} \sin \alpha x\, d\alpha = \int_0^\infty \frac{\partial}{\partial x}[g(\alpha)e^{-\alpha y} \sin \alpha x]\, d\alpha;$$

for if $|g(\alpha)| \le g_0$ and $y \ge y_0$ where y_0 is some small positive number, then the absolute value of the integrand of the integral on the far right does not exceed $g_0 \alpha \exp(-\alpha y_0)$, which is independent of x and y and integrable from $\alpha = 0$ to $\alpha = \infty$. Hence that integral is uniformly convergent. Integral (3) is

---

† See Kaplan (1973, pp. 447 ff) or Taylor and Mann (1972, pp. 722 ff), listed in the Bibliography.

then differentiable with respect to x, and similarly for the other derivatives involved in the laplacian operator $\nabla^2 = \partial^2/\partial x^2 + \partial^2/\partial y^2$. Therefore

$$(7) \qquad \nabla^2 u = \int_0^\infty g(\alpha)\nabla^2(e^{-\alpha y} \sin \alpha x)\, d\alpha = 0 \qquad (x > 0,\ y > 0).$$

Suppose now that the function (3) is also required to satisfy the nonhomogeneous boundary condition

$$(8) \qquad u(x,0) = f(x) \qquad (x > 0),$$

where f is a given function satisfying the conditions stated in Corollary 1. We need to determine the function $g(\alpha)$ in equation (3) so that

$$(9) \qquad f(x) = \int_0^\infty g(\alpha) \sin \alpha x\, d\alpha \qquad (x > 0).$$

This is easily done since representation (9) is the Fourier sine integral formula (3), Sec. 64, when

$$(10) \qquad g(\alpha) = \frac{2}{\pi} \int_0^\infty f(x) \sin \alpha x\, dx \qquad (\alpha > 0).$$

We have shown here that the function (3) with $g(\alpha)$ given by equation (10) is a solution of the boundary value problem in the quadrant $x \geq 0$, $y \geq 0$ consisting of equations (1), (2), and (8) together with the requirement that u be bounded.

68. Temperatures in a Semi-infinite Solid

The face $x = 0$ of a semi-infinite solid $x \geq 0$ is kept at temperature zero (Fig. 24). Let us find the temperatures $u(x,t)$ in the solid when the initial temperature distribution is $f(x)$, assuming at present that f and f' are piecewise continuous on each bounded interval of the positive x axis and that f is bounded and absolutely integrable from $x = 0$ to $x = \infty$.

If the solid is considered as a limiting case of a slab $0 \leq x \leq c$ as c increases, some condition corresponding to a thermal condition on the face $x = c$ seems to be needed; otherwise the temperatures on that face may be

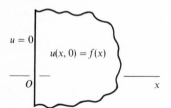

$u = 0$

$u(x, 0) = f(x)$

O

x

Figure 24

increased in any manner as c increases. We require that our function u be bounded; that condition also implies that there is no instantaneous source of heat on the face $x = 0$ at the instant $t = 0$. Then

$$(1) \qquad\qquad u_t(x,t) = k u_{xx}(x,t) \qquad\qquad (x > 0, t > 0),$$

$$u(0,t) = 0 \qquad\qquad (t > 0),$$

$$(2)$$

$$|u(x,t)| < M \qquad\qquad (x > 0, t > 0),$$

where M is some positive constant, and

$$(3) \qquad\qquad u(x,0) = f(x) \qquad\qquad (x > 0).$$

Linear combinations of functions XT will not ordinarily be bounded unless X and T themselves are bounded. Upon separating variables, we then have the conditions

$$X''(x) + \lambda X(x) = 0 \qquad\qquad (x > 0),$$

$$(4) \qquad\qquad X(0) = 0, \qquad |X(x)| < M_1,$$

and $\qquad\qquad T'(t) + \lambda k T(t) = 0, \qquad |T(t)| < M_2 \qquad\qquad (t > 0),$

where M_1 and M_2 are positive constants. As pointed out in Sec. 64, the singular eigenvalue problem (4) has continuous eigenvalues $\lambda = \alpha^2$ where α represents *all positive real numbers*; $X = \sin \alpha x$ are the eigenfunctions. In this case the corresponding functions $T = \exp(-\alpha^2 kt)$ are bounded. The generalized linear combination of the functions XT for all positive α,

$$(5) \qquad\qquad u(x,t) = \int_0^\infty g(\alpha) \exp(-\alpha^2 kt) \sin \alpha x \, d\alpha,$$

will formally satisfy all conditions of the boundary value problem if the function g can be determined so that

$$(6) \qquad\qquad f(x) = \int_0^\infty g(\alpha) \sin \alpha x \, d\alpha \qquad\qquad (x > 0).$$

As in the previous section, we note that representation (6) is the Fourier sine integral formula (3), Sec. 64, for our function f if

$$(7) \qquad\qquad g(\alpha) = \frac{2}{\pi} \int_0^\infty f(x) \sin \alpha x \, dx \qquad\qquad (\alpha > 0).$$

Our formal solution is therefore

$$(8) \qquad u(x,t) = \frac{2}{\pi} \int_0^\infty \exp(-\alpha^2 kt) \sin \alpha x \int_0^\infty f(s) \sin \alpha s \, ds \, d\alpha.$$

We can simplify this result by formally reversing the order of integration, replacing $2 \sin \alpha s \sin \alpha x$ by $\cos [\alpha(s - x)] - \cos [\alpha(s + x)]$, and then applying the integration formula (Problem 20, Sec. 69)

$$(9) \qquad \int_0^\infty \exp(-\alpha^2 a) \cos \alpha b \, d\alpha = \frac{1}{2} \sqrt{\frac{\pi}{a}} \exp\left(-\frac{b^2}{4a}\right) \qquad (a > 0).$$

Equation (8) then becomes

$$(10) \qquad u(x,t) = \frac{1}{2\sqrt{\pi k t}} \int_0^\infty f(s) \left\{ \exp\left[-\frac{(s - x)^2}{4kt}\right] - \exp\left[-\frac{(s + x)^2}{4kt}\right] \right\} ds$$

when $t > 0$. An alternative form of equation (10), obtained by introducing new variables of integration, is

$$(11) \qquad u(x,t) = \frac{1}{\sqrt{\pi}} \int_{-x/(2\sqrt{kt})}^\infty f(x + 2r\sqrt{kt}) \exp(-r^2) \, dr$$

$$- \frac{1}{\sqrt{\pi}} \int_{x/(2\sqrt{kt})}^\infty f(-x + 2r\sqrt{kt}) \exp(-r^2) \, dr.$$

Our use of the Fourier sine integral formula in obtaining the solution (8) suggests that we apply Corollary 1 in verifying that solution. The forms (10) and (11) suggest, however, that the condition in Corollary 1 that $|f(x)|$ be integrable from 0 to ∞ can be relaxed in verifying $u(x,t)$. To be precise, when s is kept fixed and $t > 0$, the functions

$$\frac{1}{\sqrt{t}} \exp\left[\frac{-(s \pm x)^2}{4kt}\right]$$

satisfy the heat equation (1). Then, under the assumption that f is continuous and bounded when $x \geq 0$, it is possible to show that the function (10) satisfies the heat equation when $x_0 < x < x_1$ and $t_0 < t < t_1$, where x_0, x_1, t_0, and t_1 are any positive numbers. Conditions (2) and (3) can be verified by using expression (11). By adding step functions (see Problem 4, Sec. 69) to f, we can permit f to have a finite number of jumps on the half line $x > 0$. Except for special cases, details in the verification of formal solutions of this problem are, however, tedious.

When $f(x) = 1$, it follows from equation (11) that

$$(12) \qquad u(x,t) = \frac{1}{\sqrt{\pi}} \left[\int_{-x/(2\sqrt{kt})}^\infty \exp(-r^2) \, dr - \int_{x/(2\sqrt{kt})}^\infty \exp(-r^2) \, dr \right].$$

In terms of the *error function*

$$(13) \qquad \text{erf}(x) = \frac{2}{\sqrt{\pi}} \int_0^x \exp(-r^2) \, dr,$$

where erf $(x) \to 1$ as $x \to \infty$ (see Problem 19, Sec. 69), equation (12) can be written

(14) $$u(x,t) = \text{erf}\left(\frac{x}{2\sqrt{kt}}\right).$$

The full verification of this result is not difficult.

69. Temperatures in an Unlimited Medium

For an application of the general Fourier integral formula, we now derive expressions for the temperatures $u(x,t)$ in a medium that occupies all space, where the initial temperature distribution is $f(x)$. We assume that f is bounded and, for the present, that it satisfies conditions under which it is represented by its Fourier integral formula. The boundary value problem consists of a boundedness condition $|u(x,t)| < M$ and the conditions

(1) $$u_t(x,t) = ku_{xx}(x,t) \qquad (-\infty < x < \infty, \, t > 0),$$

(2) $$u(x,0) = f(x) \qquad (-\infty < x < \infty).$$

Separation of variables leads to the singular eigenvalue problem

$$X''(x) + \lambda X(x) = 0, \qquad |X(x)| < M_1 \quad (-\infty < x < \infty),$$

whose eigenvalues are $\lambda = \alpha^2$ where $\alpha \geq 0$, and to two linearly independent eigenfunctions $\cos \alpha x$ and $\sin \alpha x$ corresponding to each nonzero value of α.

Our generalized linear combination of functions XT is

(3) $$u(x,t) = \int_0^\infty \exp(-\alpha^2 kt)[A(\alpha)\cos \alpha x + B(\alpha)\sin \alpha x]\, d\alpha.$$

The coefficients $A(\alpha)$ and $B(\alpha)$ are to be determined so that the integral here represents $f(x)$ $(-\infty < x < \infty)$ when $t = 0$. According to equations (8) and (9) of Sec. 61 and our Fourier integral theorem (Sec. 63), the representation is valid if

$$A(\alpha) = \frac{1}{\pi}\int_{-\infty}^\infty f(s)\cos \alpha s\, ds, \qquad B(\alpha) = \frac{1}{\pi}\int_{-\infty}^\infty f(s)\sin \alpha s\, ds.$$

Equation (3) then becomes

(4) $$u(x,t) = \frac{1}{\pi}\int_0^\infty \exp(-\alpha^2 kt)\int_{-\infty}^\infty f(s)\cos[\alpha(s-x)]\, ds\, d\alpha.$$

If we formally reverse the order of integration here, the integration formula (9) of Sec. 68 can be used to write equation (4) in the form

(5) $$u(x,t) = \frac{1}{2\sqrt{\pi kt}}\int_{-\infty}^\infty f(s)\exp\left[-\frac{(s-x)^2}{4kt}\right]ds \qquad (t > 0).$$

An alternative form of this is

(6) $$u(x,t) = \frac{1}{\sqrt{\pi}} \int_{-\infty}^{\infty} f(x + 2r\sqrt{kt}) \exp(-r^2)\, dr.$$

Forms (5) and (6) can be verified by assuming only that f is piecewise continuous over some bounded interval $|x| < c$ and continuous and bounded over the rest of the x axis, when $|x| \geq c$. If f is an odd function, $u(x,t)$ becomes the function found in the preceding section for positive values of x.

PROBLEMS

1. (a) Give details showing how the functions $\exp(-\alpha y) \sin \alpha x$ ($\alpha > 0$) arise by means of separation of variables from equations (1) and (2), Sec. 67, and the condition that the function $u(x,y)$ there is to be bounded when $x > 0$, $y > 0$.

(b) Show that if the function $u(x,y)$ in Sec. 67 satisfies the nonhomogeneous boundary condition

$$u(x,0) = e^{-x} - e^{-2x} \qquad (x \geq 0),$$

then

$$u(x,y) = \int_0^{\infty} g(\alpha)e^{-\alpha y} \sin \alpha x\, d\alpha \qquad (x \geq 0,\ y \geq 0)$$

where $g(\alpha)$ is the absolutely integrable function

$$g(\alpha) = \frac{6}{\pi} \frac{\alpha}{(\alpha^2 + 1)(\alpha^2 + 4)} \qquad (\alpha \geq 0).$$

Suggestion: Note that the singular eigenvalue problem that arises in part (a) has already been solved in Sec. 64. In part (b), refer to the result in Problem 4, Sec. 66, to write

$$e^{-x} - e^{-2x} = \frac{6}{\pi} \int_0^{\infty} \frac{\alpha \sin \alpha x}{(\alpha^2 + 1)(\alpha^2 + 4)}\, d\alpha \qquad (x \geq 0)$$

and then use this representation.

2. (a) Substitute expression (10), Sec. 67, for the function $g(\alpha)$ into equation (3) of that section. Then, by formally reversing the order of integration, show that the solution of the boundary value problem treated in Sec. 67 can be written

$$u(x,y) = \frac{y}{\pi} \int_0^{\infty} f(s) \left[\frac{1}{(s - x)^2 + y^2} - \frac{1}{(s + x)^2 + y^2} \right] ds.$$

(b) Show that when $f(x) = 1$, the form of the solution obtained in part (a) can be written in terms of the inverse tangent function as

$$u(x, y) = \frac{2}{\pi} \tan^{-1}\left(\frac{x}{y}\right).$$

3. Verify that the function $u = \mathrm{erf}[x/(2\sqrt{kt})]$ in Sec. 68 satisfies the heat equation $u_t = ku_{xx}$ when $x > 0$, $t > 0$ as well as the conditions

$$u(0+,t) = 0 \qquad\qquad (t > 0),$$

$$u(x,0+) = 1 \qquad\qquad (x > 0),$$

$$|u(x,t)| < 1 \qquad\qquad (x > 0,\, t > 0).$$

4. Show that if $f(x) = 0$ when $0 < x < c$ and $f(x) = 1$ when $x > c$, expression (11), Sec. 68, reduces to

$$u(x,t) = \frac{1}{2}\mathrm{erf}\left(\frac{c+x}{2\sqrt{kt}}\right) - \frac{1}{2}\mathrm{erf}\left(\frac{c-x}{2\sqrt{kt}}\right).$$

Verify this solution of the boundary value problem in Sec. 68 when f is this function.

5. The face $x = 0$ of a semi-infinite solid $x \geq 0$ is kept at constant temperature u_0 after its interior $x > 0$ is initially at temperature zero throughout. Obtain an expression for the temperatures $u(x,t)$ in the body.

$$Ans. \quad u(x,t) = u_0\left[1 - \mathrm{erf}\left(\frac{x}{2\sqrt{kt}}\right)\right].$$

6. The face $x = 0$ of a semi-infinite solid $x \geq 0$ is insulated, and the initial temperature distribution is $f(x)$. Derive the temperature formula

$$u(x,t) = \frac{1}{\sqrt{\pi}}\int_{-x/(2\sqrt{kt})}^{\infty} f(x + 2r\sqrt{kt})\exp(-r^2)\,dr$$

$$+ \frac{1}{\sqrt{\pi}}\int_{x/(2\sqrt{kt})}^{\infty} f(-x + 2r\sqrt{kt})\exp(-r^2)\,dr.$$

7. In Problem 6, write $f(x) = 1$ when $0 < x < c$ and $f(x) = 0$ when $x > c$. Then show that

$$u(x,t) = \frac{1}{2}\mathrm{erf}\left(\frac{c+x}{2\sqrt{kt}}\right) + \frac{1}{2}\mathrm{erf}\left(\frac{c-x}{2\sqrt{kt}}\right).$$

8. Let the initial temperature distribution $f(x)$ in the unlimited medium in Sec. 69 be $f(x) = 0$ when $x < 0$ and $f(x) = 1$ when $x > 0$. Show that

$$u(x,t) = \frac{1}{2} + \frac{1}{2}\mathrm{erf}\left(\frac{x}{2\sqrt{kt}}\right).$$

Verify this solution of the boundary value problem in Sec. 69 when f is this function.

9. Derive this solution of the wave equation $y_{tt} = a^2 y_{xx}$ $(-\infty < x < \infty,\ t > 0)$ that satisfies the conditions $y(x,0) = f(x)$, $y_t(x,0) = 0$ when $-\infty < x < \infty$:

$$y(x,t) = \frac{1}{\pi}\int_0^{\infty} \cos \alpha at \int_{-\infty}^{\infty} f(s)\cos[\alpha(s - x)]\,ds\,d\alpha.$$

Also, reduce the solution to the form obtained in Example 2, Sec. 19:

$$y(x,t) = \frac{1}{2}[f(x + at) + f(x - at)].$$

10. A semi-infinite string with one end fixed at the origin is stretched along the positive half of the x axis and released at rest from a position $y = f(x)$ $(x \geq 0)$. Derive the expression

$$y(x,t) = \frac{2}{\pi}\int_0^{\infty} \cos \alpha at \sin \alpha x \int_0^{\infty} f(s)\sin \alpha s\,ds\,d\alpha$$

for its transverse displacements. Let $F(x)$ $(-\infty < x < \infty)$ be the odd extension of $f(x)$ and show how this result reduces to the form

$$y(x,t) = \frac{1}{2}[F(x + at) + F(x - at)].$$

[Compare solution (9), Sec. 51, of the string problem treated in that section.]

11. Find the bounded harmonic function $V(x,y)$ in the semi-infinite strip $x > 0, 0 < y < 1$ that satisfies the boundary conditions (Fig. 25)

$$V_x(0,y) = 0, \qquad V_y(x,0) = 0, \qquad V(x,1) = f(x)$$

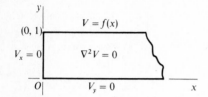

Figure 25

where (a) $f(x) = 1$ when $0 < x < 1$ and $f(x) = 0$ when $x > 1$; (b) $f(x) = e^{-x}$ $(x > 0)$.

Ans. (a) $V(x,y) = \dfrac{2}{\pi}\displaystyle\int_0^\infty \dfrac{\sin \alpha \cos \alpha x \cosh \alpha y}{\alpha \cosh \alpha}\,d\alpha;$ (b) $V(x,y) = \dfrac{2}{\pi}\displaystyle\int_0^\infty \dfrac{\cos \alpha x \cosh \alpha y}{(1 + \alpha^2)\cosh \alpha}\,d\alpha.$

12. Find $V(x,y)$ when the boundary conditions in Problem 11 are replaced by the conditions

$$V_x(0,y) = 0, \qquad V_y(x,1) = -V(x,1), \qquad V(x,0) = f(x)$$

where $f(x) = 1$ when $0 < x < 1$ and $f(x) = 0$ when $x > 1$. Interpret this problem physically.

Ans. $V(x,y) = \dfrac{2}{\pi}\displaystyle\int_0^\infty \dfrac{\alpha \cosh [\alpha(1 - y)] + \sinh [\alpha(1 - y)]}{\alpha^2 \cosh \alpha + \alpha \sinh \alpha}\sin \alpha \cos \alpha x\, d\alpha.$

13. Find the bounded harmonic function $V(x,y)$ in the semi-infinite strip $0 < x < 1, y > 0$ which satisfies the conditions

$$V_y(x,0) = 0, \qquad V(0,y) = 0, \qquad V_x(1,y) = f(y).$$

Ans. $V(x,y) = \dfrac{2}{\pi}\displaystyle\int_0^\infty \dfrac{\sinh \alpha x \cos \alpha y}{\alpha \cosh \alpha}\displaystyle\int_0^\infty f(s) \cos \alpha s\, ds\, d\alpha.$

14. Find $V(x,y)$ when the boundary conditions in Problem 13 are replaced by the conditions

$$V_y(x,0) = 0, \qquad V(0,y) = e^{-y}, \qquad V(1,y) = 0.$$

Ans. $V(x,y) = \dfrac{2}{\pi}\displaystyle\int_0^\infty \dfrac{\sinh [\alpha(1 - x)] \cos \alpha y}{(1 + \alpha^2)\sinh \alpha}\,d\alpha.$

15. Find the bounded harmonic function $V(x,y)$ in the strip $-\infty < x < \infty, 0 < y < b$ such that $V(x,0) = 0$ and $V(x,b) = f(x)$ $(-\infty < x < \infty)$, where f is bounded and represented by its Fourier integral.

Ans. $V(x,y) = \dfrac{1}{\pi}\displaystyle\int_0^\infty \dfrac{\sinh \alpha y}{\sinh \alpha b}\displaystyle\int_{-\infty}^\infty f(s) \cos [\alpha(s - x)]\, ds\, d\alpha.$

16. Let a semi-infinite solid $x \geq 0$, which is initially at a uniform temperature, be cooled or heated by keeping its boundary at a uniform constant temperature (Sec. 68). Show that the times required for two interior points to reach the same temperature are proportional to the squares of the distances of those points from the boundary plane.

17. Solve the following boundary value problem for steady temperatures $u(x,y)$ in a thin plate in the shape of a semi-infinite strip when heat transfer to the surroundings at temperature zero takes place at the faces of the plate:

$$u_{xx}(x,y) + u_{yy}(x,y) - hu(x,y) = 0 \qquad (x > 0, \, 0 < y < 1),$$

$$u_x(0,y) = 0 \qquad (0 < y < 1),$$

$$u(x,0) = 0, \qquad u(x,1) = f(x) \qquad (x > 0)$$

where h is a positive constant and $f(x) = 1$ when $0 < x < c$ and $f(x) = 0$ when $x > c$.

$$\textit{Ans.} \quad u(x,y) = \frac{2}{\pi} \int_0^\infty \frac{\sin \alpha c \, \cos \alpha x \, \sinh \, (y\sqrt{\alpha^2 + h})}{\alpha \, \sinh \, \sqrt{\alpha^2 + h}} \, d\alpha.$$

18. Verify that, for any constant C, the function

$$v(x,t) = Cxt^{-3/2} \exp\left(\frac{-x^2}{4kt}\right)$$

satisfies the heat equation $v_t = kv_{xx}$ when $x > 0$ and $t > 0$. Also verify that $v(0+,t) = 0$ when $t > 0$ and that $v(x,0+) = 0$ when $x > 0$. Thus v can be added to the function u found in Sec. 68 to form other solutions of the problem there if the temperature function is not required to be bounded. But note that v is unbounded as x and t tend to zero, as can be seen by letting x vanish while $t = x^2$.

19. Let I denote the integral of $\exp\left(-x^2\right)$ from 0 to ∞ and write

$$I^2 = \int_0^\infty e^{-x^2} \, dx \int_0^\infty e^{-y^2} \, dy = \int_0^\infty \int_0^\infty e^{-(x^2+y^2)} \, dx \, dy.$$

Evaluate the iterated integral by using polar coordinates and show that $I = \sqrt{\pi}/2$. Thus verify that erf (x), defined in equation (13), Sec. 68, tends to unity as x tends to infinity.

20. Derive the integration formula (9), Sec. 68, by first writing

$$y(x) = \int_0^\infty \exp\left(-\alpha^2 a\right) \cos \alpha x \, d\alpha \qquad (a > 0)$$

and differentiating the integral to find $y'(x)$. Then integrate the new integral by parts to show that $2ay'(x) = -xy(x)$, point out why

$$y(0) = \frac{1}{2}\sqrt{\frac{\pi}{a}}$$

(see Problem 19), and solve for $y(x)$. The desired result is the value of y when $x = b$.†

---

† Another derivation is indicated in Churchill, Brown, and Verhey (1974, p. 191), listed in the Bibliography.

EIGHT

BESSEL FUNCTIONS AND APPLICATIONS

70. Bessel's Equation

In boundary value problems that involve the laplacian $\nabla^2 u$ expressed in cylindrical coordinates, the process of separating variables often produces an equation of the form

$$(1) \qquad \rho^2 \frac{d^2 y}{d\rho^2} + \rho \frac{dy}{d\rho} + (\lambda \rho^2 - v^2) y = 0,$$

where y is a function of the cylindrical coordinate ρ. In such problems, as we shall observe in examples presented in this chapter, $-\lambda$ is the separation constant and the values of λ are eigenvalues associated with equation (1). The parameter v is a real number determined by other aspects of the boundary value problem. Since v is squared where it appears in equation (1), it is sufficient to solve that equation with the agreement that $v \geq 0$. Usually v is either zero or a positive integer.

In our applications it turns out that $\lambda \geq 0$; and, when $\lambda > 0$, the substitution

$$x = \sqrt{\lambda}\, \rho$$

can be used to transform equation (1) into a form that is free of λ:

$$(2) \qquad x^2 y''(x) + x y'(x) + (x^2 - v^2) y(x) = 0.$$

This linear homogeneous differential equation is known as *Bessel's equation*. Its solutions are called *Bessel functions*, or sometimes cylindrical functions.

Upon comparing equation (2) with the standard form

$$y''(x) + A(x) y'(x) + B(x) y(x) = 0,$$

we see that $A(x) = 1/x$ and $B(x) = 1 - (v/x)^2$. Those coefficients are both continuous except at the origin, which is a singular point of Bessel's equation. The existence and uniqueness theorem stated in Sec. 18 applies to the equation for any closed bounded interval that does not include the origin.

But for boundary value problems in regions interior to circles or cylinders $\rho = c$, the origin $x = 0$, corresponding to the center or axis $\rho = 0$, is interior to the region. The interval for the variable x then has the origin as an end point.

We limit our attention primarily to the cases $v = n$ where $n = 0, 1, 2, \ldots$. In those cases we shall see that there is a solution of Bessel's equation which is *analytic for all values of x*, including the origin; that is, one solution is represented by a power series in x which is convergent for every x. That solution, denoted by $J_n(x)$, and its derivatives of all orders are therefore everywhere continuous functions. In referring to any power series, we agree that we shall mean a Maclaurin series, or a Taylor series about the origin.

71. Bessel Functions J_n

We let n denote any fixed nonnegative integer and seek a solution of Bessel's equation

$$(1) \qquad x^2 y''(x) + xy'(x) + (x^2 - n^2)y(x) = 0 \qquad (n = 0, 1, 2, \ldots)$$

in the form of a power series multiplied by x^p, where the first term in that series is nonzero and p is some constant. That is, we attempt to determine p and the coefficients a_j so that the function

$$(2) \qquad y = x^p \sum_{j=0}^{\infty} a_j x^j = \sum_{j=0}^{\infty} a_j x^{p+j} \qquad (a_0 \neq 0)$$

satisfies equation (1).†

Assume for the present that the series is differentiable. Then, upon substituting the function (2) and its derivatives into equation (1), we obtain the equation

$$\sum_{j=0}^{\infty} [(p + j)(p + j - 1) + (p + j) + (x^2 - n^2)]a_j x^{p+j} = 0,$$

or

$$\sum_{j=0}^{\infty} [(p + j)^2 - n^2]a_j x^j + \sum_{j=0}^{\infty} a_j x^{j+2} = 0.$$

The last sum can be written

$$\sum_{j=2}^{\infty} a_{j-2} x^j;$$

---

† The series method used here to solve equation (1) is often referred to as the method of Frobenius and is treated in introductory texts on ordinary differential equations, such as the one by Boyce and DiPrima (1977) or the one by Rainville and Bedient (1974). Both of these are listed in the Bibliography.

thus

$$(3) \quad (p - n)(p + n)a_0 + (p - n + 1)(p + n + 1)a_1 x$$

$$+ \sum_{j=2}^{\infty} [(p - n + j)(p + n + j)a_j + a_{j-2}]x^j = 0.$$

Equation (3) is an identity in x if the coefficient of each power of x vanishes. This condition is satisfied if $p = n$ or $p = -n$, so that the constant term vanishes, and if $a_1 = 0$ and

$$(p - n + j)(p + n + j)a_j + a_{j-2} = 0 \qquad (j = 2, 3, \ldots).$$

We make the choice $p = n$. Then the *recurrence relation*

$$(4) \qquad\qquad a_j = \frac{-1}{j(2n + j)} a_{j-2} \qquad\qquad (j = 2, 3, \ldots)$$

is obtained, giving each coefficient in terms of the second one preceding it in the series. Note that $2n + j \neq 0$ for all j in relation (4); when $n \neq 0$, the choice $p = -n$ fails to give such a recurrence relation.

Since $a_1 = 0$, relation (4) requires that $a_3 = 0$; then $a_5 = 0$, and so on. That is,

$$(5) \qquad\qquad a_{2k+1} = 0 \qquad\qquad (k = 0, 1, 2, \ldots).$$

To obtain the remaining coefficients, we let k denote any positive integer and use relation (4) to write the following k equations:

$$a_2 = \frac{-1}{1(n + 1)2^2} a_0,$$

$$a_4 = \frac{-1}{2(n + 2)2^2} a_2,$$

$$\cdots$$

$$a_{2k} = \frac{-1}{k(n + k)2^2} a_{2k-2}.$$

Upon equating the product of the left-hand sides of these equations to the product of their right-hand sides, we find that

$$(6) \qquad a_{2k} = \frac{(-1)^k}{k!(n + 1)(n + 2) \cdots (n + k)2^{2k}} a_0 \qquad (k = 1, 2, \ldots).$$

The coefficient a_0 may have any nonzero value. To simplify expression (6), we write

$$a_0 = \frac{1}{n! \, 2^n}.$$

The nonvanishing coefficients are then

$$(7) \qquad a_{2k} = \frac{(-1)^k}{k!\,(n+k)!} \frac{1}{2^{n+2k}} \qquad (k = 0, 1, 2, \ldots),$$

where we use the convention that $0! = 1$.

Our proposed solution (2) of Bessel's equation (1) can now be written, in view of expressions (5) and (7), as

$$y = J_n(x)$$

where J_n is the *Bessel function of the first kind of order, or index, n* $(n = 0, 1, 2, \ldots)$:

$$(8) \qquad J_n(x) = \sum_{k=0}^{\infty} \frac{(-1)^k}{k!\,(n+k)!} \left(\frac{x}{2}\right)^{n+2k}$$

$$= \frac{x^n}{2^n n!}\left[1 - \frac{x^2}{2^2 1!\,(n+1)} + \frac{x^4}{2^4 2!\,(n+1)(n+2)}\right.$$

$$\left. - \frac{x^6}{2^6 3!\,(n+1)(n+2)(n+3)} + \cdots \right].$$

The power series (8) is absolutely convergent for all x $(-\infty < x < \infty)$, according to the ratio test. It therefore represents a continuous function and is differentiable with respect to x any number of times. Since it is differentiable and its coefficients satisfy the recurrence relation needed to make its sum satisfy Bessel's equation (1), $y = J_n(x)$ is indeed a solution of that equation.

Theorem 1. *For all x, the analytic function $J_n(x)$ is a solution of Bessel's equation* (1).

Since Bessel's equation is linear and homogeneous, the function $CJ_n(x)$, where C is any constant, is also a solution.

From expression (8) we note that

$$(9) \qquad J_n(-x) = (-1)^n J_n(x);$$

that is, J_n is an even function if $n = 0, 2, 4, \ldots$, but odd if $n = 1, 3, 5, \ldots$.

The series representation

$$(10) \qquad J_0(x) = 1 - \frac{x^2}{2^2} + \frac{x^4}{2^2 4^2} - \frac{x^6}{2^2 4^2 6^2} + \cdots$$

bears some resemblance to the power series for cos x. There is also a similarity between the power series representations of the odd functions $J_1(x)$ and sin x. Similarities between the properties of those functions include, as we shall see, the differentiation formula $J_0'(x) = -J_1(x)$, corresponding to the formula for the derivative of cos x. Graphs of J_0 and J_1 will be shown in Sec. 77.

72. Some Other Bessel Functions

Functions linearly independent of J_n that satisfy Bessel's equation

$$(1) \qquad x^2 y''(x) + xy'(x) + (x^2 - n^2)y(x) = 0 \qquad (n = 0, 1, 2, \ldots)$$

can be obtained by various methods of a fairly elementary nature.

The singular point $x = 0$ of equation (1) belongs to the class of such points known as *regular* singular points. The power series procedure, extended so as to give general solutions near regular singular points, applies to Bessel's equation. We do not give further details here but only state the results.

When $n = 0$, the general solution is found to be

$$(2) \quad y = AJ_0(x)$$

$$+ B\left[J_0(x) \log x + \frac{x^2}{2^2} - \frac{x^4}{2^2 4^2}\left(1 + \frac{1}{2}\right) + \frac{x^6}{2^2 4^2 6^2}\left(1 + \frac{1}{2} + \frac{1}{3}\right) - \cdots\right],$$

where A and B are arbitrary constants and $x > 0$. Observe that as long as $B \neq 0$, any choice of A and B yields a solution which is unbounded as x tends to zero through positive values. Such a solution cannot therefore be expressed as a constant times $J_0(x)$, which tends to unity as x tends to zero. So $J_0(x)$ and the solution (2) are linearly independent when $B \neq 0$. It is most common to use *Euler's constant* $\gamma = 0.5772\ldots$, which is defined as the limit of the sequence

$$(3) \qquad s_n = 1 + \frac{1}{2} + \frac{1}{3} + \cdots + \frac{1}{n} - \log n \qquad (n = 1, 2, \ldots),$$

and write $A = (2/\pi)(\gamma - \log 2)$, $B = 2/\pi$. When A and B are assigned those values, the second solution that arises is *Weber's Bessel function of the second kind of order zero:*

$$(4) \qquad Y_0(x) = \frac{2}{\pi}\left[\left(\log\frac{x}{2} + \gamma\right)J_0(x)\right.$$

$$\left. + \frac{x^2}{2^2} - \frac{x^4}{2^2 4^2}\left(1 + \frac{1}{2}\right) + \frac{x^6}{2^2 4^2 6^2}\left(1 + \frac{1}{2} + \frac{1}{3}\right) - \cdots\right].\dagger$$

---

† There are other Bessel functions, and the notation varies widely throughout the literature. The treatise by Watson (1952) that is listed in the Bibliography is, however, usually regarded as the standard reference.

More generally, when n has any one of the values $n = 0, 1, 2, \ldots$, equation (1) has a solution $Y_n(x)$ which is valid when $x > 0$ and is unbounded as x tends to zero. Since $J_n(x)$ is continuous at $x = 0$, $J_n(x)$ and $Y_n(x)$ are linearly independent; and when $x > 0$, the general solution of equation (1) is

$$(5) \qquad\qquad y = CJ_n(x) + DY_n(x) \qquad\qquad (n = 0, 1, 2, \ldots),$$

where C and D are arbitrary constants. The theory of the second solution $Y_n(x)$ is considerably more involved than that of $J_n(x)$, and we shall limit our applications to problems where it is only necessary to know that $Y_n(x)$ is unbounded on any interval $(0,c)$.

To write the general solution of Bessel's equation

$$(6) \qquad x^2 y''(x) + xy'(x) + (x^2 - v^2)y(x) = 0 \quad (v > 0, \, v \neq 1, 2, \ldots),$$

where v is any positive number other than $1, 2, \ldots$, it is necessary to know some elementary properties of the *gamma function*, defined when $v > 0$ by means of the equation

$$(7) \qquad\qquad \Gamma(v) = \int_0^\infty e^{-t} t^{v-1} \, dt \qquad\qquad (v > 0).$$

An integration by parts shows that

$$(8) \qquad\qquad \Gamma(v + 1) = v\Gamma(v)$$

when $v > 0$. That property is *assigned* to the function when $v < 0$, so that $\Gamma(v) = \Gamma(v + 1)/v$ when $-1 < v < 0$, $-2 < v < -1$, and so on. Thus equations (7) and (8) together define $\Gamma(v)$ for all v except $v = 0, -1, -2, \ldots$. We find from equation (7) that $\Gamma(1) = 1$; also, it can be shown that Γ is continuous when $v > 0$. It then follows from property (8) that $\Gamma(0+) = \infty$ and consequently that $|\Gamma(v)|$ becomes infinite as $v \to -n$ $(n = 1, 2, \ldots)$.

When $v = 1, 2, 3, \ldots$, $\Gamma(v)$ reduces to a factorial; specifically,

$$(9) \qquad\qquad \Gamma(n + 1) = n! \qquad\qquad (n = 0, 1, 2, \ldots).$$

The proof of property (9) and the further property that

$$\Gamma(\tfrac{1}{2}) = \sqrt{\pi}$$

is left to the problems.

We define the *Bessel function of the first kind of order* v $(v \geq 0)$ as

$$(10) \qquad\qquad J_v(x) = \sum_{k=0}^\infty \frac{(-1)^k}{k!\,\Gamma(v + k + 1)}\left(\frac{x}{2}\right)^{v+2k},$$

and we note that this becomes expression (8), Sec. 71, for $J_n(x)$ when $v = n = 0, 1, 2, \ldots$. *The Bessel function $J_{-v}(x)$ $(v > 0)$ is also well-defined* when v is replaced by $-v$ in equation (10); if v is a positive integer, the terms in the expression for $J_{-v}(x)$ where the argument $(-v + k + 1)$ of Γ is zero or a negative integer are to be dropped from the series. It is not difficult to verify by direct substitution into equation (6) that J_v and J_{-v} are solutions of that equation. These solutions are arrived at by a modification, involving property (8) of the gamma function, of the procedure used in Sec. 71.

When $v > 0$ and $v \neq 1, 2, \ldots$, the Bessel function $J_{-v}(x)$ is the product of $1/x^v$ and a power series in x whose initial term $(k = 0)$ is nonzero; hence $J_{-v}(x)$ is unbounded as $x \to 0$. Since $J_v(x)$ tends to zero as $x \to 0$, it is evident that J_v and J_{-v} are linearly independent functions. The general solution of Bessel's equation (6) can therefore be written

$$(11) \qquad\qquad y = CJ_v(x) + DJ_{-v}(x) \qquad\qquad (v > 0, \ v \neq 1, 2, \ldots),$$

where C and D are arbitrary constants. [Contrast this with solution (5) of equation (1).]

It can be shown that J_n and J_{-n} are linearly dependent because

$$(12) \qquad\qquad J_{-n}(x) = (-1)^n J_n(x) \qquad\qquad (n = 0, 1, 2, \ldots)$$

(see Problem 7, Sec. 73). So if $v = n = 0, 1, 2, \ldots$, solution (11) cannot be the *general* solution of equation (6).

73. Recurrence Relations

Starting with the equation

$$x^{-n} J_n(x) = \frac{1}{2^n} \sum_{k=0}^{\infty} \frac{(-1)^k}{k!\,(n+k)!} \left(\frac{x}{2}\right)^{2k} \qquad (n = 0, 1, 2, \ldots),$$

let us write

$$\frac{d}{dx}[x^{-n} J_n(x)] = \frac{1}{2^n} \sum_{k=1}^{\infty} \frac{k(-1)^k}{k(k-1)!\,(n+k)!} \left(\frac{x}{2}\right)^{2k-1}.$$

If we replace k by $k + 1$ here, so that k runs from 0 to ∞ again, it follows that

$$\frac{d}{dx}[x^{-n} J_n(x)] = x^{-n} \left(\frac{x}{2}\right)^n \sum_{k=0}^{\infty} \frac{(-1)^{k+1}}{k!\,(n+k+1)!} \left(\frac{x}{2}\right)^{2k+1}$$

$$= -x^{-n} \sum_{k=0}^{\infty} \frac{(-1)^k}{k!\,(n+1+k)!} \left(\frac{x}{2}\right)^{n+1+2k},$$

or

$$(1) \qquad\qquad \frac{d}{dx}[x^{-n} J_n(x)] = -x^{-n} J_{n+1}(x) \qquad\qquad (n = 0, 1, 2, \ldots).$$

The special case

(2) $$J_0'(x) = -J_1(x)$$

was mentioned at the end of Sec. 71.

Similarly, from the power series representation of $x^n J_n(x)$, we can show that

(3) $$\frac{d}{dx}[x^n J_n(x)] = x^n J_{n-1}(x) \qquad (n = 1, 2, \ldots).$$

Relations (1) and (3), which are called *recurrence relations*, can be written

$$x J_n'(x) = n J_n(x) - x J_{n+1}(x),$$
$$x J_n'(x) = -n J_n(x) + x J_{n-1}(x).$$

Eliminating $J_n'(x)$ from these equations, we find that

(4) $$x J_{n+1}(x) = 2n J_n(x) - x J_{n-1}(x) \qquad (n = 1, 2, \ldots).$$

This recurrence relation expresses J_{n+1} in terms of the functions J_n and J_{n-1} of lower orders.

From equation (3) we get the integration formula

(5) $$\int_0^x s^n J_{n-1}(s) \, ds = x^n J_n(x) \qquad (n = 1, 2, \ldots).$$

An important special case is

(6) $$\int_0^x s J_0(s) \, ds = x J_1(x).$$

Relations (1), (3), and (4) are valid when n is replaced by the unrestricted parameter v. Modifications of the derivations simply consist of writing $\Gamma(v + k + 1)$ or $(v + k)\Gamma(v + k)$ in place of $(n + k)!$.

PROBLEMS

1. Show that $J_0(0) = 1$, $J_n(0) = 0$ $(n = 1, 2, \ldots)$.

2. From the series representation (8), Sec. 71, for J_n show that

$$i^{-n} J_n(ix) = \sum_{k=0}^{\infty} \frac{1}{k!(n+k)!} \left(\frac{x}{2}\right)^{n+2k} \qquad (n = 0, 1, 2, \ldots).$$

The function $I_n(x) = i^{-n} J_n(ix)$ is the *modified Bessel function of the first kind of order n*. Show that the series here converges for all x, that $I_n(x) > 0$ when $x > 0$, and that $I_n(-x) = (-1)^n I_n(x)$. Also show that since $J_n(x)$ satisfies Bessel's equation (1), Sec. 71, $I_n(x)$ is a solution of the modified equation

$$x^2 y''(x) + x y'(x) - (x^2 + n^2) y(x) = 0.$$

3. (a) Derive the factorial property $\Gamma(v + 1) = v\Gamma(v)$ of the gamma function when $v > 0$. (b) Show that $\Gamma(1) = 1$ and hence that $\Gamma(n + 1) = n!$ when $n = 0, 1, 2, \ldots$.

4. Refer to the result obtained in Problem 19, Sec. 69, and show that

$$\Gamma(\tfrac{1}{2}) = 2 \int_0^\infty e^{-x^2} \, dx = \sqrt{\pi}.$$

5. Verify that the function J_v $(v \geq 0)$ defined by equation (10), Sec. 72, satisfies Bessel's equation (6) in that section. Point out how it follows that J_{-v} is also a solution.

6. From definition (10), Sec. 72, of J_v show that

$$(a) \quad J_{1/2}(x) = \sqrt{\frac{2}{\pi x}} \sin x; \qquad (b) \quad J_{-1/2}(x) = \sqrt{\frac{2}{\pi x}} \cos x.$$

7. Using series (10), Sec. 72, and recalling that certain terms are to be dropped, show that

$$J_{-n}(x) = (-1)^n J_n(x) \qquad (n = 0, 1, 2, \ldots)$$

and hence that the functions J_n and J_{-n} are linearly dependent.

8. Derive relation (3), Sec. 73.

9. Establish the differentiation formula

$$x^2 J_n''(x) = (n^2 - n - x^2)J_n(x) + xJ_{n+1}(x) \qquad (n = 0, 1, 2, \ldots).$$

10. Derive the differentiation formula

$$\frac{d}{dx}[x^{-v} J_v(x)] = -x^{-v} J_{v+1}(x),$$

where $v \geq 0$, and point out why it is also valid when v is replaced by $-v$ $(v > 0)$. [Compare relation (1), Sec. 73.]

11. Use results in Problems 6 and 10 to show that

$$J_{3/2}(x) = \sqrt{\frac{2}{\pi x}} \left(\frac{\sin x}{x} - \cos x \right).$$

12. Let $y(x)$ denote a solution of Bessel's equation (1), Sec. 72, when $n = 0$.

(a) Put that special case of Bessel's equation in self-adjoint form $L[y(x)] = 0$ (see Sec. 31) and, by writing $X = J_0$ and $Y = y$ in Lagrange's identity [Problem 17(a), Sec. 34], show that

$$\frac{d}{dx}\left[\frac{y(x)}{J_0(x)} \right] = \frac{B}{x[J_0(x)]^2}$$

where B is some constant.

(b) Assuming that the function $1/[J_0(x)]^2$ has a Maclaurin series expansion of the form

$$\frac{1}{[J_0(x)]^2} = 1 + \sum_{k=1}^\infty c_k x^{2k}\dagger$$

---

† This valid assumption is most easily justified by methods from the theory of functions of a complex variable. See, for instance, Churchill, Brown, and Verhey (1974, pp. 164–165), listed in the Bibliography.

and that the resulting expansion obtained by multiplying each side by $1/x$ can be integrated term by term, use the result in part (a) to show formally that y can be written in the form

$$y = AJ_0(x) + B\left[J_0(x) \log x + \sum_{k=1}^{\infty} d_k x^{2k}\right]$$

where A, B, and d_k ($k = 1, 2, \ldots$) are constants. [Compare equation (2), Sec. 72.]

13. Since the two Bessel functions $J_0(x)$ and $Y_0(x)$ are linearly independent, their *Wronskian*

$$W[J_0(x),\ Y_0(x)] = \begin{vmatrix} J_0 & Y_0 \\ J_0' & Y_0' \end{vmatrix} = J_0 Y_0' - J_0' Y_0$$

never vanishes when $x > 0$.† Find a simple expression for that Wronskian in the following way.

(a) By writing $y(x) = Y_0(x)$ in the result obtained in Problem 12(a), show that

$$W[J_0(x),\ Y_0(x)] = \frac{C}{x} \qquad (x > 0)$$

where C is some constant.

(b) Use expression (4), Sec. 72, for $Y_0(x)$ to show that

$$W[J_0(x),\ Y_0(x)] = \frac{2}{\pi x}[J_0(x)]^2 + \frac{2}{\pi}[J_0(x)S'(x) - J_0'(x)S(x)]$$

where

$$S(x) = \frac{x^2}{2^2} - \frac{x^4}{2^2 4^2}\left(1 + \frac{1}{2}\right) + \frac{x^6}{2^2 4^2 6^2}\left(1 + \frac{1}{2} + \frac{1}{3}\right) - \cdots \qquad (x > 0).$$

Thus show that

$$\lim_{\substack{x \to 0 \\ x > 0}} x W[J_0(x),\ Y_0(x)] = \frac{2}{\pi}.$$

(c) Conclude from part (b) that the constant C in part (a) has the value $C = 2/\pi$ and hence that

$$W[J_0(x),\ Y_0(x)] = \frac{2}{\pi x} \qquad (x > 0).$$

14. Let s_n ($n = 1, 2, \ldots$) be the sequence defined in equation (3), Sec. 72. Show that $s_n > 0$ and $s_n - s_{n+1} > 0$ for each n. Thus show that the sequence is bounded and decreasing and that it therefore converges to some number γ. Also, point out how it follows that $0 \le \gamma < 1$.

Suggestion: Observe from the graph of the function $y = 1/x$ that

$$\sum_{k=1}^{n-1} \frac{1}{k} > \int_1^n \frac{dx}{x} = \log n \qquad (n \ge 2)$$

and

$$\frac{1}{n+1} < \int_n^{n+1} \frac{dx}{x} = \log (n+1) - \log n \qquad (n \ge 1).$$

† See, for example, one of the texts mentioned in the footnote in Sec. 71 for a discussion of the Wronskian, which plays an important role in the theory of ordinary differential equations.

15. (a) Derive the *reduction formula*

$$\int_0^x s^n J_0(s)\, ds = x^n J_1(x) + (n-1)x^{n-1}J_0(x) - (n-1)^2 \int_0^x s^{n-2}J_0(s)\, ds \qquad (n = 2, 3, \ldots)$$

by using integration by parts twice and using the relations (Sec. 73)

$$\frac{d}{ds}[sJ_1(s)] = sJ_0(s), \qquad \frac{d}{ds}J_0(s) = -J_1(s)$$

in the first and second of those integrations, respectively.

(b) Note that, in view of equation (6), Sec. 73, the identity obtained in part (a) can be applied successively to evaluate the integral on the left-hand side of that identity when the integer n is *odd*.† Illustrate this by showing that

$$\int_0^x s^5 J_0(s)\, ds = x(x^2 - 8)[4xJ_0(x) + (x^2 - 8)J_1(x)].$$

74. Bessel's Integral Form of J_n

Consider two power series converging for every real number x, with sums $f(x)$ and $g(x)$:

$$(1) \qquad \sum_{m=0}^{\infty} a_m x^m = f(x), \qquad \sum_{m=0}^{\infty} b_m x^m = g(x).$$

The *Cauchy product*

$$a_0 b_0 + (a_0 b_1 + a_1 b_0)x + (a_0 b_2 + a_1 b_1 + a_2 b_0)x^2 + \cdots$$

of those series also converges for every x and represents the product $f(x)g(x)$; that is,

$$(2) \qquad \sum_{m=0}^{\infty} c_m x^m = f(x)g(x) \quad \text{where} \quad c_m = \sum_{j=0}^{m} a_j b_{m-j}.$$

This is true, moreover, when the coefficients a_m and b_m are complex numbers.‡

In particular, since

$$\sum_{m=0}^{\infty} \frac{(i/2)^m}{m!} e^{im\theta} x^m = \exp\left(\frac{ix}{2} e^{i\theta}\right)$$

† Note, too, that when n is even, the reduction formula can be used to transform the problem of evaluating $\int_0^x s^n J_0(s)\, ds$ into that of evaluating $\int_0^x J_0(s)\, ds$, which is tabulated for various values of x on, for example, pp. 492–493 of the book edited by Abramowitz and Stegun (1965) that is listed in the Bibliography. Further references are given on pp. 490–491 of that book.

‡ See, for example, Taylor and Mann (1972, pp. 635 and 662) and also Churchill, Brown, and Verhey (1974, pp. 164–165), both listed in the Bibliography.

and

$$\sum_{m=0}^{\infty} \frac{(i/2)^m}{m!} e^{i(n-m)\theta} x^m = \exp\left(\frac{ix}{2} e^{-i\theta}\right) e^{in\theta},$$

where θ is independent of x and where n is a fixed nonnegative integer, it follows that

(3)
$$\sum_{m=0}^{\infty} c_m(\theta) x^m = \exp\,(ix \cos\,\theta) e^{in\theta}$$

where

(4)
$$c_m(\theta) = \frac{i^m}{2^m} \sum_{j=0}^{m} \frac{1}{j!\,(m-j)!}\, e^{i(n-m+2j)\theta}.$$

We shall now show that, except for a multiplicative constant, the integral from $-\pi$ to π with respect to θ of the right-hand side of equation (3) is the Bessel function $J_n(x)$. To do this, we observe first that the series on the left-hand side of that equation is uniformly convergent with respect to θ on the interval $-\pi \le \theta \le \pi$. For expression (4) and the binomial expansion formula allow us to write

$$|c_m(\theta)x^m| \le \left(\frac{|x|}{2}\right)^m \frac{1}{m!} \sum_{j=0}^{m} \frac{m!}{j!\,(m-j)!} = \left(\frac{|x|}{2}\right)^m \frac{(1+1)^m}{m!} = \frac{|x|^m}{m!},$$

and the Weierstrass M-test can be applied. Term by term integration of the series in equation (3) over the interval $-\pi \le \theta \le \pi$ is therefore justified. That is,

(5)
$$\sum_{m=0}^{\infty} \int_{-\pi}^{\pi} c_m(\theta)\,d\theta\,x^m = \int_{-\pi}^{\pi} \exp\,(ix \cos\,\theta) e^{in\theta}\,d\theta.$$

The integrals on the left-hand side of this last equation can be evaluated as follows. Using expression (4), write

(6)
$$\int_{-\pi}^{\pi} c_m(\theta)\,d\theta = \frac{i^m}{2^m} \sum_{j=0}^{m} \frac{1}{j!\,(m-j)!} \int_{-\pi}^{\pi} e^{i(n-m+2j)\theta}\,d\theta.$$

Since

$$\int_{-\pi}^{\pi} e^{iN\theta}\,d\theta = \begin{cases} 0 & \text{if } N = \pm 1, \pm 2, \ldots, \\ 2\pi & \text{if } N = 0, \end{cases}$$

it is evident that all the terms on the right-hand side of equation (6) vanish unless $m = n + 2j$. So the only integers m for which it is possible that the integral on the left-hand side of equation (6) is nonzero are those of the form

$m = n + 2k$ $(k = 0, 1, 2, \ldots)$. When m is such an integer, all the terms on the right-hand side of equation (6) vanish except when $j = k$. To be precise,

$$(7) \qquad \int_{-\pi}^{\pi} c_{n+2k}(\theta) \, d\theta = \frac{2\pi}{k!(n+k)!} \left(\frac{i}{2}\right)^{n+2k} \qquad (k = 0, 1, 2, \ldots).$$

Equation (5) therefore reduces to

$$2\pi i^n \sum_{k=0}^{\infty} \frac{(-1)^k}{k!(n+k)!} \left(\frac{x}{2}\right)^{n+2k} = \int_{-\pi}^{\pi} \exp{(ix \cos \theta)} e^{in\theta} \, d\theta,$$

or

$$(8) \qquad J_n(x) = \frac{i^{-n}}{2\pi} \int_{-\pi}^{\pi} \exp{(ix \cos \theta)} e^{in\theta} \, d\theta \qquad (n = 0, 1, 2, \ldots).$$

Now $i = \exp{(i\pi/2)}$. Hence

$$J_n(x) = \frac{1}{2\pi} \int_{-\pi}^{\pi} \exp{(ix \cos \theta)} \exp\left[in\left(\theta - \frac{\pi}{2}\right)\right] d\theta.$$

If we make the substitution $\phi = (\pi/2) - \theta$ and note that the resulting integrand is periodic in ϕ with period 2π (see Problem 15, Sec. 42), we find that

$$(9) \qquad J_n(x) = \frac{1}{2\pi} \int_{-\pi}^{\pi} \exp{[i(x \sin \phi - n\phi)]} \, d\phi.$$

The imaginary part of this integral vanishes, of course, since $J_n(x)$ is real when x is real. Therefore

$$J_n(x) = \frac{1}{2\pi} \int_{-\pi}^{\pi} \cos{(x \sin \phi - n\phi)} \, d\phi.$$

Finally, since the integrand here is an even function of ϕ,

$$(10) \qquad J_n(x) = \frac{1}{\pi} \int_{0}^{\pi} \cos{(n\phi - x \sin \phi)} \, d\phi \qquad (n = 0, 1, 2, \ldots).$$

This is *Bessel's integral form* of $J_n(x)$.

75. Consequences of Bessel's Integral Form

From the integral representation

$$(1) \qquad J_n(x) = \frac{1}{\pi} \int_{0}^{\pi} \cos{(n\phi - x \sin \phi)} \, d\phi \qquad (n = 0, 1, 2, \ldots)$$

just obtained, it follows that

$$J_n'(x) = \frac{1}{\pi} \int_{0}^{\pi} \sin{(n\phi - x \sin \phi)} \sin \phi \, d\phi.$$

Continued differentiation yields integral representations for $J_n''(x)$, and so on. In each case the absolute value of the integrand that arises does not exceed unity. The following *boundedness properties* are then consequences of Bessel's integral form (1):

$$(2) \qquad |J_n(x)| \le 1, \qquad \left| \frac{d^k}{dx^k} J_n(x) \right| \le 1 \qquad (-\infty < x < \infty, \; k = 1, 2, \ldots).$$

For example, the first inequality is true because

$$|J_n(x)| \le \frac{1}{\pi} \int_0^\pi |\cos (n\phi - x \sin \phi)| \, d\phi \le \frac{1}{\pi} \int_0^\pi d\phi = 1.$$

The integrand in expression (1) can be written

$$\cos (n\phi - x \sin \phi) = f_n(x,\phi) + g_n(x,\phi)$$

where

$$f_n(x,\phi) = \cos n\phi \cos (x \sin \phi), \qquad g_n(x,\phi) = \sin n\phi \sin (x \sin \phi).$$

For each fixed x the graphs of f_n and g_n have these properties of symmetry with respect to the line $\phi = \pi/2$:

$$(3) \qquad \begin{aligned} f_n(x,\pi - \phi) &= (-1)^n f_n(x,\phi), \\ g_n(x,\pi - \phi) &= -(-1)^n g_n(x,\phi). \end{aligned}$$

Consequently

$$\int_0^\pi f_{2n}(x, \phi) \, d\phi = 2 \int_0^{\pi/2} f_{2n}(x, \phi) \, d\phi, \qquad \int_0^\pi g_{2n}(x, \phi) \, d\phi = 0,$$

and therefore

$$J_{2n}(x) = \frac{1}{\pi} \int_0^\pi \cos 2n\phi \cos (x \sin \phi) \, d\phi,$$

or

$$(4) \qquad J_{2n}(x) = \frac{2}{\pi} \int_0^{\pi/2} \cos 2n\phi \cos (x \sin \phi) \, d\phi \qquad (n = 0, 1, 2, \ldots).$$

In like manner, we see that

$$J_{2n-1}(x) = \frac{1}{\pi} \int_0^\pi \sin (2n - 1)\phi \sin (x \sin \phi) \, d\phi,$$

or

$$(5) \qquad J_{2n-1}(x) = \frac{2}{\pi} \int_0^{\pi/2} \sin (2n - 1)\phi \sin (x \sin \phi) \, d\phi \qquad (n = 1, 2, \ldots).$$

Note the special case

$$(6) \qquad J_0(x) = \frac{2}{\pi} \int_0^{\pi/2} \cos (x \sin \phi) \, d\phi$$

of representation (4), which can also be written

$$(7) \qquad J_0(x) = \frac{2}{\pi} \int_0^{\pi/2} \cos (x \cos \theta) \, d\theta$$

by means of the substitution $\theta = (\pi/2) - \phi$.

For each fixed x the Riemann-Lebesgue lemma (Lemma 1, Sec. 40) applies to the integrals in expressions (4) and (5) to show that $J_{2n}(x)$ and $J_{2n-1}(x)$ tend to zero as n increases. Thus

$$(8) \qquad \lim_{n \to \infty} J_n(x) = 0 \qquad\qquad (n = 0, 1, 2, \ldots).$$

Another important property is that, for each fixed n,

$$(9) \qquad \lim_{x \to \infty} J_n(x) = 0 \qquad\qquad (n = 0, 1, 2, \ldots).$$

To indicate a method of proving this, we consider the special case $n = 0$. We substitute $u = \sin \phi$ in equation (6) to write

$$\frac{\pi}{2} J_0(x) = \int_0^c \frac{\cos xu}{\sqrt{1 - u^2}} \, du + \int_c^1 \frac{\cos xu}{\sqrt{1 - u^2}} \, du,$$

where $0 < c < 1$. The second integral here is improper but uniformly convergent with respect to x. Corresponding to any positive number ε, the absolute value of that integral can be made less than $\varepsilon/2$, uniformly for all x, by selecting c so that $1 - c$ is sufficiently small and positive. The Riemann-Lebesgue lemma then applies to the first integral with that value of c. That is, there is a number x_ε such that the absolute value of the first integral is less than $\varepsilon/2$ whenever $x > x_\varepsilon$. Therefore

$$\frac{\pi}{2} |J_0(x)| < \varepsilon \qquad\qquad \text{whenever } x > x_\varepsilon,$$

and this proves property (9) when $n = 0$. The proof for the other cases is left to the problems.

PROBLEMS

1. Use integral representations for J_n to verify that (a) $J_0(0) = 1$; (b) $J_n(0) = 0$ when $n = 1, 2, \ldots$; (c) $J_0'(x) = -J_1(x)$.

2. Deduce from expression (4), Sec. 75, that

$$J_{2n}(x) = (-1)^n \frac{2}{\pi} \int_0^{\pi/2} \cos 2n\theta \cos (x \cos \theta) \, d\theta$$

when $n = 0, 1, 2, \ldots$.

3. Deduce from expression (5), Sec. 75, that

$$J_{2n-1}(x) = -(-1)^n \frac{2}{\pi} \int_0^{\pi/2} \cos(2n-1)\theta \sin(x\cos\theta)\, d\theta$$

when $n = 1, 2, \ldots$.

4. Complete the proof of property (9), Sec. 75, that

$$\lim_{x \to \infty} J_n(x) = 0 \qquad\qquad (n = 0, 1, 2, \ldots).$$

5. Verify directly from the representation

$$J_0(x) = \frac{2}{\pi} \int_0^{\pi/2} \cos(x\sin\phi)\, d\phi$$

that $J_0(x)$ satisfies Bessel's equation (1), Sec. 71, when $n = 0$ there.

6. Apply integration by parts to representations (4) and (5) in Sec. 75 and then use the Riemann-Lebesgue lemma (Sec. 40) to show that

$$\lim_{n \to \infty} n J_n(x) = 0 \qquad\qquad (n = 0, 1, 2, \ldots)$$

for each fixed x.

7. Recall from equation (4), Sec. 75, that when $n = 0, 1, 2, \ldots$,

$$J_{2n}(x) = \frac{2}{\pi} \int_0^{\pi/2} \cos(x\sin\phi)\cos 2n\phi\, d\phi.$$

Then, by using this expression and Corollary 2 in Sec. 43 applied to Fourier cosine series on the interval $0 < \phi < \pi/2$, prove that

$$\cos(x\sin\phi) = J_0(x) + 2\sum_{n=1}^{\infty} J_{2n}(x)\cos 2n\phi$$

for all real x and ϕ. Observe that this is also a Fourier cosine series on the interval $0 < \phi < \pi$.

8. Show that when $n = 1, 2, \ldots$,

$$\int_0^{\pi} \sin(x\sin\phi)\sin 2n\phi\, d\phi = 0.$$

Then, using the Fourier sine series for the function $\sin(x\sin\phi)$ on the interval $0 < \phi < \pi$, prove that

$$\sin(x\sin\phi) = 2\sum_{n=1}^{\infty} J_{2n-1}(x)\sin(2n-1)\phi$$

for all real x and ϕ.

9. According to Sec. 46, if a function f and its derivative f' are both continuous on the interval $-\pi \le x \le \pi$ and if $f(-\pi) = f(\pi)$, then Parseval's equation

$$\frac{1}{\pi} \int_{-\pi}^{\pi} [f(x)]^2\, dx = \frac{1}{2}a_0^2 + \sum_{n=1}^{\infty} (a_n^2 + b_n^2)$$

holds, where the numbers a_n $(n = 0, 1, 2, \ldots)$ and b_n $(n = 1, 2, \ldots)$ are the Fourier coefficients

$$a_n = \frac{1}{\pi} \int_{-\pi}^{\pi} f(x)\cos nx\, dx, \qquad b_n = \frac{1}{\pi} \int_{-\pi}^{\pi} f(x)\sin nx\, dx.$$

(a) By applying that result to $f(\phi) = \cos(x \sin \phi)$, an even function of ϕ, and referring to the Fourier (cosine) series for $f(\phi)$ in Problem 7, show that

$$\frac{1}{\pi} \int_0^\pi \cos^2(x \sin \phi)\, d\phi = [J_0(x)]^2 + 2 \sum_{n=1}^\infty [J_{2n}(x)]^2 \qquad (-\infty < x < \infty).$$

(b) Similarly, by writing $f(\phi) = \sin(x \sin \phi)$ and referring to the Fourier (sine) series expansion in Problem 8, show that

$$\frac{1}{\pi} \int_0^\pi \sin^2(x \sin \phi)\, d\phi = 2 \sum_{n=1}^\infty [J_{2n-1}(x)]^2 \qquad (-\infty < x < \infty).$$

(c) Combine the results in parts (a) and (b) to show that

$$[J_0(x)]^2 + 2 \sum_{n=1}^\infty [J_n(x)]^2 = 1 \qquad (-\infty < x < \infty),$$

and point out how it follows from this identity that $|J_0(x)| \le 1$ and $|J_n(x)| \le 1/\sqrt{2}$ ($n = 1, 2, \ldots$) for all x.

10. Since the series in the expansions

$$\exp\left(\frac{xt}{2}\right) = \sum_{j=0}^\infty \frac{x^j}{j!\,2^j}\, t^j, \qquad \exp\left(\frac{-x}{2t}\right) = \sum_{k=0}^\infty \frac{(-1)^k x^k}{k!\,2^k}\, t^{-k}$$

are absolutely convergent when x is any real number and $t \ne 0$, the product of the two functions represented is itself represented by the series formed by multiplying each term in one series by every term in the other and then arranging the resulting terms *in any order*.† By noting that each of the terms in the resulting series can be written as a function of x times either t^n ($n = 0, 1, 2, \ldots$) or t^{-n} ($n = 1, 2, \ldots$) and arranging those terms according to powers of t, show that

$$\exp\left[\frac{x}{2}\left(t - \frac{1}{t}\right)\right] = J_0(x) + \sum_{n=1}^\infty [J_n(x)t^n + (-1)^n J_n(x)t^{-n}] \qquad (t \ne 0).$$

Also, write this result in the form

$$\exp\left[\frac{x}{2}\left(t - \frac{1}{t}\right)\right] = \lim_{m \to \infty} \sum_{n=-m}^m J_n(x)t^n \qquad (t \ne 0).$$

The exponential function here is said to be a *generating function* for the Bessel functions $J_n(x)$ ($n = 0, \pm 1, \pm 2, \ldots$).

Suggestion: Observe that the coefficient of t^n ($n = 0, 1, 2, \ldots$) can be obtained by multiplying the kth term of the second series by the term of the first series whose index is $j = n + k$ and then summing over k from 0 to ∞. Similarly, the coefficient of t^{-n} ($n = 1, 2, \ldots$) can be obtained by multiplying the jth term of the first series by the term of the second series with index $k = n + j$ and summing over j from 0 to ∞.

11. By writing $t = e^{i\phi}$ in the identity found in Problem 10 and noting that its left-hand side becomes $\exp(ix \sin \phi)$, obtain the Fourier series expansions in Problems 7 and 8.

---

† For a justification of this procedure, see, for example, Taylor and Mann (1972, pp. 635–636), listed in the Bibliography.

76. The Zeros of $J_0(x)$

A modified form of Bessel's equation

(1) $$x^2 y''(x) + x y'(x) + (x^2 - v^2) y(x) = 0$$

in which the term containing the first derivative is absent is sometimes useful. That form is easily found (Problem 1, Sec. 78) by making the substitution $y(x) = x^c u(x)$ in equation (1) and observing that the coefficient of $u'(x)$ in the resulting differential equation in u,

(2) $$x^2 u''(x) + (1 + 2c) x u'(x) + (x^2 - v^2 + c^2) u(x) = 0,$$

is zero if $c = -\frac{1}{2}$. The desired modified form of equation (1) is then

(3) $$x^2 u''(x) + \left(x^2 - v^2 + \frac{1}{4} \right) u(x) = 0,$$

and the function $u = \sqrt{x} J_v(x)$ is evidently a solution of it. In particular, when $v = 0$, we see that the function $u = \sqrt{x} J_0(x)$ satisfies the differential equation

(4) $$u''(x) + \left(1 + \frac{1}{4x^2} \right) u(x) = 0.$$

Observe now that the differential operator $L = d^2/dx^2$ is self-adjoint (Sec. 31) and that Lagrange's identity [Problem 17(a), Sec. 34] for that operator is

(5) $$U(x) V''(x) - V(x) U''(x) = \frac{d}{dx} (U V' - U' V),$$

where $U(x)$ and $V(x)$ are any functions whose first and second derivatives exist. We write

(6) $$U(x) = \sqrt{x} J_0(x)$$

and note that if $V(x)$ is chosen so that $V''(x) = -V(x)$, identity (5) becomes

$$-[U''(x) + U(x)] V(x) = \frac{d}{dx} (U V' - U' V).$$

But, since the function $u = U(x)$ satisfies equation (4), we know that $-[U''(x) + U(x)] = U(x)/(4x^2)$. Hence if V satisfies the condition $V'' = -V$ and if $0 < a < b$, then

(7) $$\int_a^b \frac{U(x) V(x)}{4x^2} dx = [U(x) V'(x) - U'(x) V(x)]_a^b .$$

With the aid of identity (7), we shall now show that *the positive zeros of* $J_0(x)$, *or roots of the equation* $J_0(x) = 0$, *form an infinite sequence of numbers* x_j $(j = 1, 2, \ldots)$ *such that* $x_j \to \infty$ *as* $j \to \infty$.†

First we show that our function $U(x) = \sqrt{x} J_0(x)$, and hence $J_0(x)$, has at least one zero in each interval $2k\pi \leq x \leq 2k\pi + \pi$ $(k = 1, 2, \ldots)$. We do this by assuming that $U(x) \neq 0$ anywhere in the interval $2k\pi \leq x \leq 2k\pi + \pi$ and obtaining a contradiction. According to that assumption, either $U(x) > 0$ for all x in the interval or $U(x) < 0$ for all such x, since $U(x)$ is continuous and it cannot therefore change sign without having a zero value at some point in the interval.

Suppose that $U(x) > 0$ when $2k\pi \leq x \leq 2k\pi + \pi$. In identity (7) write $a = 2k\pi$, $b = 2k\pi + \pi$, and $V(x) = \sin x$. Since $V(a) = V(b) = 0$, $V'(a) = 1$, and $V'(b) = -1$, that identity becomes

$$(8) \qquad \int_{2k\pi}^{2k\pi + \pi} U(x) \frac{\sin x}{4x^2} \, dx = -[U(2k\pi) + U(2k\pi + \pi)].$$

The integrand here is positive when $2k\pi < x < 2k\pi + \pi$. Hence the left-hand side of this equation has a positive value while the right-hand side is negative, giving a contradiction.

If, on the other hand, $U(x) < 0$ when $2k\pi \leq x \leq 2k\pi + \pi$, those two sides of equation (8) are negative and positive, respectively. This is again a contradiction. Thus $J_0(x)$ has at least one zero in each interval $2k\pi \leq x \leq 2k\pi + \pi$ $(k = 1, 2, \ldots)$.

Actually, $J_0(x)$ can have *at most* a finite number of zeros in any closed bounded interval $a \leq x \leq b$. To see that this is so, we assume that the interval $a \leq x \leq b$ does contain an infinite number of zeros. From advanced calculus we know that if a given infinite set of points lies in a closed bounded interval, there is always a sequence of distinct points in that set which converges to a point in the interval.‡ In particular, then, our assumption that the interval $a \leq x \leq b$ contains an infinite number of zeros of $J_0(x)$ implies that there exists a sequence x_m $(m = 1, 2, \ldots)$ of distinct zeros such that $x_m \to c$ as $m \to \infty$, where c is a point which also lies in the interval. Since the function $J_0(x)$ is continuous, $J_0(c) = 0$; and, by the definition of the limit of a sequence, every interval centered about c contains other zeros of $J_0(x)$. But the fact that $J_0(x)$ is not identically zero and has a Maclaurin series representation which is valid for all x means that there exists some interval centered at c which contains no other zeros.§ Since this is contrary to what

† Our method is a modification of the one used by A. Czarnecki, *Amer. Math. Monthly,* vol. 71, no. 4, pp. 403–404, 1964, who considers Bessel functions $J_\nu(x)$ where $-\frac{1}{2} \leq \nu \leq \frac{1}{2}$.

‡ See, for example, Taylor and Mann (1972, pp. 542–547), listed in the Bibliography.

§ That is, the zeros of such a function are *isolated*. The argument for this is given in a somewhat more general setting by Churchill, Brown, and Verhey (1974, p. 167), listed in the Bibliography.

Table 1. $J_0(x_j) = 0$

j	1	2	3	4	5
x_j	2.405	5.520	8.654	11.79	14.93
$J_1(x_j)$	0.5191	-0.3403	0.2715	-0.2325	0.2065

has just been shown, the number of zeros in the interval $a \leq x \leq b$ cannot then be infinite.

It is now evident that the positive zeros of $J_0(x)$ can, in fact, be arranged as an infinite sequence of numbers tending to infinity. Table 1 gives the values, correct to four significant figures, of the first five zeros of $J_0(x)$ and the corresponding values of $J_1(x)$. Extensive tables of numerical values of Bessel and related functions, together with their zeros, will be found in books listed in the Bibliography.†

77. Zeros of Related Functions

If for some pair of positive numbers a and b it is true that $J_n(a) = 0$ and $J_n(b) = 0$, then $x^{-n}J_n(x)$ also vanishes when $x = a$ and when $x = b$. It follows from Rolle's theorem that the derivative of $x^{-n}J_n(x)$ vanishes for at least one value of x between a and b. But (Sec. 73)

$$\frac{d}{dx}[x^{-n}J_n(x)] = -x^{-n}J_{n+1}(x) \qquad (n = 0, 1, 2, \ldots);$$

and so, when $n = 0, 1, 2, \ldots$, there is at least one zero of $J_{n+1}(x)$ between two positive zeros of $J_n(x)$. Also, just as in the case of $J_0(x)$ (Sec. 76), $J_{n+1}(x)$ can have at most a finite number of zeros on each bounded interval.

We have already shown that the positive zeros of $J_0(x)$ constitute an unbounded infinite sequence of numbers. It now follows that the zeros of $J_1(x)$ must form such a set. The same is then true for $J_2(x)$, and so on. That is, for each fixed nonnegative integer n, the set of all positive roots of the equation $J_n(x) = 0$ forms an infinite sequence $x = x_{nj}$ $(j = 1, 2, \ldots)$ where $x_{nj} \to \infty$ as $j \to \infty$.

Graphs of $J_0(x)$ and $J_1(x)$ are shown in Fig. 26.

The function $y = J_n(x)$ satisfies Bessel's equation, which is a linear homogeneous differential equation of the second order with the origin as a singular point. According to the uniqueness theorem stated in Sec. 18, there is just one solution that satisfies conditions $y(c) = y'(c) = 0$ where $c > 0$;

---

† See the book edited by Abramowitz and Stegun (1965) and the ones by Jahnke, Emde, and Lösch (1960), Gray and Mathews (1966), and Watson (1952), all listed there.

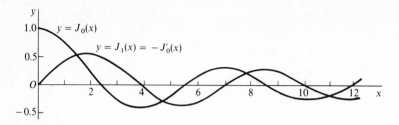

Figure 26

that solution is identically zero. Consequently, there is no positive number c such that $J_n(c) = J_n'(c) = 0$. That is, $J_n'(x)$ cannot vanish at a positive zero of $J_n(x)$; thus $J_n(x)$ must change its sign at that point.

Let a and b $(0 < a < b)$ be two consecutive zeros of $J_n(x)$. If $J_n'(a) > 0$, then $J_n(x) > 0$ when $a < x < b$ and $J_n(x)$ is decreasing at its zero b; that is, $J_n'(b) < 0$. Similarly, if $J_n'(a) < 0$, then $J_n'(b) > 0$. So the values of J_n' alternate in sign at consecutive positive zeros of J_n.

We now consider the function $hJ_n(x) + xJ_n'(x)$, where $h \geq 0$ and constant. Its zeros will also arise in certain boundary value problems. If a and b are consecutive positive zeros of $J_n(x)$, the function $hJ_n(x) + xJ_n'(x)$ has the values $aJ_n'(a)$ and $bJ_n'(b)$ at the points $x = a$ and $x = b$, respectively. Since one of those values is positive and the other negative, the function vanishes at some point, or at some finite number of points, between a and b. It therefore has an infinite sequence of positive zeros.† We collect our principal results as follows.

Theorem 2. *For each fixed n $(n = 0, 1, 2, \ldots)$, the set of all positive roots of the equation*

$$(1) \qquad\qquad J_n(x) = 0$$

consists of an infinite sequence $x = x_{nj}$ $(j = 1, 2, \ldots)$ such that $x_{nj} \to \infty$ as $j \to \infty$; also, the set of all positive roots of the equation

$$(2) \qquad\qquad hJ_n(x) + xJ_n'(x) = 0 \qquad\qquad (h \geq 0),$$

where h is a constant, is always a sequence of that type.

Observe that $x = 0$ is a root of both equations (1) and (2) if n is a positive integer. It is also a root of the equation $J_0'(x) = 0$.

---

† In the important special case when $n = 0$, the first six zeros are tabulated for various positive values of h in, for example, Appendix IV of the book on heat conduction by Carslaw and Jaeger (1959) that is listed in the Bibliography.

If $x = c$ is a root of equation (1), then $x = -c$ is also a root since $J_n(-c) = (-1)^n J_n(c)$. That statement is true of equation (2) as well; for, in view of the recurrence relation (Sec. 73)

$$x J_n'(x) = n J_n(x) - x J_{n+1}(x),$$

equation (2) can be written

(3) $$(h + n)J_n(x) - x J_{n+1}(x) = 0,$$

and we note that

$$(h + n)J_n(-x) - (-x)J_{n+1}(-x) = (-1)^n[(h + n)J_n(x) - x J_{n+1}(x)].$$

Finally, although our discussion leading up to Theorem 2 need not have excluded the possibility that h be negative, those values of h will not arise in our applications.

78. Orthogonal Sets of Bessel Functions

As indicated in Sec. 70, where somewhat different notation was used, the physical applications in this chapter will involve solutions of the differential equation

(1) $$x^2 \frac{d^2 X}{dx^2} + x \frac{dX}{dx} + (\lambda x^2 - n^2)X = 0 \qquad (n = 0, 1, 2, \ldots).$$

Equation (1) is readily expressed in the self-adjoint form (Sec. 31)

(2) $$\frac{d}{dx}\left(x \frac{dX}{dx}\right) + \left(\lambda x - \frac{n^2}{x}\right)X = 0.$$

For each fixed n, this form of equation (1) is a special case of the Sturm-Liouville equation (4), Sec. 31, where $r(x) = p(x) = x$ and $q(x) = -n^2/x$.

In particular, we shall need to solve the *singular* (Sec. 32) Sturm-Liouville problem, on an interval $(0,c)$, consisting of equation (1) and a boundary condition

(3) $$b_1 X(c) + b_2 X'(c) = 0,$$

where it is understood that X and X' are to be continuous on the closed interval $0 \leq x \leq c$. The constants b_1 and b_2 are real and not both zero. The end point $x = 0$ is the singular point of the differential equation; also, q is discontinuous there unless $n = 0$.

In the important special case when $b_2 = 0$, the boundary condition (3) is

(4) $$X(c) = 0.$$

When $b_2 \neq 0$, we may multiply through condition (3) by c/b_2 and write it as

(5) $$hX(c) + cX'(c) = 0$$

where $h = cb_1/b_2$. In solving our Sturm-Liouville problem, we shall find it convenient to use the boundary condition (3) in its separate forms (4) and (5); and, when using condition (5), *we shall always assume that $h \geq 0$*.

Cases (*a*) of Theorems 3 and 4 in Sec. 32 give conditions ensuring that our singular Sturm-Liouville problem has orthogonal eigenfunctions and *real eigenvalues λ*. We note that the proofs of those theorems do not depend on the continuity of $q(x)$ at the end point $x = a$ ($a = 0$ in the present case) as long as X and X' are continuous on the closed interval $a \leq x \leq b$. We now consider the three possibilities that λ be zero, positive, or negative.

When $\lambda = 0$, equation (1) is a Cauchy-Euler equation (see Problem 13, Sec. 34):

(6)
$$x^2 \frac{d^2 X}{dx^2} + x \frac{dX}{dx} - n^2 X = 0.$$

To solve it, we write $x = e^s$ and put it into the form

(7)
$$\frac{d^2 X}{ds^2} - n^2 X = 0.$$

If $n = 1, 2, \ldots$, it follows that $X = Ae^{ns} + Be^{-ns}$, or $X(x) = Ax^n + Bx^{-n}$, where A and B are constants. Since our solution must be continuous, and therefore bounded, on the interval $0 \leq x \leq c$, we require that $B = 0$; that is, $X(x) = Ax^n$. It is now easy to see that $A = 0$ if either of conditions (4) or (5) is to be satisfied, and we arrive at only the trivial solution $X(x) \equiv 0$. Thus zero is not an eigenvalue if $n = 1, 2, \ldots$.

If, on the other hand, $n = 0$, equation (7) has the general solution $X = As + B$; and the general solution of equation (6) when $n = 0$ is therefore $X(x) = A \log x + B$. According to the continuity requirements, then, $X(x) = B$. When condition (4) is imposed, $B = 0$; the same is true of condition (5) when $h > 0$. But when $h = 0$, condition (5) becomes simply $X'(c) = 0$, and B can remain arbitrary. So *if $n = 0$ and condition (5) is used when $h = 0$, we have the eigenfunction*

$$X(x) = 1 \qquad corresponding\ to\ \lambda = 0.$$

This is the only case in which $\lambda = 0$ is an eigenvalue.

We consider next the case when $\lambda > 0$ and write $\lambda = \alpha^2$ ($\alpha > 0$). Equation (1) is then

(8)
$$x^2 \frac{d^2 X}{dx^2} + x \frac{dX}{dx} + (\alpha^2 x^2 - n^2)X = 0.$$

Except for the notation used, it has already been pointed out in Sec. 70 that the substitution $s = \alpha x$ may be used to transform equation (8) into Bessel's equation

$$(9) \qquad s^2 \frac{d^2 X}{ds^2} + s \frac{dX}{ds} + (s^2 - n^2)X = 0;$$

and the general solution of equation (8) is evidently (Sec. 72)

$$(10) \qquad X(x) = C_1 J_n(\alpha x) + C_2 Y_n(\alpha x).$$

Our continuity requirements imply that $C_2 = 0$, since $Y_n(\alpha x)$ is unbounded on the interval $(0,c)$. Hence any nontrivial solution of equation (8) which meets those requirements must be a nonzero constant multiple of the function $X(x) = J_n(\alpha x)$.

In applying the boundary condition at $x = c$, let us agree that *the symbol $J_n'(\alpha x)$ stands for the derivative of J_n with respect to the argument of J_n. That is, $J_n'(\alpha x) = d/ds[J_n(s)]$ where $d/ds[J_n(s)]$ is to be evaluated at $s = \alpha x$.* Then $d/dx[J_n(\alpha x)] = \alpha J_n'(\alpha x)$; and conditions (4) and (5) require that

$$(11) \qquad J_n(\alpha c) = 0$$

and

$$(12) \qquad hJ_n(\alpha c) + (\alpha c)J_n'(\alpha c) = 0,$$

respectively. Note that since equation (2), Sec. 77, can be written in the form (3) in that section, equation (12) can also be written

$$(h + n)J_n(\alpha c) - (\alpha c)J_{n+1}(\alpha c) = 0.$$

According to Theorem 2, there is an infinite sequence of positive values of αc satisfying equation (11) or (12). If we designate such a sequence by x_j $(j = 1, 2, \ldots)$, but keep in mind that *the numbers x_j depend on the value of n and also on the value of h in the case of equation (12)*, it follows that equation (11) or (12) is satisfied when $\alpha c = x_j$. That is, the positive roots of either of those equations have the form

$$(13) \qquad \alpha_j = \frac{x_j}{c} \qquad\qquad (j = 1, 2, \ldots).$$

Our Sturm-Liouville problem thus has eigenvalues $\lambda_j = \alpha_j^2$ $(j = 1, 2, \ldots)$, and the corresponding eigenfunctions are

$$(14) \qquad X_j(x) = J_n(\alpha_j x) \qquad\qquad (j = 1, 2, \ldots).$$

We note that if the numbers α_j are the positive roots of equation (12) when $n = h = 0$, which is the only case when $\lambda = 0$ was found to be an eigenvalue, that equation can be written

$$(15) \qquad J_1(\alpha c) = 0.$$

The numbers α_j are then given more directly as the positive roots of equation (15). Also, making a minor exception in our notation, we let the subscript j range over the values $j = 2, 3, \ldots$, instead of starting from unity. The subscript $j = 1$ is reserved for writing $\alpha_1 = 0$ and $\lambda_1 = \alpha_1^2 = 0$. This allows us to include the eigenvalue $\lambda_1 = 0$ together with the eigenfunction $X_1(x) = J_0(\alpha_1 x) = 1$, obtained earlier for the case $n = h = 0$. Note that it is also possible to describe the numbers α_j $(j = 1, 2, \ldots)$ here as the *nonnegative* roots of equation (15).

Finally, we consider the case when $\lambda < 0$, or $\lambda = -\beta^2$ $(\beta > 0)$, and write equation (1) as

$$(16) \qquad x^2 \frac{d^2 X}{dx^2} + x \frac{dX}{dx} - (\beta^2 x^2 + n^2)X = 0.$$

The substitution $s = \beta x$, which is like the one used to transform equation (8) into Bessel's equation (9), can be used here to put equation (16) into the form

$$(17) \qquad s^2 \frac{d^2 X}{ds^2} + s \frac{dX}{ds} - (s^2 + n^2)X = 0.$$

From Problem 2, Sec. 73, we know that the modified Bessel function $X = I_n(s) = i^{-n} J_n(is)$ satisfies equation (17); and, since it has a power series representation which converges for all x, it satisfies the continuity requirements in our problem. As was the case with equation (9), equation (17) has a second solution which is *discontinuous at* $s = 0$, that solution being analogous to Y_n.† Thus we know that, except for an arbitrary constant factor, $X(x) = I_n(\beta x)$.

We now show that for each positive value of β the function $X(x) = I_n(\beta x)$ fails to satisfy either of the conditions (4) or (5). In each case our proof rests on the fact that $I_n(x) > 0$ when $x > 0$, as demonstrated in Problem 2, Sec. 73.

Since $I_n(\beta c) > 0$ when $\beta > 0$, it is obvious that condition (4), which requires that $I_n(\beta c) = 0$, fails to be satisfied by any positive number β. Also, in view of the alternative form (3), Sec. 77, of equation (2) in that section, condition (5), when applied to our function $X(x) = I_n(\beta x) = i^{-n} J_n(i\beta x)$, becomes

$$(h + n)i^{-n} J_n(i\beta c) + \beta c i^{-(n+1)} J_{n+1}(i\beta c) = 0,$$

or

$$(18) \qquad (h + n)I_n(\beta c) + \beta c I_{n+1}(\beta c) = 0.$$

---

† For a detailed discussion of this, see, for example, pp. 16 ff of the book by Tranter (1969) listed in the Bibliography.

Since $\beta > 0$, the left-hand side of this last equation is positive; and, once again, no positive values of β can occur as roots. We conclude, then, that there are no negative eigenvalues.

We have now completely solved our singular Sturm-Liouville problem consisting of equations (1) and (3), where, since we have assumed that the constant $h = cb_1/b_2$ is nonnegative, it is understood that the constants b_1 and b_2 in equation (3) have the same sign when neither is zero.

The eigenvalues are all represented by the numbers $\lambda_j = \alpha_j^2$ where the α_j are given by equation (13) and where $\lambda_j > 0$, except that $\lambda_1 = 0$ in the case $n = h = 0$. Let us agree to arrange the numbers α_j in ascending order of magnitude so that $\alpha_j < \alpha_{j+1}$, and therefore $\lambda_j < \lambda_{j+1}$.

Theorem 3, Sec. 32, gives the orthogonality property

$$(19) \qquad \int_0^c xJ_n(\alpha_j x)J_n(\alpha_k x)\,dx = 0 \qquad\qquad (j \neq k).$$

Note that this orthogonality of the eigenfunctions with weight function x, on the interval $(0,c)$, is the same as ordinary orthogonality of the functions $\sqrt{x}\,J_n(\alpha_j x)$ on that same interval. Also note that many orthogonal sets are represented here, depending on the values of n, c, and h.

We summarize our results in the following theorem.

Theorem 3. *Let n have one of the values $n = 0, 1, 2, \ldots$. The set $\{J_n(\alpha_j x)\}$ $(j = 1, 2, \ldots)$ is orthogonal on the interval $(0,c)$ with weight function x when the numbers $\alpha = \alpha_j$ $(j = 1, 2, \ldots)$ are either*

(a) the positive roots of the equation

$$J_n(\alpha c) = 0,$$

(b) the positive roots of the equation

$$hJ_n(\alpha c) + (\alpha c)J_n'(\alpha c) = 0 \qquad (h \geq 0, h + n > 0),$$

or (c) (when $n = 0$) zero and the positive roots of the equation

$$J_0'(\alpha c) = 0.$$

The numbers $\lambda_j = \alpha_j^2$ are the eigenvalues, and $X_j = J_n(\alpha_j x)$ are the corresponding eigenfunctions, of the singular Sturm-Liouville problem consisting of the equation

$$(20) \qquad x^2 \frac{d^2 X}{dx^2} + x \frac{dX}{dx} + (\lambda x^2 - n^2)X = 0,$$

whose self-adjoint form is

$$\frac{d}{dx}\left(x \frac{dX}{dx}\right) + \left(\lambda x - \frac{n^2}{x}\right)X = 0,$$

together with the boundary condition

(21) $$X(c) = 0$$

in case (a), the condition

(22) $$hX(c) + cX'(c) = 0 \qquad (h \geq 0, \, h + n > 0)$$

in case (b), or the condition

(23) $$X'(c) = 0$$

in case (c), the value of n in equation (20) for this last case being zero.

PROBLEMS

1. By means of the substitution $y(x) = x^c u(x)$, transform Bessel's equation

$$x^2 y''(x) + xy'(x) + (x^2 - v^2)y(x) = 0$$

into the differential equation

$$x^2 u''(x) + (1 + 2c)xu'(x) + (x^2 - v^2 + c^2)u(x) = 0,$$

which becomes equation (3), Sec. 76, when $c = -\frac{1}{2}$.

2. Use equation (3), Sec. 76, to obtain a general solution of Bessel's equation when $v = \frac{1}{2}$. Then, using the expressions in Problem 6, Sec. 73, point out how $J_{1/2}(x)$ and $J_{-1/2}(x)$ are special cases of that solution.

3. By referring to Theorem 3, show that the eigenvalues of the singular Sturm-Liouville problem

$$(xX')' + \lambda xX = 0 \qquad (0 < x < 2),$$

$$X(2) = 0$$

are the numbers $\lambda_j = \alpha_j^2$ $(j = 1, 2, \ldots)$ where α_j are the positive roots of the equation $J_0(2\alpha) = 0$ and that the corresponding eigenfunctions are $X_j = J_0(\alpha_j x)$ $(j = 1, 2, \ldots)$. With the aid of Table 1 in Sec. 76, obtain the numerical values $\alpha_1 = 1.2, \alpha_2 = 2.8, \alpha_3 = 4.3$, valid to one decimal place.

4. Write $U(x) = \sqrt{x} J_n(\alpha x)$ where n has any one of the values $n = 0, 1, 2, \ldots$ and α is a positive constant.

(a) Use equation (3), Sec. 76, to show that

$$U''(x) + \left(\alpha^2 + \frac{1 - 4n^2}{4x^2}\right)U(x) = 0.$$

(b) Let c denote any fixed positive number and write $U_j(x) = \sqrt{x} J_n(\alpha_j x)$ $(j = 1, 2, \ldots)$ where α_j are the positive roots of the equation $J_n(\alpha c) = 0$. Use the result in part (a) to show that

$$(\alpha_j^2 - \alpha_k^2)U_j(x)U_k(x) = U_j U_k'' - U_k U_j''.$$

(c) Use the result in part (b) and Lagrange's identity [Problem 17(a), Sec. 34] for the self-adjoint operator $L = d^2/dx^2$ to show that the set $\{U_j(x)\}$ $(j = 1, 2, \ldots)$ is orthogonal on $(0,c)$ with weight function unity. Thus give a new proof that the set $\{J_n(\alpha_j x)\}$ $(j = 1, 2, \ldots)$ in case (a) of Theorem 3 is orthogonal on $(0,c)$ with weight function x.

5. Let n have any one of the fixed values $n = 0, 1, 2, \ldots$.

(a) Suppose that $J_n(ib) = 0$ $(b \neq 0)$ and use results in Problem 2, Sec. 73, to reach a contradiction. Thus show that the function $J_n(z)$ has no pure imaginary zeros $z = ib$ $(b \neq 0)$.

(b) Since our series representation of $J_n(x)$ converges when x is replaced by any complex number z and since the coefficients of the powers of z in that representation are all real, it is true that $J_n(\bar{z}) = \overline{J_n(z)}$, where $\bar{z}$ denotes the complex conjugate $x - iy$ of the number $z = x + iy$.† Also, the proof of orthogonality in Problem 4 above remains valid when α is a nonzero complex number and the set of roots α_j there is allowed to include any nonzero complex roots of the equation $J_n(\alpha c) = 0$ that may occur. Use these facts to show that if the complex number $a + ib$ $(a \neq 0, b \neq 0)$ is a zero of $J_n(z)$, then $a - ib$ is also a zero and that

$$0 = \int_0^1 x J_n[(a + ib)x] J_n[(a - ib)x]\, dx = \int_0^1 x |J_n[(a + ib)x]|^2\, dx.$$

Point out why the value of the integral on the far right is actually positive and, with this contradiction, deduce that $J_n(z)$ has no zeros of the form $a + ib$ $(a \neq 0, b \neq 0)$. Conclude that if $z = x_j$ $(j = 1, 2, \ldots)$ are the positive zeros of $J_n(z)$, the only other zeros, real or complex, are the numbers $z = -x_j$ $(j = 1, 2, \ldots)$, and also $z = 0$ when $n \geq 1$.

79. The Orthonormal Functions

From the previous section we know that if α is a positive constant, the function

$$X(x) = J_n(\alpha x) \qquad\qquad (n = 0, 1, 2, \ldots)$$

satisfies the equation

(1) $$(xX')' + \left(\alpha^2 x - \frac{n^2}{x}\right) X = 0 \qquad\qquad (n = 0, 1, 2, \ldots).$$

We multiply each side by $2xX'$ and write

$$\frac{d}{dx}(xX')^2 + (\alpha^2 x^2 - n^2)\frac{d}{dx}(X^2) = 0.$$

After integrating both terms here and using integration by parts in the second term, we find that

$$[(xX')^2 + (\alpha^2 x^2 - n^2)X^2]_0^c - 2\alpha^2 \int_0^c xX^2\, dx = 0,$$

where c is any positive number. When $n = 0$, the quantity inside the brackets clearly vanishes at $x = 0$; and the same is true when $n = 1, 2, \ldots$, since $X(0) = J_n(0) = 0$ then. We thus arrive at the expression

(2) $$2\alpha^2 \int_0^c x[J_n(\alpha x)]^2\, dx = \alpha^2 c^2 [J_n'(\alpha c)]^2 + (\alpha^2 c^2 - n^2)[J_n(\alpha c)]^2,$$

---

† The ratio test applies even when the terms in a series are complex numbers. Also, see Churchill, Brown, and Verhey (1974, p. 291), listed in the Bibliography.

which we now use to find the norms of all the eigenfunctions in Theorem 3 except for the one corresponding to the zero eigenvalue in case (c) there. That norm is treated separately.

(a) When α_j $(j = 1, 2, \ldots)$ are the positive roots of the equation

$$(3) \qquad J_n(\alpha c) = 0,$$

equation (2) becomes

$$2 \int_0^c x[J_n(\alpha_j x)]^2 \, dx = c^2[J'_n(\alpha_j c)]^2.$$

The integral here is the square of the norm of $J_n(\alpha_j x)$ on the interval $(0,c)$ with weight function x. Also (Sec. 73)

$$xJ'_n(x) = nJ_n(x) - xJ_{n+1}(x),$$

and therefore $\alpha_j cJ'_n(\alpha_j c) = -\alpha_j cJ_{n+1}(\alpha_j c)$. Hence

$$(4) \qquad \|J_n(\alpha_j x)\|^2 = \frac{c^2}{2}[J_{n+1}(\alpha_j c)]^2 \qquad (j = 1, 2, \ldots).$$

(b) When α_j $(j = 1, 2, \ldots)$ are the positive roots of the equation

$$(5) \qquad hJ_n(\alpha c) + (\alpha c)J'_n(\alpha c) = 0 \qquad (h \geq 0, \, h + n > 0),$$

we find from equation (2) that

$$(6) \qquad \|J_n(\alpha_j x)\|^2 = \frac{\alpha_j^2 c^2 - n^2 + h^2}{2\alpha_j^2}[J_n(\alpha_j c)]^2 \qquad (j = 1, 2, \ldots).$$

(c) Suppose now that $\alpha_1 = 0$ and that α_j $(j = 2, 3, \ldots)$ are the positive roots of the equation

$$(7) \qquad J'_0(\alpha c) = 0.$$

Since $J_0(\alpha_1 x) = 1$,

$$(8) \qquad \|J_0(\alpha_1 x)\|^2 = \int_0^c x \, dx = \frac{c^2}{2}.$$

Expressions for $\|J_0(\alpha_j c)\|^2$ $(j = 2, 3, \ldots)$ are obtained by writing $n = h = 0$ in equation (6):

$$(9) \qquad \|J_0(\alpha_j x)\|^2 = \frac{c^2}{2}[J_0(\alpha_j c)]^2 \qquad (j = 2, 3, \ldots).$$

For equation (7) is simply equation (5) when $n = h = 0$, and the restriction $h + n > 0$ is not actually needed in deriving expression (6).

The orthogonal eigenfunctions $X_j(x) = J_n(\alpha_j x)$ of our singular Sturm-Liouville problem can now be written in normalized form as

$$(10) \qquad\qquad \phi_j(x) = \frac{J_n(\alpha_j x)}{\|J_n(\alpha_j x)\|} \qquad\qquad (j = 1, 2, \ldots).$$

The norms here are given for the eigenfunctions in cases (a) and (b) of Theorem 3 by equations (4) and (6), respectively. In case (c) they are given by equations (8) and (9). The fact that the set (10) is orthonormal on the interval $(0,c)$ with weight functions x is, of course, expressed by the equations

$$\int_0^c x\phi_j(x)\phi_k(x)\,dx = \begin{cases} 0 & \text{if } j \neq k, \\ 1 & \text{if } j = k. \end{cases}$$

80. Fourier-Bessel Series

Since the functions $\phi_j(x)$ constitute all the normalized eigenfunctions of our singular Sturm-Liouville problem, we can anticipate representations of functions $f(x)$ on the interval $(0,c)$ by means of generalized Fourier series (Sec. 26) involving those orthonormal functions.

Let c_j $(j = 1, 2, \ldots)$ denote the Fourier constants of a function $f(x)$ with respect to the set $\{\phi_j(x)\}$ $(j = 1, 2, \ldots)$ on the interval $(0,c)$. Then

$$c_j = \int_0^c xf(x)\phi_j(x)\,dx = \frac{1}{\|J_n(\alpha_j x)\|}\int_0^c xf(x)J_n(\alpha_j x)\,dx,$$

and the generalized Fourier series corresponding to f is

$$(1) \qquad \sum_{j=1}^{\infty} c_j\phi_j(x) = \sum_{j=1}^{\infty} \frac{J_n(\alpha_j x)}{\|J_n(\alpha_j x)\|^2}\int_0^c sf(s)J_n(\alpha_j s)\,ds.$$

In view of the expressions for the norms found in the preceding section, we thus have the correspondence

$$(2) \qquad\qquad f(x) \sim \sum_{j=1}^{\infty} A_j J_n(\alpha_j x) \qquad\qquad (0 < x < c)$$

where the coefficients A_j have the following values.

 (a) When α_j $(j = 1, 2, \ldots)$ are the positive roots of the equation

$$(3) \qquad\qquad J_n(\alpha c) = 0,$$

$$(4) \qquad\qquad A_j = \frac{2}{c^2[J_{n+1}(\alpha_j c)]^2}\int_0^c xf(x)J_n(\alpha_j x)\,dx \qquad (j = 1, 2, \ldots).$$

(b) *When* α_j $(j = 1, 2, \ldots)$ *are the positive roots of the equation*

(5) $$hJ_n(\alpha c) + (\alpha c)J'_n(\alpha c) = 0 \qquad (h \geq 0,\ h + n > 0),$$

(6) $$A_j = \frac{2\alpha_j^2}{(\alpha_j^2 c^2 - n^2 + h^2)[J_n(\alpha_j c)]^2} \int_0^c xf(x)J_n(\alpha_j x)\,dx \qquad (j = 1, 2, \ldots).$$

(c) *When* $n = 0$ *in series* (2) *and when* $\alpha_1 = 0$ *and* α_j $(j = 2, 3, \ldots)$ *are the positive roots of the equation*

(7) $$J'_0(\alpha c) = 0,$$

then

(8) $$A_1 = \frac{2}{c^2} \int_0^c xf(x)\,dx$$

and

(9) $$A_j = \frac{2}{c^2[J_0(\alpha_j c)]^2} \int_0^c xf(x)J_0(\alpha_j x)\,dx \qquad (j = 2, 3, \ldots).$$

Note that equation (9) reduces to equation (8) if $j = 1$, since $\alpha_1 = 0$ in case (c).

Proofs that correspondence (2) is actually an equality, under conditions similar to those used to ensure the representation of a function by its Fourier cosine or sine series, usually involve the theory of functions of a complex variable. We state without proof one form of such a representation theorem and refer the reader to the Bibliography.†

Theorem 4. *If f is a function which, together with its derivative f', is piecewise continuous on the interval $0 < x < c$, then series (2) converges to the mean value of the one-sided limits of f at each point in that interval. That is,*

(10) $$\frac{1}{2}[f(x+) + f(x-)] = \sum_{j=1}^{\infty} A_j J_n(\alpha_j x) \qquad (0 < x < c)$$

where the coefficients A_j are defined by equation (4) *or* (6) *or the pair of equations* (8) *and* (9), *depending on the particular equation that determines the roots α_j.*

The expansion (10) is called a *Fourier-Bessel series* representation of *f*.

An example is the expansion of the function $f(x) = 1$ $(0 < x < c)$ into a series of the functions $J_0(\alpha_j x)$ $(j = 1, 2, \ldots)$ when α_j are the positive roots of

---

† This theorem is proved in the book by Watson (1952). Also see pp. 84–86 in part I of the work by Titchmarsh (1962), as well as the books by Gray and Mathews (1966) and Bowman (1958). These are all listed in the Bibliography.

the equation $J_0(\alpha c) = 0$. In view of the integration formula (Sec. 73)

$$\int_0^x s J_0(s)\, ds = x J_1(x),$$

it is easy to evaluate the integral in expression (4) that is needed:

$$\int_0^c x J_0(\alpha_j x)\, dx = \frac{1}{\alpha_j^2} \int_0^{\alpha_j c} s J_0(s)\, ds = \frac{c}{\alpha_j} J_1(\alpha_j c).$$

Consequently,

$$(11) \qquad\qquad 1 = \frac{2}{c} \sum_{j=1}^{\infty} \frac{J_0(\alpha_j x)}{\alpha_j J_1(\alpha_j c)} \qquad\qquad (0 < x < c).$$

The results stated in Theorems 2, 3, and 4 are also valid when n is replaced by an arbitrary positive number ν, although we have not developed properties of the functions J_ν far enough to establish this fact.

For functions on the unbounded interval $x > 0$ there is an integral representation in terms of J_ν, corresponding to the Fourier cosine or sine integral formula.† The representation, for a fixed ν $(\nu \geq -\tfrac{1}{2})$, is

$$f(x) = \int_0^{\infty} \alpha J_\nu(\alpha x) \int_0^{\infty} s f(s) J_\nu(\alpha s)\, ds\, d\alpha \qquad\qquad (x > 0)$$

and is known as *Hankel's integral formula*. It is valid if f and f' are piecewise continuous on each bounded interval and if $\sqrt{x} f(x)$ is absolutely integrable from zero to infinity and $f(x)$ is defined as its mean value at each point of discontinuity.

If the interval $(0,c)$ is replaced by some interval (a,b) where $0 < a < b$, the Sturm-Liouville problem treated in Sec. 78 is no longer singular when the same differential equation is used and boundary conditions of type (3) in that section are applied at *each* end point. The eigenfunctions involve both of the Bessel functions J_n and Y_n.

PROBLEMS

1. Find the coefficients A_j $(j = 1, 2, \ldots)$ in the expansion

$$1 = \sum_{j=1}^{\infty} A_j J_0(\alpha_j x) \qquad\qquad (0 < x < c)$$

when $\alpha_1 = 0$ and α_j $(j = 2, 3, \ldots)$ are the positive roots of the equation $J_0'(\alpha c) = 0$.

Ans. $A_1 = 1,\ A_j = 0$ $(j = 2, 3, \ldots)$

---

† See Chap. 2 of the book by Sneddon (1951) that is listed in the Bibliography. For a summary of representations in terms of Bessel functions, see Chap. 7 in vol. 2 of the work edited by Erdélyi (1953) which is also listed.

2. (*a*) Obtain the representation

$$1 = 2c \sum_{j=1}^{\infty} \frac{\alpha_j J_1(\alpha_j c) J_0(\alpha_j x)}{(\alpha_j^2 c^2 + h^2)[J_0(\alpha_j c)]^2} \qquad (0 < x < c)$$

where α_j ($j = 1, 2, \ldots$) are the positive roots of the equation

$$h J_0(\alpha c) + (\alpha c) J_0'(\alpha c) = 0 \qquad (h > 0).$$

[Contrast this representation with the representation (11), Sec. 80, and also the one in Problem 1.]
(*b*) Show how the result in part (*a*) can be written in the form

$$1 = \frac{2}{c} \sum_{j=1}^{\infty} \left(\frac{1}{\alpha_j}\right) \frac{J_1(\alpha_j c) J_0(\alpha_j x)}{[J_0(\alpha_j c)]^2 + [J_1(\alpha_j c)]^2} \qquad (0 < x < c).$$

3. Show that if

$$f(x) = \begin{cases} 1 & \text{when } 0 < x < 1, \\ 0 & \text{when } 1 < x < 2, \end{cases}$$

and $f(1) = \frac{1}{2}$ then

$$f(x) = \frac{1}{2} \sum_{j=1}^{\infty} \frac{J_1(\alpha_j)}{\alpha_j [J_1(2\alpha_j)]^2} J_0(\alpha_j x) \qquad (0 < x < 2)$$

where α_j ($j = 1, 2, \ldots$) are the positive roots of the equation $J_0(2\alpha) = 0$.

4. Let α_j ($j = 1, 2, \ldots$) denote the positive roots of the equation $J_0(\alpha c) = 0$, where c is a fixed positive number.

(*a*) With the aid of the reduction formula found in Problem 15, Sec. 73, obtain the expansion

$$x^2 = \frac{2}{c} \sum_{j=1}^{\infty} \frac{(\alpha_j c)^2 - 4}{\alpha_j^3 J_1(\alpha_j c)} J_0(\alpha_j x) \qquad (0 < x < c).$$

(*b*) Combine expansion (11), Sec. 80, with the one in part (*a*) to show that

$$c^2 - x^2 = \frac{8}{c} \sum_{j=1}^{\infty} \frac{J_0(\alpha_j x)}{\alpha_j^3 J_1(\alpha_j c)} \qquad (0 < x < c).$$

5. Show that

$$x = 2 \sum_{j=1}^{\infty} \left[1 - \frac{1}{\alpha_j^2 J_1(\alpha_j)} \int_0^{\alpha_j} J_0(s) \, ds \right] \frac{J_0(\alpha_j x)}{\alpha_j J_1(\alpha_j)} \qquad (0 < x < 1)$$

where α_j ($j = 1, 2, \ldots$) are the positive roots of the equation $J_0(\alpha) = 0$.†

6. Let n have any one of the positive values $n = 1, 2, \ldots$. With the aid of the integration formula (5), Sec. 73, show that

$$x^n = 2 \sum_{j=1}^{\infty} \frac{\alpha_j J_{n+1}(\alpha_j)}{(\alpha_j^2 - n^2)[J_n(\alpha_j)]^2} J_n(\alpha_j x) \qquad (0 < x < 1)$$

where α_j ($j = 1, 2, \ldots$) are the positive roots of the equation $J_n'(\alpha) = 0$.

† See the footnote with Problem 15(*b*) Sec. 73.

7. Point out why the eigenvalues of the singular Sturm-Liouville problem

$$(xX')' + \left(\lambda x - \frac{1}{x}\right)X = 0 \qquad (0 < x < 1),$$

$$X(1) = 0$$

are the numbers $\lambda_j = \alpha_j^2 (j = 1, 2, \ldots)$ where α_j are the positive roots of the equation $J_1(\alpha) = 0$ and why the corresponding eigenfunctions are $X_j = J_1(\alpha_j x)$ $(j = 1, 2, \ldots)$. Then obtain the representation

$$x = 2 \sum_{j=1}^{\infty} \frac{J_1(\alpha_j x)}{\alpha_j J_2(\alpha_j)} \qquad (0 \le x < 1)$$

in terms of those eigenfunctions.

81. Temperatures in a Long Cylinder

Let the lateral surface $\rho = c$ of an infinitely long circular cylinder (Fig. 27), or a circular cylinder of finite altitude with insulated bases, be kept at temperature zero; and let the initial temperature distribution be a given function of *only* the variable ρ, the distance from the axis of the cylinder. We shall derive an expression for the temperatures $u(\rho, t)$ in the cylinder, assuming that the material of that solid is homogeneous.

With the cylinder situated as shown in Fig. 27, the heat equation and boundary conditions are, in cylindrical coordinates,

(1) $$\frac{\partial u}{\partial t} = k\left(\frac{\partial^2 u}{\partial \rho^2} + \frac{1}{\rho}\frac{\partial u}{\partial \rho}\right) \qquad (0 < \rho < c, t > 0),$$

(2) $$u(c-, t) = 0 \qquad (t > 0),$$

(3) $$u(\rho, 0+) = f(\rho) \qquad (0 < \rho < c).$$

Also, when $t > 0$, the function u is to be continuous throughout the cylinder and, in particular, on the axis $\rho = 0$. We assume that f and f' are piecewise

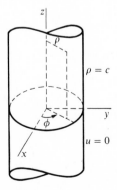

Figure 27

continuous on the interval $(0,c)$ and, for convenience, that f is defined as the mean value of its one-sided limits at each point of discontinuity.

Any solutions of the homogeneous equations (1) and (2) that are of the type $R(\rho)T(t)$ must satisfy the equations

$$RT' = kT\left(R'' + \frac{1}{\rho}R'\right), \qquad R(c)T(t) = 0.$$

Separating variables in the first equation here, we have

$$\frac{T'}{kT} = \frac{1}{R}\left(R'' + \frac{1}{\rho}R'\right) = -\lambda,$$

where $-\lambda$ is some constant yet to be specified. Thus

(4)
$$\rho R''(\rho) + R'(\rho) + \lambda\rho R(\rho) = 0 \qquad (0 < \rho < c),$$
$$R(c) = 0,$$

and

(5)
$$T'(t) + \lambda k T(t) = 0 \qquad (t > 0).$$

The differential equation in R is Bessel's equation with the parameter λ, in which $n = 0$. Problem (4), together with continuity conditions on R and R' on the interval $0 \leq \rho \leq c$, is a special case of the singular Sturm-Liouville problem consisting of equations (20) and (21) in Theorem 3 of Sec. 78. According to that theorem, the eigenvalues λ_j of problem (4) are the numbers $\lambda_j = \alpha_j^2$ $(j = 1, 2, ...)$ where α_j are the positive roots of the equation

(6)
$$J_0(\alpha c) = 0;$$

and $R = J_0(\alpha_j\rho)$ are the corresponding eigenfunctions.

When $\lambda = \lambda_j$, we see from equation (5) that $T = \exp(-\alpha_j^2 kt)$. So, except for a constant factor,

$$R(\rho)T(t) = J_0(\alpha_j\rho)\exp(-\alpha_j^2 kt) \qquad (j = 1, 2, ...).$$

The extended linear combination of those functions,

(7)
$$u(\rho,t) = \sum_{j=1}^{\infty} A_j J_0(\alpha_j\rho)\exp(-\alpha_j^2 kt),$$

formally satisfies the homogeneous equations (1) and (2) in our boundary value problem. It also satisfies the nonhomogeneous initial condition (3) if $u(\rho, 0+) = u(\rho,0)$ and if the coefficients A_j can be determined so that

$$f(\rho) = \sum_{j=1}^{\infty} A_j J_0(\alpha_j\rho) \qquad (0 < \rho < c).$$

This is a valid Fourier-Bessel series representation (Theorem 4) if the coefficients have the values

$$
(8) \qquad A_j = \frac{2}{c^2[J_1(\alpha_j c)]^2} \int_0^c \rho f(\rho) J_0(\alpha_j \rho)\, d\rho \qquad (j = 1, 2, \ldots),
$$

obtained by writing $n = 0$ in equation (4), Sec. 80.

The formal solution of the boundary value problem is therefore given by equation (7) with the coefficients (8), where α_j are the positive roots of equation (6). That is, our temperature formula can be written as

$$
u(\rho,t) = \frac{2}{c^2} \sum_{j=1}^{\infty} \frac{J_0(\alpha_j \rho)}{[J_1(\alpha_j c)]^2} \exp\left(-\alpha_j^2 kt\right) \int_0^c sf(s) J_0(\alpha_j s)\, ds.
$$

Verification. We can verify formula (7), with coefficients (8), as a solution of our problem by the procedure followed in Sec. 56 if we use two additional properties of the zeros of Bessel functions, namely that $\alpha_{j+1} - \alpha_j \to \pi/c$ as $j \to \infty$ and that the sequence of numbers $1/[\sqrt{\alpha_j} J_1(\alpha_j c)]$ $(j = 1, 2, \ldots)$ is bounded.†

Since the sequence $1/[\sqrt{\alpha_j} J_1(\alpha_j c)]$ $(j = 1, 2, \ldots)$ is bounded and since J_0 and f are bounded functions, it follows from expression (8) that the sequence of numbers A_j/α_j $(j = 1, 2, \ldots)$ is also bounded. Therefore a positive number B exists such that, for each positive number t_0, the absolute values of the terms in series (7) are less than the constant terms

$$
(9) \qquad B\alpha_j \exp\left(-\alpha_j^2 kt_0\right)
$$

when $0 \le \rho \le c$ and $t \ge t_0$. The series of those constant terms converges because $\alpha_{j+1} - \alpha_j \to \pi/c$ as $j \to \infty$ (see Problem 5, Sec. 82). Series (7) therefore converges uniformly with respect to ρ and t $(t \ge t_0)$, and its sum $u(\rho,t)$ is a continuous function of its two variables when $0 \le \rho \le c$ and $t > 0$. But $u(c,t) = 0$; hence condition (2) is satisfied.

The derivatives of $J_0(x)$ are also bounded (Sec. 75), and the series of constants $B\alpha_j^m \exp\left(-\alpha_j^2 kt_0\right)$, where $m = 2, 3$, converge. So it follows in the same way that the differentiated series converge uniformly when $t \ge t_0$, and hence that the function (7) satisfies the heat equation (1).

---

† These properties follow from an asymptotic representation of $J_n(x)$, for large values of x,

$$
\sqrt{\pi x}\, J_n(x) = \sqrt{2} \cos\left(x - \frac{1}{2}n\pi - \frac{1}{4}\pi\right) + \frac{1}{x}\theta_n(x)
$$

where $\theta_n(x)$ is bounded as x tends to infinity. For a derivation, see vol. 1, p. 526, of the work by Courant and Hilbert (1953) or the book by Watson (1952), both listed in the Bibliography.

Finally, in view of the convergence of series (7) to $f(\rho)$ when $t = 0$, Abel's test, which is to be proved in Chap. 10, applies to show that $u(\rho,0+) = u(\rho,0)$ when $0 < \rho < c$. Condition (3) is therefore satisfied, and our solution is verified.

82. Heat Transfer at the Surface of the Cylinder

Let us replace the condition that the surface of the infinite cylinder is at temperature zero by a condition that heat transfer takes place there into surroundings at temperature zero. As in Sec. 7, where Newton's law for surface heat transfer was discussed, the flux through the surface is assumed to be proportional to the difference between the temperature of the surface and that of its surroundings. That is,

$$-Ku_\rho(c,t) = H[u(c,t) - 0] \qquad (K > 0, H > 0),$$

where K is the thermal conductivity of the material of the cylinder and H is its surface conductance.

The boundary value problem for the temperature function $u(\rho,t)$ is

$$(1) \qquad u_t(\rho,t) = k\left[u_{\rho\rho}(\rho,t) + \frac{1}{\rho}u_\rho(\rho,t)\right] \qquad (0 < \rho < c, t > 0),$$

$$(2) \qquad cu_\rho(c,t) = -hu(c,t) \qquad\qquad (t > 0),$$

$$(3) \qquad u(\rho,0) = f(\rho) \qquad\qquad (0 < \rho < c).$$

We have written $h = cH/K$; and, for convenience, we allow the possibility that the constant h be zero. In that case condition (2) simply states that the surface $\rho = c$ is insulated.

When $u = R(\rho)T(t)$, separation of variables produces the eigenvalue problem

$$\rho R''(\rho) + R'(\rho) + \lambda\rho R(\rho) = 0 \qquad (0 < \rho < c),$$
$$(4)$$
$$hR(c) + cR'(c) = 0.$$

If $h > 0$, the eigenvalues are, according to Theorem 3, $\lambda_j = \alpha_j^2$ ($j = 1$, $2, \ldots$) where α_j are the positive roots of the equation

$$(5) \qquad hJ_0(\alpha c) + (\alpha c)J_0'(\alpha c) = 0;$$

and, since $T'(t) + \lambda k T(t) = 0$, we find that

$$R(\rho)T(t) = J_0(\alpha_j\rho) \exp{(-\alpha_j^2 kt)} \qquad (j = 1, 2, \ldots).$$

The formal solution of our problem is then

$$(6) \qquad u(\rho,t) = \sum_{j=1}^{\infty} A_j J_0(\alpha_j\rho) \exp{(-\alpha_j^2 kt)}$$

where, according to the initial condition (3) and Theorem 4,

$$(7) \qquad A_j = \frac{2\alpha_j^2}{(\alpha_j^2 c^2 + h^2)[J_0(\alpha_j c)]^2} \int_0^c \rho f(\rho) J_0(\alpha_j \rho) \, d\rho \quad (j = 1, 2, \ldots).$$

If $h = 0$, the boundary condition in our eigenvalue problem becomes $R'(c) = 0$. In that case, $\lambda_j = \alpha_j^2$ $(j = 1, 2, \ldots)$ where $\alpha_1 = 0$ and α_j $(j = 2, 3, \ldots)$ are the positive roots of the equation $J_0'(\alpha c) = 0$, or

$$(8) \qquad J_1(\alpha c) = 0.$$

Noting that $J_0(\alpha_1 \rho) = 1$, we may now write

$$(9) \qquad u(\rho,t) = A_1 + \sum_{j=2}^{\infty} A_j J_0(\alpha_j \rho) \exp\left(-\alpha_j^2 kt\right)$$

where (Theorem 4)

$$(10) \qquad A_1 = \frac{2}{c^2} \int_0^c \rho f(\rho) \, d\rho,$$

$$(11) \qquad A_j = \frac{2}{c^2[J_0(\alpha_j c)]^2} \int_0^c \rho f(\rho) J_0(\alpha_j \rho) \, d\rho \qquad (j = 2, 3, \ldots).$$

PROBLEMS

1. In Sec. 81, use expression (7), with coefficients (8), to find the temperatures $u(\rho,t)$ in an infinite cylinder $\rho \le 1$ under the conditions $u(1,t) = 0$, $u(\rho,0) = u_0$, where u_0 is a constant. Give approximate numerical values of the first three coefficients in the series.

Ans. $u(\rho,t) = 2u_0[0.80 J_0(2.4\rho) \exp\left(-5.8kt\right) - 0.53 J_0(5.5\rho) \exp\left(-30kt\right)$
$$+ 0.43 J_0(8.7\rho) \exp\left(-76kt\right) - \cdots].$$

2. Over a long solid cylinder $\rho \le 1$, at uniform temperature A, there is tightly fitted a long hollow cylinder $1 \le \rho \le 2$ of the same material at temperature B. The outer surface $\rho = 2$ is then kept at temperature B. Derive this expression for the temperatures in the cylinder of radius 2 so formed:

$$u(\rho,t) = B + \frac{A - B}{2} \sum_{j=1}^{\infty} \frac{J_1(\alpha_j)}{\alpha_j[J_1(2\alpha_j)]^2} J_0(\alpha_j \rho) \exp\left(-\alpha_j^2 kt\right)$$

where α_j are the positive roots of $J_0(2\alpha) = 0$. This is a temperature problem in shrunken fittings.

3. A function $V(\rho,z)$ is harmonic interior to the cylinder bounded by the three surfaces $\rho = c$, $z = 0$, and $z = b$. Assuming that $V = 0$ on the first two of those surfaces and that $V(\rho,b) = f(\rho)$ $(0 < \rho < c)$, derive the expression

$$V(\rho,z) = \sum_{j=1}^{\infty} A_j J_0(\alpha_j \rho) \frac{\sinh \alpha_j z}{\sinh \alpha_j b}$$

where α_j are the positive roots of $J_0(\alpha c) = 0$ and the coefficients A_j are given by equation (8), Sec. 81.

4. Derive an expression for the steady temperatures $u(\rho,z)$ in the solid cylinder bounded by the three surfaces $\rho = 1$, $z = 0$, and $z = 1$ when $u = 0$ on the side, the bottom is insulated, and $u = 1$ on the top.

$$Ans. \quad u(\rho,z) = 2 \sum_{j=1}^{\infty} \frac{J_0(\alpha_j \rho)}{a_j J_1(\alpha_j)} \frac{\cosh \alpha_j z}{\cosh \alpha_j} \qquad [J_0(\alpha_j) = 0, \, \alpha_j > 0].$$

5. Apply the ratio test to show that the series of constant terms (9), Sec. 81, converges. *Suggestion:* First show that $(\alpha_{j+1}/\alpha_j) - 1 \to 0$, or $(\alpha_{j+1}/\alpha_j) \to 1$, as $j \to \infty$.

6. A solid cylinder is bounded by the three surfaces $\rho = 1$, $z = 0$, and $z = b$. The side is insulated, the bottom kept at temperature zero, and the top at temperatures $f(\rho)$. Derive this expression for the steady temperatures $u(\rho,z)$ in the cylinder:

$$u(\rho,z) = \frac{2z}{b} \int_0^1 sf(s) \, ds + 2 \sum_{j=2}^{\infty} \frac{J_0(\alpha_j \rho) \sinh \alpha_j z}{[J_0(\alpha_j)]^2 \sinh \alpha_j b} \int_0^1 sf(s) J_0(\alpha_j s) \, ds$$

where $\alpha_2, \alpha_3, \ldots$ are the positive roots of $J_1(\alpha) = 0$.

7. Show that when $f(\rho) = 1$ $(0 < \rho < 1)$ in Problem 6, $u(\rho,z) = z/b$.

8. Find the bounded steady temperatures $u(\rho,z)$ in the semi-infinite cylinder $\rho \leq 1$, $z \geq 0$ when $u = 1$ on the base and there is heat transfer into surroundings at temperature zero, according to the linear law used in Sec. 82, at the surface $\rho = 1$, $z > 0$.

$$Ans. \quad u(\rho,z) = 2h \sum_{j=1}^{\infty} \frac{J_0(\alpha_j \rho) \exp(-\alpha_j z)}{J_0(\alpha_j)(\alpha_j^2 + h^2)} \qquad [hJ_0(\alpha_j) = \alpha_j J_1(\alpha_j), \, \alpha_j > 0].$$

9. Solve this boundary value problem for $u(x,t)$:

$$xu_t = (xu_x)_x - (n^2/x)u \qquad (0 < x < c, \, t > 0),$$

$$u(c,t) = 0 \qquad (t > 0),$$

$$u(x,0) = f(x) \qquad (0 < x < c),$$

where u is continuous when $0 \leq x \leq c$, $t > 0$ and where n is a nonnegative integer.

$$Ans. \quad u(x,t) = \sum_{j=1}^{\infty} A_j J_n(\alpha_j x) \exp(-\alpha_j^2 t) \text{ where } \alpha_j \text{ and } A_j \text{ are given by equations (3) and (4)}$$

in Sec. 80.

10. Let $V(\rho,z)$ denote a function which is harmonic interior to the cylinder bounded by the three surfaces $\rho = c$, $z = 0$, and $z = b$. Given that $V = 0$ on both the top and bottom of the cylinder and that $V(c,z) = f(z)$ $(0 < z < b)$, derive the expression

$$V(\rho,z) = \sum_{n=1}^{\infty} b_n \frac{I_0(n\pi\rho/b)}{I_0(n\pi c/b)} \sin \frac{n\pi z}{b}$$

where

$$b_n = \frac{2}{b} \int_0^b f(z) \sin \frac{n\pi z}{b} \, dz.$$

[See Problem 2, Sec. 73, as well as the comments immediately following equation (17), Sec. 78, regarding the solutions of that modified form of Bessel's equation.]

83. Vibration of a Circular Membrane

A membrane, stretched over a fixed circular frame $\rho = c$ in the plane $z = 0$, is given an initial displacement $z = f(\rho,\phi)$ and released from rest in that position. Its transverse displacements $z(\rho,\phi,t)$, where ρ, ϕ, and z are cylindrical coordinates, will be found as the continuous function that satisfies this boundary value problem:

$$(1) \qquad z_{tt} = a^2 \left(z_{\rho\rho} + \frac{1}{\rho} z_\rho + \frac{1}{\rho^2} z_{\phi\phi} \right),$$

$$(2) \qquad z(c,\phi,t) = 0 \qquad\qquad (-\pi \le \phi \le \pi, \, t \ge 0),$$

$$(3) \qquad z(\rho,\phi,0) = f(\rho,\phi), \quad z_t(\rho,\phi,0) = 0 \quad (0 \le \rho \le c, \, -\pi \le \phi \le \pi),$$

where the function $z(\rho,\phi,t)$ is periodic with period 2π in the variable ϕ.

A function $z = R(\rho)\Phi(\phi)T(t)$ satisfies equation (1) if

$$(4) \qquad \frac{T''}{a^2 T} = \frac{1}{R}\left(R'' + \frac{1}{\rho} R' \right) + \frac{1}{\rho^2}\frac{\Phi''}{\Phi} = -\lambda,$$

where $-\lambda$ is any constant. We separate variables again in the second of equations (4) and write $\Phi''/\Phi = -\mu$. Then we find that the function $R\Phi T$ satisfies the homogeneous equations and has the necessary periodicity with respect to ϕ if R and Φ are eigenfunctions of the Sturm-Liouville problems

$$(5) \qquad \rho^2 R''(\rho) + \rho R'(\rho) + (\lambda\rho^2 - \mu)R(\rho) = 0, \qquad R(c) = 0,$$

$$(6) \qquad \Phi''(\phi) + \mu\Phi(\phi) = 0, \quad \Phi(-\pi) = \Phi(\pi), \quad \Phi'(-\pi) = \Phi'(\pi)$$

and T is such that

$$T''(t) + \lambda a^2 T(t) = 0, \qquad\qquad T'(0) = 0.$$

If μ has one of the values

$$\mu = n^2 \qquad\qquad (n = 0, 1, 2, \ldots),$$

Theorem 3 can be applied to problem (5); and if we consider problem (6) first, we see that the constant μ must, in fact, have one of those values. For, according to Sec. 33, they are eigenvalues of problem (6). To be precise, $\Phi(\phi) = 1$ when $n = 0$; and when $n = 1, 2, \ldots, \Phi(\phi)$ can be any linear combination of $\cos n\phi$ and $\sin n\phi$. From Theorem 3 we now see that the eigenvalues of problem (5) are the numbers $\lambda_{nj} = \alpha_{nj}^2$ $(j = 1, 2, \ldots)$ where α_{nj} are the positive roots of the equation

$$(7) \qquad J_n(\alpha c) = 0 \qquad\qquad (n = 0, 1, 2, \ldots),$$

the corresponding eigenfunctions being $R(\rho) = J_n(\alpha_{nj}\rho)$. Then $T(t) = \cos(\alpha_{nj} at)$.

The generalized linear combination of our functions $R\Phi T$,

$$(8) \qquad z(\rho,\phi,t) = \sum_{j=1}^{\infty} A_{0j} J_0(\alpha_{0j}\rho) \cos(\alpha_{0j}at)$$

$$+ \sum_{n=1}^{\infty} \sum_{j=1}^{\infty} J_n(\alpha_{nj}\rho)(A_{nj} \cos n\phi + B_{nj} \sin n\phi) \cos(\alpha_{nj}at),$$

formally satisfies all the homogeneous equations. It also satisfies the condition $z(\rho,\phi,0) = f(\rho,\phi)$ if the coefficients A_{0j}, A_{nj}, and B_{nj} are such that

$$(9) \qquad f(\rho,\phi) = \sum_{j=1}^{\infty} A_{0j} J_0(\alpha_{0j}\rho)$$

$$+ \sum_{n=1}^{\infty} \left\{ \left[\sum_{j=1}^{\infty} A_{nj} J_n(\alpha_{nj}\rho) \right] \cos n\phi + \left[\sum_{j=1}^{\infty} B_{nj} J_n(\alpha_{nj}\rho) \right] \sin n\phi \right\}$$

when $0 \le \rho \le c$, $-\pi \le \phi \le \pi$.

For each fixed value of ρ, series (9) is the Fourier series for $f(\rho,\phi)$ on the interval $-\pi \le \phi \le \pi$ if

$$\sum_{j=1}^{\infty} A_{0j} J_0(\alpha_{0j}\rho) = \frac{1}{2\pi} \int_{-\pi}^{\pi} f(\rho,\phi) \, d\phi,$$

$$\sum_{j=1}^{\infty} A_{nj} J_n(\alpha_{nj}\rho) = \frac{1}{\pi} \int_{-\pi}^{\pi} f(\rho,\phi) \cos n\phi \, d\phi \qquad (n = 1, 2, \ldots),$$

$$\sum_{j=1}^{\infty} B_{nj} J_n(\alpha_{nj}\rho) = \frac{1}{\pi} \int_{-\pi}^{\pi} f(\rho,\phi) \sin n\phi \, d\phi \qquad (n = 1, 2, \ldots).$$

For each fixed n, the equations here are Fourier-Bessel series representations of the functions of ρ appearing on the right-hand sides of those equations, on the interval $(0,c)$, provided that (Theorem 4)

$$(10) \qquad A_{0j} = \frac{1}{\pi c^2 [J_1(\alpha_{0j}c)]^2} \int_0^c \rho J_0(\alpha_{0j}\rho) \int_{-\pi}^{\pi} f(\rho,\phi) \, d\phi \, d\rho,$$

and

$$(11) \qquad A_{nj} = \frac{2}{\pi c^2 [J_{n+1}(\alpha_{nj}c)]^2} \int_0^c \rho J_n(\alpha_{nj}\rho) \int_{-\pi}^{\pi} f(\rho,\phi) \cos n\phi \, d\phi \, d\rho,$$

$$(12) \qquad B_{nj} = \frac{2}{\pi c^2 [J_{n+1}(\alpha_{nj}c)]^2} \int_0^c \rho J_n(\alpha_{nj}\rho) \int_{-\pi}^{\pi} f(\rho,\phi) \sin n\phi \, d\phi \, d\rho$$

when $n = 1, 2, \ldots$.

The displacements $z(\rho,\phi,t)$ are then given by equation (8) when the coefficients have the values (10), (11), and (12). We assume, of course, that the function f is such that the series in expression (8) has adequate properties of convergence and differentiability.

PROBLEMS

1. Suppose that in Sec. 83 the initial displacement function $f(\rho,\phi)$ is a linear combination of a finite number of the functions $J_0(\alpha_{0j}\rho)$ and $J_n(\alpha_{nj}\rho)\cos n\phi$, $J_n(\alpha_{nj}\rho)\sin n\phi$ ($n = 1, 2, \ldots$). Point out why the iterated series in expression (8) of that section then contains only a finite number of terms and represents a rigorous solution of the boundary value problem.

2. Let the initial displacement of the membrane in Sec. 83 be $f(\rho)$, a function of ρ only, and derive the expression

$$z(\rho,t) = \frac{2}{c^2} \sum_{j=1}^{\infty} \frac{J_0(\alpha_j\rho)\cos(\alpha_j at)}{[J_1(\alpha_j c)]^2} \int_0^c sf(s)J_0(\alpha_j s)\,ds,$$

where α_j are the positive roots of $J_0(\alpha c) = 0$, for the displacements when $t > 0$.

3. Show that if the initial displacement of the membrane in Sec. 83 is $AJ_0(\alpha_k\rho)$, where A is a constant and α_k is some positive root of $J_0(\alpha c) = 0$, the subsequent displacements are

$$z(\rho,t) = AJ_0(\alpha_k\rho)\cos(\alpha_k at).$$

Observe that these displacements are all periodic in t with a common period; thus the membrane gives a musical note.

4. Replace the initial conditions (3), Sec. 83, by the conditions that $z = 0$ and $z_t = 1$ when $t = 0$. This is the case if the membrane and its frame are moving with unit velocity in the z direction and the frame is brought to rest at the instant $t = 0$. Derive the expression

$$z(\rho,t) = \frac{2}{ac} \sum_{j=1}^{\infty} \frac{\sin(\alpha_j at)}{\alpha_j^2 J_1(\alpha_j c)} J_0(\alpha_j\rho),$$

where α_j are the positive roots of $J_0(\alpha c) = 0$, for the displacements when $t > 0$.

5. Derive the following expression for the temperatures $u(\rho,\phi,t)$ in an infinite cylinder $\rho \leq c$ when $u = 0$ on the surface $\rho = c$ and $u = f(\rho,\phi)$ at time $t = 0$:

$$u(\rho,\phi,t) = \sum_{j=1}^{\infty} A_{0j}J_0(\alpha_{0j}\rho)\exp\left(-\alpha_{0j}^2 kt\right)$$

$$+ \sum_{n=1}^{\infty} \sum_{j=1}^{\infty} J_n(\alpha_{nj}\rho)(A_{nj}\cos n\phi + B_{nj}\sin n\phi)\exp\left(-\alpha_{nj}^2 kt\right)$$

where α_{nj}, A_{0j}, A_{nj}, and B_{nj} are the numbers defined in Sec. 83.

6. Derive an expression for the temperatures $u(\rho,z,t)$ in a solid cylinder $\rho \leq c, 0 \leq z \leq \pi$ whose entire surface is kept at temperature zero and whose initial temperature is a constant A. Show that it can be written

$$u(\rho,z,t) = Av(z,t)w(\rho,t)$$

where

$$v(z,t) = \frac{4}{\pi} \sum_{n=1}^{\infty} \frac{\sin(2n-1)z}{2n-1} \exp\left[-(2n-1)^2 kt\right]$$

and

$$w(\rho,t) = \frac{2}{c} \sum_{j=1}^{\infty} \frac{J_0(\alpha_j\rho)}{\alpha_j J_1(\alpha_j c)} \exp\left(-\alpha_j^2 kt\right) \qquad [J_0(\alpha_j c) = 0, \ \alpha_j > 0].$$

Show that $v(z,t)$ represents temperatures in a slab $0 \leq z \leq \pi$ and $w(\rho,t)$ temperatures in an infinite cylinder $\rho \leq c$, both with zero boundary temperature and unit initial temperature (see Secs. 56 and 81).

7. Derive the following expression for temperatures $u(\rho,\phi,t)$ in the long right-angled cylindrical wedge formed by the surface $\rho = 1$ and the planes $\phi = 0$ and $\phi = \pi/2$ when $u = 0$ on its entire surface and $u = f(\rho,\phi)$ at time $t = 0$:

$$u(\rho,\phi,t) = \sum_{n=1}^{\infty} \sum_{j=1}^{\infty} B_{nj} J_{2n}(\alpha_{nj}\rho) \, \sin 2n\phi \exp\left(-\alpha_{nj}^2 kt\right)$$

where α_{nj} are the positive roots of $J_{2n}(\alpha) = 0$ and

$$B_{nj}[J_{2n+1}(\alpha_{nj})]^2 = \frac{8}{\pi} \int_0^{\pi/2} \sin 2n\phi \int_0^1 \rho f(\rho,\phi) J_{2n}(\alpha_{nj}\rho) \, d\rho \, d\phi.$$

8. Show that if the plane $\phi = \pi/2$ in Problem 7 is replaced by a plane $\phi = \phi_0$, the expression for the temperatures in the wedge will in general involve Bessel functions J_ν of *nonintegral* orders.

9. Solve Problem 7 when the entire surface of the wedge is insulated, instead of being kept at temperature zero.

10. Solve the following problem for temperatures $u(\rho,t)$ in a thin circular plate with heat transfer from its faces into surroundings at temperature zero:

$$u_t = u_{\rho\rho} + \frac{1}{\rho}u_\rho - hu \qquad\qquad (\rho < 1, t > 0; h > 0),$$

$$u(1,t) = 0, \qquad u(\rho,0) = 1.$$

$$\text{Ans.} \quad u(\rho,t) = 2e^{-ht} \sum_{j=1}^{\infty} \frac{J_0(\alpha_j\rho)}{\alpha_j J_1(\alpha_j)} \exp\left(-\alpha_j^2 t\right) \qquad [J_0(\alpha_j) = 0, \, \alpha_j > 0].$$

11. Solve Problem 10 after replacing the condition $u(1,t) = 0$ by this surface heat transfer condition at the edge:

$$u_\rho(1,t) = -h_0 u(1,t) \qquad\qquad (h_0 > 0).$$

12. Solve this Dirichlet problem for $V(\rho,z)$:

$$\nabla^2 V = 0 \qquad\qquad (\rho < 1, z > 0),$$

$$V(1,z) = 0 \qquad\qquad (z > 0),$$

$$V(\rho,0) = 1 \qquad\qquad (\rho < 1),$$

and V is to be bounded in the domain $\rho < 1$, $z > 0$.

$$\text{Ans.} \quad V(\rho,z) = 2 \sum_{j=1}^{\infty} \frac{J_0(\alpha_j\rho)}{\alpha_j J_1(\alpha_j)} \exp\left(-\alpha_j z\right) \qquad [J_0(\alpha_j) = 0, \, \alpha_j > 0].$$

13. Let the steady temperatures $u(\rho,z)$ in a semi-infinite cylinder $\rho \leq 1$, $z \geq 0$ whose base is insulated be such that

$$u(1,z) = \begin{cases} 1 & \text{when } 0 < z < 1, \\ 0 & \text{when } z > 1. \end{cases}$$

Derive the expression

$$u(\rho,z) = \frac{2}{\pi} \int_0^\infty \frac{I_0(\alpha\rho)}{\alpha I_0(\alpha)} \cos \alpha z \, \sin \alpha \, d\alpha$$

for those temperatures.

14. Given a function $f(z)$ that is represented by its Fourier integral formula for all real z, derive the following expression for the harmonic function $V(\rho,z)$ inside the cylinder $\rho = c$ such that $V(c,z) = f(z)$ $(-\infty < z < \infty)$:

$$V(\rho,z) = \frac{1}{\pi} \int_0^\infty \frac{I_0(\alpha\rho)}{I_0(\alpha c)} \int_{-\infty}^\infty f(s) \cos\left[\alpha(s - z)\right] \, ds \, d\alpha.$$

NINE

LEGENDRE POLYNOMIALS AND APPLICATIONS

84. Solutions of Legendre's Equation

As we shall see later on (Sec. 91), separation of variables in Laplace's equation written in terms of the spherical coordinates r and θ leads, after the substitution $x = \cos \theta$ is made, to *Legendre's equation*

$$(1) \qquad (1 - x^2)y''(x) - 2xy'(x) + \lambda y(x) = 0.$$

The points $x = 1$ and $x = -1$, corresponding to $\theta = 0$ and $\theta = \pi$, are singular points of that linear homogeneous differential equation. We show now that there is a certain infinite set of values of the parameter λ for each of which equation (1) has a polynomial as a solution.

The point $x = 0$ is an *ordinary point* of equation (1); and so, to determine a solution, we substitute

$$(2) \qquad y = \sum_{j=0}^{\infty} a_j x^j$$

into that equation.† This yields the identity

$$\sum_{j=0}^{\infty} j(j - 1)a_j x^{j-2} - \sum_{j=0}^{\infty} [j(j - 1) + 2j - \lambda]a_j x^j = 0.$$

Since the first two terms in the first series here are actually zero and since $j(j - 1) + 2j = j(j + 1)$, we may write

$$\sum_{j=2}^{\infty} j(j - 1)a_j x^{j-2} - \sum_{j=0}^{\infty} [j(j + 1) - \lambda]a_j x^j = 0.$$

Modifying the first series once again, we write it in the form

$$\sum_{j=0}^{\infty} (j + 2)(j + 1)a_{j+2} x^j$$

---

† For a discussion of ordinary points and a justification for this substitution, see, for example, the books referred to earlier in the footnote in Sec. 71.

so that

$$(3) \qquad \sum_{j=0}^{\infty} \{(j+2)(j+1)a_{j+2} - [j(j+1) - \lambda]a_j\}x^j = 0.$$

Equation (3) is an identify in x if the coefficients a_j satisfy the recurrence relation

$$(4) \qquad a_{j+2} = \frac{j(j+1) - \lambda}{(j+2)(j+1)} a_j \qquad (j = 0, 1, 2, \ldots).$$

The power series (2) then represents a solution of Legendre's equation within its interval of convergence if its coefficients satisfy relation (4). This leaves a_0 and a_1 as arbitrary constants.

If $a_1 = 0$, it follows from relation (4) that $a_3 = a_5 = \cdots = 0$. Thus one nontrivial solution of Legendre's equation is

$$(5) \qquad y_1 = a_0 + \sum_{k=1}^{\infty} a_{2k} x^{2k} \qquad (a_0 \neq 0)$$

where a_0 is an arbitrary nonzero constant and where the remaining coefficients $a_2, a_4, \ldots$ are expressed in terms of a_0 by successive applications of relation (4). Another such solution is obtained by writing $a_0 = 0$ and letting a_1 be arbitrary. To be precise, the series

$$(6) \qquad y_2 = a_1 x + \sum_{k=1}^{\infty} a_{2k+1} x^{2k+1} \qquad (a_1 \neq 0)$$

satisfies Legendre's equation for any nonzero value of a_1 when $a_3, a_5, \ldots$ are written in terms of a_1 in accordance with relation (4). These two solutions are, of course, linearly independent since they are not constant multiples of each other.

From relation (4) it is clear that the value of λ affects the values of all but the first coefficients in series (5) and (6). In particular, if λ has any one of the integral values

$$(7) \qquad \lambda = n(n+1) \qquad (n = 0, 1, 2, \ldots),$$

so that

$$(8) \qquad a_{j+2} = \frac{j(j+1) - n(n+1)}{(j+2)(j+1)} a_j \qquad (j = 0, 1, 2, \ldots),$$

one of those series always terminates since $a_{n+2} = 0$ and, consequently, $a_{n+4} = a_{n+6} = \cdots = 0$.

Note that if $n = 0$, then $a_2 = a_4 = a_6 = \cdots = 0$; and series (5) becomes simply $y_1 = a_0$. If, moreover, n is any one of the even integers $2, 4, \ldots$, so that $n = 2m$ ($m = 1, 2, \ldots$), it is true that $a_{2m} \neq 0$ and $a_{2(m+1)} = a_{2(m+2)} = \cdots = 0$. Series (5) thus reduces to a polynomial whose degree is $2m$, or n. On the other hand, if $n = 1$, we see that $y_2 = a_1 x$; and if n is any one of the odd integers $n = 2m + 1$ ($m = 1, 2, \ldots$), then $a_{2m+1} \neq 0$ and we may write $a_{2(m+1)+1} = a_{2(m+2)+1} = \cdots = 0$. Hence series (6) becomes a polynomial of degree n if n is odd.

To summarize our discussion of solutions of Legendre's equation (1) when $\lambda = n(n + 1)$ ($n = 0, 1, 2, \ldots$), *solution* (5) *reduces to the polynomial*

$$(9) \qquad\qquad y_1 = a_0 + a_2 x^2 + \cdots + a_n x^n \qquad\qquad (a_n \neq 0)$$

when n is even; and solution (6) *becomes*

$$(10) \qquad\qquad y_2 = a_1 x + a_3 x^3 + \cdots + a_n x^n \qquad\qquad (a_n \neq 0)$$

when n is odd. The coefficients a_0 and a_1 are arbitrary nonzero constants, and the others are determined by successive applications of relation (8). Observe that when n is even, solution (6) remains an infinite series and that when n is odd, the same is true of solution (5).

If n is even, it is customary to assign a value to a_0 such that when the coefficients $a_2, \ldots, a_n$ in expression (9) are determined by means of relation (8), the final coefficient a_n has the value

$$(11) \qquad\qquad a_n = \frac{(2n)!}{2^n (n!)^2}.$$

The precise value of a_0 that is required is not important to us. Using the convention that $0! = 1$, we note that $a_0 = 1$ if $n = 0$. In that case, $y_1 = 1$. If n is odd, let us choose a_1 so that the final coefficient in expression (10) is also given by equation (11). Note that $y_2 = x$ if $n = 1$, since $a_1 = 1$ for that value of n.

When $n = 2, 3, \ldots$, relation (8) can be used to write all the coefficients that precede a_n in expressions (9) and (10) in terms of a_n. To accomplish this, we first observe that the numerator on the right-hand side of relation (8) can be written

$$j(j + 1) - n(n + 1) = -[(n^2 - j^2) + (n - j)] = -(n - j)(n + j + 1).$$

We then solve for a_j; the result is

$$(12) \qquad\qquad a_j = -\frac{(j + 2)(j + 1)}{(n - j)(n + j + 1)} a_{j+2}.$$

To express a_{n-2k} in terms of a_n, we now use relation (12) to write the following k equations:

$$a_{n-2} = -\frac{(n)(n-1)}{(2)(2n-1)}a_n,$$

$$a_{n-4} = -\frac{(n-2)(n-3)}{(4)(2n-3)}a_{n-2},$$

$$\cdots$$

$$a_{n-2k} = -\frac{(n-2k+2)(n-2k+1)}{(2k)(2n-2k+1)}a_{n-2k+2}.$$

Equating the product of the left-hand sides of these equations to the product of their right-hand sides, we find that

$$(13) \qquad a_{n-2k} = \frac{(-1)^k}{2^k k!}\frac{n(n-1)\cdots(n-2k+1)}{(2n-1)(2n-3)\cdots(2n-2k+1)}a_n.$$

Then, upon substituting expression (11) for a_n into equation (13) and combining various terms into the appropriate factorials (see Problem 1, Sec. 86), we arrive that the desired expression:

$$(14) \qquad a_{n-2k} = \frac{1}{2^n}\frac{(-1)^k}{k!}\frac{(2n-2k)!}{(n-2k)!(n-k)!}.$$

As usual, we agree that $0! = 1$.

In view of equation (14), the polynomials (9) and (10), with the values of a_0 and a_1 assigned in the manner described above, can be written

$$(15) \qquad P_n(x) = \frac{1}{2^n}\sum_{k=0}^{m}\frac{(-1)^k}{k!}\frac{(2n-2k)!}{(n-2k)!(n-k)!}x^{n-2k} \qquad (n = 0, 1, 2, \ldots)$$

where $m = n/2$ if n is even and $m = (n-1)/2$ if n is odd. Another expression for $P_n(x)$ will be given in Sec. 86.

The polynomial $P_n(x)$ is called the *Legendre polynomial of degree n.* For the first several values of n expression (15) becomes

$$P_0(x) = 1, \qquad P_1(x) = x, \qquad P_2(x) = \frac{1}{2}(3x^2 - 1),$$

$$P_3(x) = \frac{1}{2}(5x^3 - 3x), \qquad P_4(x) = \frac{1}{8}(35x^4 - 30x^2 + 3),$$

$$P_5(x) = \frac{1}{8}(63x^5 - 70x^3 + 15x).$$

We have just seen that Legendre's equation

(16) $\qquad (1 - x^2)y''(x) - 2xy'(x) + n(n + 1)y(x) = 0 \qquad (n = 0, 1, 2, \ldots)$

always has the polynomial solution $y = P_n(x)$, which is solution (9) (n even) or solution (10) (n odd) when appropriate values are assigned to the arbitrary constants a_0 and a_1 in those solutions. Details regarding the standard form of the accompanying series solution, which is denoted by $Q_n(x)$ and is called a *Legendre function of the second kind*, are left to the problems. The series representation for $Q_n(x)$ is shown there to be convergent on the interval $-1 < x < 1$, and each $Q_n(x)$ is therefore continuous on that open interval. Although it is somewhat more difficult to show, the series diverges at the end points $x = \pm 1$.† It will, however, be sufficient for us to know that $Q_n(x)$ and $Q_n'(x)$ fail to be a pair of continuous functions on the *closed* interval $-1 \le x \le 1$ (Problem 13, Sec. 90). Since $P_n(x)$ and $Q_n(x)$ are linearly independent, the general solution of equation (16) is

(17) $\qquad\qquad y = CP_n(x) + DQ_n(x),$

where C and D are arbitrary constants.

Finally, we note that when λ is unrestricted, two fundamental solutions of Legendre's equation (1) can be written as power series in x by means of relation (4).

85. Orthogonality of Legendre Polynomials

The self-adjoint form of Legendre's equation (1), Sec. 84, is

(1) $\qquad\qquad \dfrac{d}{dx}\left[(1 - x^2)\dfrac{dy}{dx}\right] + \lambda y = 0.$

It is a special case of the Sturm-Liouville equation (4), Sec. 31, in which $r(x) = 1 - x^2$, $p(x) = 1$, and $q(x) = 0$. Since $r(-1) = 0$ and $r(1) = 0$, no boundary conditions at the ends of the interval $(-1,1)$ are needed in order to complete the singular Sturm-Liouville problem on that interval (see Sec. 32). The problem consists of equation (1) and the condition that y and y' be continuous when $-1 \le x \le 1$.

Equation (17) in the preceding section gives the general solution of equation (1) here if $\lambda = n(n + 1)$ where $n = 0, 1, 2, \ldots$. Since the polynomial $P_n(x)$ and its derivative are continuous on the entire interval $-1 \le x \le 1$ and since this is not true of the Legendre function $Q_n(x)$, it is clear that the continuity requirements on y and y' are met only when y is a constant

† See, for example, pp. 230–231 of the book by Bell (1968) that is listed in the Bibliography.

multiple of $P_n(x)$. Our singular Sturm-Liouville problem on the interval $(-1,1)$ therefore has eigenvalues

$$(2) \qquad\qquad \lambda = n(n+1) \qquad\qquad (n = 0, 1, 2, \ldots)$$

and corresponding eigenfunctions

$$(3) \qquad\qquad y = P_n(x) \qquad\qquad (n = 0, 1, 2, \ldots).$$

Since these eigenfunctions all correspond to different eigenvalues, we may conclude from Theorem 3, Sec. 32, that *the set $\{P_n(x)\}$ $(n = 0, 1, 2, \ldots)$ of Legendre polynomials is orthogonal on the interval $(-1,1)$ with weight function $p(x) = 1$*. That is,

$$(4) \qquad\qquad \int_{-1}^{1} P_m(x)P_n(x)\,dx = 0 \qquad\qquad (m \neq n).$$

In the notation used for inner products, property (4) reads $(P_m, P_n) = 0$ $(m \neq n)$.

Later on (Secs. 89 and 90) we shall prove that whenever a function f and its derivative f' are piecewise continuous on the interval $-1 < x < 1$, the generalized Fourier series for f with respect to the normalized form of the orthogonal set $\{P_n(x)\}$ converges to $f(x)$ at each point in the interval where f is continuous. That orthonormal set is therefore closed in the sense of pointwise convergence, and this means that it is complete in the class of all such functions f (see Sec. 28). It follows that *there can be no other eigenvalues than the set* (2). For, suppose $y_0(x)$ is an eigenfunction corresponding to another eigenvalue λ_0. Then $(y_0, P_n) = 0$ for each n; and the completeness of the set $\{P_n(x)\}$ requires that y_0, which is continuous on the entire interval $-1 \le x \le 1$, have value zero for each x in that interval. Hence y_0 cannot be an eigenfunction.

Since $P_n(x)$ is a polynomial containing only even powers of x if n is even and only odd powers if n is odd, it is an even or an odd function, depending on whether n is even or odd; that is,

$$(5) \qquad\qquad P_n(-x) = (-1)^n P_n(x) \qquad\qquad (n = 0, 1, 2, \ldots).$$

Hence if m and n are both even or both odd, the product $P_m(x)P_n(x)$ is an even function. It then follows from property (4) that

$$(6) \qquad\qquad \int_{0}^{1} P_{2m}(x)P_{2n}(x)\,dx = 0 \qquad\qquad (m \neq n)$$

and

$$(7) \qquad\qquad \int_{0}^{1} P_{2m+1}(x)P_{2n+1}(x)\,dx = 0 \qquad\qquad (m \neq n),$$

where $m, n = 0, 1, 2, \ldots$.

Condition (6) states that *the set* $\{P_{2n}(x)\}$ $(n = 0, 1, 2, \ldots)$ *of Legendre polynomials of even degree is orthogonal on the interval* $(0,1)$. Since $P'_{2n}(0) = 0$, those polynomials are the eigenfunctions of the singular Sturm-Liouville problem consisting of Legendre's equation (1) on the interval $(0,1)$ and the condition

(8) $$y'(0) = 0,$$

together with the condition that y and y' be continuous when $0 \le x \le 1$. The eigenvalues are $\lambda = 2n(2n + 1)$ where $n = 0, 1, 2, \ldots$.

Similarly, according to condition (7), *the set* $\{P_{2n+1}(x)\}$ $(n = 0, 1, 2, \ldots)$ *of Legendre polynomials of odd degree is orthogonal on the interval* $(0,1)$. Those polynomials are the eigenfunctions of the problem consisting of equation (1), the boundary condition

(9) $$y(0) = 0,$$

and the same continuity conditions on the closed interval $0 \le x \le 1$. The eigenvalues are $\lambda = (2n + 1)(2n + 2)$ where $n = 0, 1, 2, \ldots$.

The Fourier cosine series and sine series representations on half intervals followed from the basic Fourier series of sines and cosines (Sec. 37). In just the same way, representations of functions on the interval $(0,1)$ in terms of either P_{2n} or P_{2n+1} follow from representations in terms of P_n on the interval $(-1,1)$.

Further properties of $P_n(x)$ must be developed before we can establish the expansion theorem and certain order properties that are needed in our applications of those polynomials.†

86. Rodrigues' Formula and Norms

According to Sec. 84,

(1) $$P_n(x) = \frac{1}{2^n n!} \sum_{k=0}^{m} (-1)^k \frac{n!}{k!(n-k)!} \frac{(2n-2k)!}{(n-2k)!} x^{n-2k}$$

where $m = n/2$ if n is even and $m = (n-1)/2$ if n is odd. Since

$$\frac{d^n}{dx^n} x^{2n-2k} = \frac{(2n-2k)!}{(n-2k)!} x^{n-2k} \qquad (0 \le k \le m)$$

and because of the linearity of the differential operator d^n/dx^n, expression (1) can be written

(2) $$P_n(x) = \frac{1}{2^n n!} \frac{d^n}{dx^n} \sum_{k=0}^{m} (-1)^k \frac{n!}{k!(n-k)!} x^{2n-2k}.$$

† Readers who wish to pass to the applications at this time may read Theorem 1 (Sec. 88) and Sec. 89 and then proceed to Sec. 91.

The powers of x in the sum here decrease in steps of two as the index k increases; and the lowest power is $2n - 2m$, which is n if n is even and $n + 1$ if n is odd. Evidently, then, the sum can be extended so that k ranges from 0 to n; for the additional polynomial that is introduced is of degree less than n, and its nth derivative is therefore zero. Since the resulting sum is the binomial expansion of $(x^2 - 1)^n$, it follows from equation (2) that

$$(3) \qquad\qquad P_n(x) = \frac{1}{2^n n!} \frac{d^n}{dx^n} (x^2 - 1)^n \qquad\qquad (n = 0, 1, 2, \ldots).$$

This is *Rodrigues' formula* for the Legendre polynomials.

Various useful properties of Legendre polynomials are readily obtained from Rodrigues' formula with the aid of *Leibnitz' rule* for the nth derivative $D^n[f(x)g(x)]$ of the product of two functions:

$$(4) \qquad\qquad D^n(fg) = \sum_{k=0}^{n} \frac{n!}{k!(n-k)!} D^k(f) D^{n-k}(g),$$

where it is understood that all the required derivatives exist and that the zero-order derivative of a function is the function itself.

We note, for example, that if we write $u = x^2 - 1$, so that

$$u^n = (x^2 - 1)^n = (x + 1)^n (x - 1)^n,$$

it follows from Leibnitz' rule that

$$D^n u^n = \sum_{k=0}^{n} \frac{n!}{k!(n-k)!} D^k[(x + 1)^n] D^{n-k}[(x - 1)^n].$$

Since only the first term, occurring when $k = 0$, in this sum is free of the factor $x - 1$, we find that the value of the sum when $x = 1$ is $2^n n!$. According to Rodrigues' formula, then,

$$(5) \qquad\qquad P_n(1) = 1 \qquad\qquad (n = 0, 1, 2, \ldots).$$

(See Problem 13 of this section for another derivation of this result.) Note how it follows from this and equation (5), Sec. 85, that $P_n(-1) = (-1)^n$.

For another application of Rodrigues' formula, we write

$$2^{n+1}(n + 1)! P_{n+1}(x) = D^{n+1} u^{n+1} = D^{n-1}(D^2 u^{n+1}),$$

where $u = x^2 - 1$. But

$$Du^{n+1} = 2(n + 1)xu^n,$$

and so

$$D^2 u^{n+1} = 2(n + 1)(u^n + 2nx^2 u^{n-1})$$
$$= 2(n + 1)[u^n + 2n(x^2 - 1)u^{n-1} + 2nu^{n-1}]$$
$$= 2(n + 1)[(2n + 1)u^n + 2nu^{n-1}].$$

Consequently,

$$2^n n! \, P_{n+1}(x) = (2n + 1)D^{n-1}u^n + 2nD^{n-1}u^{n-1}.$$

Substituting $2^{n-1}(n-1)! \, P_{n-1}(x)$ for $D^{n-1}u^{n-1}$ here, we find that

$$(6) \qquad P_{n+1}(x) - P_{n-1}(x) = \frac{2n+1}{2^n n!} D^{n-1}u^n.$$

On the other hand, Leibnitz' rule allows us to write

$$P_{n+1}(x) = \frac{D^n(Du^{n+1})}{2^{n+1}(n+1)!} = \frac{D^n(xu^n)}{2^n n!} = \frac{xD^n u^n + nD^{n-1}u^n}{2^n n!};$$

and since $D^n u^n = 2^n n! \, P_n(x)$,

$$(7) \qquad P_{n+1}(x) - xP_n(x) = \frac{n}{2^n n!} D^{n-1}u^n.$$

Elimination of $D^{n-1}u^n$ between this and equation (6) gives the *recurrence relation*

$$(8) \qquad (n+1)P_{n+1}(x) + nP_{n-1}(x) = (2n+1)xP_n(x) \qquad (n = 1, 2, \ldots).$$

Note, too, that the relation

$$(9) \qquad P'_{n+1}(x) - P'_{n-1}(x) = (2n+1)P_n(x) \qquad (n = 1, 2, \ldots)$$

is an immediate consequence of equation (6).

We now show how relation (8) and its form

$$(10) \qquad nP_n(x) + (n-1)P_{n-2}(x) = (2n-1)xP_{n-1}(x) \quad (n = 2, 3, \ldots),$$

obtained by replacing n by $n-1$, can be used to find the norms $\|P_n\| = (P_n, P_n)^{1/2}$ of the orthogonal polynomials P_n. Keeping in mind that $(P_{n+1}, P_{n-1}) = 0$ and $(P_{n-2}, P_n) = 0$, we find from equations (8) and (10), respectively, that

$$(11) \qquad n(P_{n-1}, P_{n-1}) = (2n+1)(xP_n, P_{n-1})$$

and

$$(12) \qquad n(P_n, P_n) = (2n-1)(xP_{n-1}, P_n).$$

The integrals representing (xP_n, P_{n-1}) and (xP_{n-1}, P_n) are identical, and we need only eliminate those quantities from equations (11) and (12) to see that

$$(2n+1)(P_n, P_n) = (2n-1)(P_{n-1}, P_{n-1}),$$

or

$$(13) \qquad (2n+1)\|P_n\|^2 = (2n-1)\|P_{n-1}\|^2 \qquad (n = 2, 3, \ldots).$$

It is easy to verify directly that equation (13) is also valid when $n = 1$.

Next let n be any fixed positive integer and use equation (13) to write the following n equations:

$$(2n + 1)\|P_n\|^2 = (2n - 1)\|P_{n-1}\|^2,$$

$$(2n - 1)\|P_{n-1}\|^2 = (2n - 3)\|P_{n-2}\|^2,$$

$$\cdots$$

$$(5)\|P_2\|^2 = (3)\|P_1\|^2,$$

$$(3)\|P_1\|^2 = (1)\|P_0\|^2.$$

Setting the product of the left-hand sides of these equations equal to the product of their right-hand sides and canceling appropriately, we arrive at the result

$$(2n + 1)\|P_n\|^2 = \|P_0\|^2 \qquad (n = 1, 2, \ldots).$$

Since $\|P_0\|^2 = 2$, this means that

(14) $$\|P_n\| = \sqrt{\frac{2}{2n + 1}} \qquad (n = 0, 1, 2, \ldots).$$

The set

(15) $$\left\{ \sqrt{\frac{2n + 1}{2}} \, P_n(x) \right\} \qquad (n = 0, 1, 2, \ldots)$$

is therefore *orthonormal on the interval* $(-1, 1)$.

PROBLEMS

1. Obtain expression (14), Sec. 84, for the coefficients a_{n-2k} from equations (11) and (13) in that section.

Suggestion: Observe that the factorials in equation (11), Sec. 84, can be written

$$(2n)! = (2n)(2n - 1)(2n - 2) \cdots (2n - 2k + 1)(2n - 2k)!,$$

$$n! = n(n - 1) \cdots (n - 2k + 1)(n - 2k)!,$$

$$n! = n(n - 1) \cdots (n - k + 1)(n - k)!.$$

2. Establish these properties of Legendre polynomials:

(a) $P_{2n+1}(0) = 0$; (b) $P_{2n}(0) = (-1)^n \dfrac{(2n)!}{2^{2n}(n!)^2}$; (c) $P'_{2n}(0) = 0$.

3. Verify directly that the set of functions $P_0(x)$, $P_1(x)$, and $P_2(x)$ is orthogonal on the interval $(-1, 1)$. Draw the graphs of these functions.

4. Use induction on the integer n to verify Leibnitz' rule (4), Sec. 86.

5. From relation (9), Sec. 86, obtain the integration formula

$$\int_x^1 P_n(s)\, ds = \frac{1}{2n+1}[P_{n-1}(x) - P_{n+1}(x)] \qquad (n = 1, 2, \ldots).$$

6. From the orthogonality of the set $\{P_n(x)\}$, state why

(a) $\displaystyle\int_{-1}^1 P_n(x)\, dx = 0$ when $n = 1, 2, \ldots$;

(b) $\displaystyle\int_{-1}^1 (Ax + B)P_n(x)\, dx = 0$ when $n = 2, 3, \ldots$ (A, B constant).

7. Why is the set $\{\sqrt{4n+1}\, P_{2n}(x)\}$ $(n = 0, 1, 2, \ldots)$ *orthonormal on the interval* $(0,1)$?

8. Why is the set $\{\sqrt{4n+3}\, P_{2n+1}(x)\}$ $(n = 0, 1, 2, \ldots)$ *orthonormal on the interval* $(0,1)$?

9. With the aid of expression (15), Sec. 84, for $P_n(x)$, show that when $n = 2, 3, \ldots$, the constants a_0 and a_1 in equations (9) and (10) in that section must have the following values in order for the final constant a_n to have the value specified in equation (11) there:

$$a_0 = (-1)^{n/2} \frac{(1)(3)(5) \cdots (n-1)}{(2)(4)(6) \cdots (n)} \qquad (n = 2, 4, \ldots),$$

$$a_1 = (-1)^{(n-1)/2} \frac{(1)(3)(5) \cdots (n)}{(2)(4)(6) \cdots (n-1)} \qquad (n = 3, 5, \ldots).$$

10. When the value of the constant λ in Legendre's equation (1), Sec. 84, is unrestricted, it is customary to write that equation as

$$(1 - x^2)y''(x) - 2xy'(x) + v(v + 1)y(x) = 0,$$

where v is an unrestricted complex number.

(a) Show that relation (4), Sec. 84, takes the form

$$a_{j+2} = -\frac{(v - j)(v + j + 1)}{(j + 2)(j + 1)} a_j \qquad (j = 0, 1, 2, \ldots),$$

and then use it to obtain the following nontrivial solutions of Legendre's equation:

$$y_1 = a_0\left\{1 + \sum_{k=1}^{\infty} (-1)^k \frac{[v(v-2) \cdots (v - 2k + 2)][(v+1)(v+3) \cdots (v + 2k - 1)]}{(2k)!} x^{2k}\right\},$$

$$y_2 = a_1\left\{x + \sum_{k=1}^{\infty} (-1)^k \frac{[(v-1)(v-3) \cdots (v - 2k + 1)][(v+2)(v+4) \cdots (v + 2k)]}{(2k+1)!} x^{2k+1}\right\},$$

where a_0 and a_1 are arbitrary nonzero constants.

(b) Apply the ratio test to the series obtained in part (a) to show that they are convergent when $-1 < x < 1$.

11. Note that the solutions y_1 and y_2 obtained in Problem 10 are solutions (5) and (6) in Sec. 84 when $\lambda = v(v + 1)$. They remain infinite series when $v = n = 1, 3, 5, \ldots$ and $v = n = 0, 2, 4, \ldots$, respectively. When $v = n = 2m$ $(m = 0,1,2,\ldots)$, the Legendre function Q_n of the second kind is defined as y_2 where

$$a_1 = \frac{(-1)^m 2^{2m}(m!)^2}{(2m)!};$$

and when $v = n = 2m + 1$ $(m = 0,1,2,...)$, it is defined as y_1 where

$$a_0 = -\frac{(-1)^m 2^{2m}(m!)^2}{(2m+1)!}.$$

Using the fact that

$$\log\left(\frac{1+x}{1-x}\right) = 2 \sum_{k=0}^{\infty} \frac{x^{2k+1}}{2k+1} \qquad (-1 < x < 1),$$

show that

$$Q_0(x) = \frac{1}{2}\log\left(\frac{1+x}{1-x}\right)$$

and

$$Q_1(x) = \frac{x}{2}\log\left(\frac{1+x}{1-x}\right) - 1 = xQ_0(x) - 1.$$

12. Write

$$F(x,t) = (1 - 2xt + t^2)^{-1/2},$$

where $|x| \le 1$ and t is as yet unrestricted.

(a) Note that $x = \cos\theta$ for some uniquely determined value of θ between 0 and π inclusive, and show that

$$F(x,t) = (1 - e^{i\theta}t)^{-1/2}(1 - e^{-i\theta}t)^{-1/2}.$$

Then, using the fact that $(1 - z)^{-1/2}$ has a valid Maclaurin series expansion when $|z| < 1$, point out why it is true that the functions $(1 - e^{\pm i\theta}t)^{-1/2}$, considered as functions of t, can be represented by Maclaurin series which are valid when $|t| < 1$. It follows that the product of those two functions also has such a representation when $|t| < 1$.[†] That is, there are functions $f_n(x)$ $(n = 0, 1, 2, ...)$ such that

$$F(x,t) = \sum_{n=0}^{\infty} f_n(x)t^n \qquad (|t| < 1).$$

(b) Show that the function $F(x,t)$ satisfies the identity

$$(1 - 2xt + t^2)\frac{\partial F}{\partial t} = (x - t)F,$$

and use this result to show that the functions $f_n(x)$ in part (a) satisfy the recurrence relation

$$(n + 1)f_{n+1}(x) + nf_{n-1}(x) = (2n + 1)xf_n(x) \qquad (n = 1, 2, ...).$$

(c) Show that the first two functions $f_0(x)$ and $f_1(x)$ in part (a) are 1 and x, respectively, and notice that the recurrence relation obtained in part (b) can then be used to determine $f_n(x)$ when $n = 2, 3, ...$ Compare that relation with relation (8), Sec. 86, and conclude that the functions $f_n(x)$ are, in fact, the Legendre polynomials $P_n(x)$; that is, show that

$$(1 - 2xt + t^2)^{-1/2} = \sum_{n=0}^{\infty} P_n(x)t^n \qquad (|x| \le 1, |t| < 1).$$

The function F is thus a *generating function* for the Legendre polynomials.

13. Use the generating function obtained in Problem 12 to show that $P_n(1) = 1$ $(n = 0, 1, 2, ...)$.

[†] For a discussion of this point, see, for example, Churchill, Brown, and Verhey (1974, p. 164), listed in the Bibliography.

87. Laplace's Integral Form of P_n

Useful orders of magnitude of $P_n(x)$ with respect to x and n follow from a certain integral representation for those polynomials.

To derive the representation, we start with the fact that

$$\int_{-\pi}^{\pi} e^{iN\phi}\, d\phi = \begin{cases} 0 & \text{when } N = 1, 2, \ldots, \\ 2\pi & \text{when } N = 0. \end{cases}$$

Because of the nature of the binomial expansion of $(x + ye^{i\phi})^k$ $(k = 0, 1, 2, \ldots)$, where x and y $(y \neq 0)$ are independent of ϕ, it follows that

$$\frac{1}{2\pi} \int_{-\pi}^{\pi} (x + ye^{i\phi})^k\, d\phi = x^k \qquad (k = 0, 1, 2, \ldots)$$

and hence that any polynomial

$$q(x) = \sum_{k=0}^{n} a_k x^k$$

satisfies the identity

(1) $$q(x) = \frac{1}{2\pi} \int_{-\pi}^{\pi} q(x + ye^{i\phi})\, d\phi.$$

Details are left to the problems.

In particular, let $q(x)$ be the Legendre polynomial $P_n(x)$. In view of Rodrigues' formula (3) in the previous section, this choice of $q(x)$ can also be written as the nth derivative $p^{(n)}(x)$ of the polynomial

(2) $$p(x) = \frac{1}{2^n n!} (x^2 - 1)^n.$$

Equation (1) then becomes

(3) $$P_n(x) = \frac{1}{2\pi} \int_{-\pi}^{\pi} p^{(n)}(x + ye^{i\phi})\, d\phi.$$

Now it is readily shown by means of integration by parts that

(4) $$\int_{-\pi}^{\pi} e^{-ik\phi} p^{(n-k)}(x + ye^{i\phi})\, d\phi = \frac{k+1}{y} \int_{-\pi}^{\pi} e^{-i(k+1)\phi} p^{(n-k-1)}(x + ye^{i\phi})\, d\phi$$

where k is any integer such that $0 \leq k \leq n - 1$ (see Problem 9, Sec. 90). Successive applications of this reduction formula enable us to write equation (3) in the form

(5) $$P_n(x) = \frac{n!}{2\pi y^n} \int_{-\pi}^{\pi} e^{-in\phi} p(x + ye^{i\phi})\, d\phi.$$

If at this point we require that $x \neq \pm 1$ and write $y = \sqrt{x^2 - 1}$, definition (2) of the polynomial $p(x)$ and some elementary simplifications reveal that

$$\frac{n!}{y^n} e^{-in\phi} p(x + ye^{i\phi}) = (x + \sqrt{x^2 - 1} \cos \phi)^n.$$

With the above choice of y, then, expression (5) reduces to

$$P_n(x) = \frac{1}{2\pi} \int_{-\pi}^{\pi} (x + \sqrt{x^2 - 1} \cos \phi)^n \, d\phi,$$

or

(6) $$P_n(x) = \frac{1}{\pi} \int_0^{\pi} (x + \sqrt{x^2 - 1} \cos \phi)^n \, d\phi \qquad (n = 0, 1, 2, \ldots).$$

This is called *Laplace's integral form of $P_n(x)$*. Since $P_n(1) = 1$ and $P_n(-1) = (-1)^n$ (Sec. 86), it is also valid when $x = \pm 1$. Thus it is valid for all x.

Suppose that $-1 \leq x \leq 1$. Since $x = \cos \theta$ for some value of θ between 0 and π inclusive, equation (6) can be written

(7) $$P_n(\cos \theta) = \frac{1}{\pi} \int_0^{\pi} (\cos \theta + i \sin \theta \cos \phi)^n \, d\phi \qquad (0 \leq \theta \leq \pi).$$

Furthermore,

$$|\cos \theta + i \sin \theta \cos \phi| = (1 - \sin^2 \theta \sin^2 \phi)^{1/2};$$

and so, if we also note that $\sin (\pi - \phi) = \sin \phi$, we have the inequality

(8) $$|P_n(\cos \theta)| \leq \frac{2}{\pi} \int_0^{\pi/2} (1 - \sin^2 \theta \sin^2 \phi)^{n/2} \, d\phi.$$

Since the nonnegative quantity $1 - \sin^2 \theta \sin^2 \phi$ is always less than or equal to unity, it follows from inequality (8) that $|P_n(\cos \theta)| \leq 1$, or

(9) $$|P_n(x)| \leq 1 \qquad \text{when } -1 \leq x \leq 1 \qquad (n = 0, 1, 2, \ldots).$$

To obtain another important consequence of inequality (8), we observe from the graph of $\sin \phi$ that $\sin \phi > 2\phi/\pi$ when $0 < \phi < \pi/2$. Thus if $u = (2\phi/\pi)^2 \sin^2 \theta$, it is true that

$$1 - \sin^2 \theta \sin^2 \phi < 1 - u$$

when $0 < \phi < \pi/2$ and $0 < \theta < \pi$. Also $1 - u \leq e^{-u}$, as can be seen graphically or from the Maclaurin series for e^{-u}. According to inequality (8), then,

$$|P_n(\cos \theta)| < \frac{2}{\pi} \int_0^{\pi/2} \exp\left(-\frac{2n \sin^2 \theta}{\pi^2} \phi^2\right) d\phi \qquad (0 < \theta < \pi, n = 1, 2, \ldots).$$

Changing the variable of integration in this last integral by means of the substitution $r = (\sqrt{2n} \sin \theta)\phi/\pi$ and noting that the resulting upper limit of integration can be replaced by ∞, we find that

$$|P_n(\cos \theta)| < \sqrt{\frac{2}{n} \frac{1}{\sin \theta}} \int_0^\infty \exp(-r^2)\, dr \qquad (0 < \theta < \pi, n = 1, 2, \ldots).$$

Finally, since the value of the integral here is $\sqrt{\pi}/2$ (Problem 19, Sec. 69), we arrive at the inequality

$$(10) \qquad |P_n(x)| < \sqrt{\frac{\pi}{2n(1 - x^2)}} \qquad (-1 < x < 1, n = 1, 2, \ldots).$$

That is, *for each fixed x where $-1 < x < 1$, $P_n(x)$ is of the order of $n^{-1/2}$ as n increases*.

88. Further Order Properties

Since $P_k(x)$ is a polynomial of degree k containing only alternate powers of x, we know that

$$x^k = cP_k(x) + c_{k-2}x^{k-2} + c_{k-4}x^{k-4} + \cdots$$

where the coefficients are constants. Similarly, x^{k-2} is a linear combination of $P_{k-2}(x)$ and a polynomial of degree $k - 4$, and so on. Thus x^k is a finite linear combination of the polynomials $P_k(x)$, $P_{k-2}(x)$, $P_{k-4}(x)$, ...

The derivative $P_n'(x)$ $(n = 1,2,\ldots)$ is of degree $n - 1$ and contains only alternate powers of x; and each of those powers can be written in the above manner as a linear combination of Legendre polynomials. Therefore $P_n'(x)$ itself can be written as a linear combination

$$(1) \qquad P_n'(x) = C_{n-1}P_{n-1}(x) + C_{n-3}P_{n-3}(x) + \cdots \qquad (n = 1, 2, \ldots)$$

of Legendre polynomials. To find any one of the constants C_j $(j = n - 1, n - 3, \ldots)$, we take inner products of both sides of equation (1) with $P_j(x)$:

$$\int_{-1}^1 P_j(x)P_n'(x)\, dx = C_j(P_j, P_j).$$

When integrated by parts, the integral on the left becomes

$$[P_j(x)P_n(x)]_{-1}^1 - \int_{-1}^1 P_n(x)P_j'(x)\, dx;$$

and this last integral vanishes because $P_j'(x)$ is a linear combination of Legendre polynomials of degree less than n. Also $P_k(1) = 1$, $P_k(-1) = (-1)^k$, and $(P_j, P_j) = 2/(2j + 1)$. Therefore

$$C_j = 2j + 1 \qquad (j = n - 1, n - 3, \ldots).$$

Representation (1), which is valid for all x, then becomes

$$(2) \quad P_n'(x) = (2n - 1)P_{n-1}(x) + (2n - 5)P_{n-3}(x) + \cdots \quad (n = 1, 2, \ldots),$$

ending with $3P_1(x)$ if n is even and with $P_0(x)$ if n is odd.

Assume now that $-1 \leq x \leq 1$. Then $|P_n(x)| \leq 1$ (Sec. 87), and it follows from equation (2) that

$$|P_{2n}'(x)| \leq (4n - 1) + (4n - 5) + \cdots + 3 = n(2n + 1),$$

$$|P_{2n+1}'(x)| \leq (4n + 1) + (4n - 3) + \cdots + 1 = (n + 1)(2n + 1).$$

But $n(2n + 1) \leq (2n)^2$ and $(n + 1)(2n + 1) \leq (2n + 1)^2$; and so

$$(3) \qquad |P_n'(x)| \leq n^2 \qquad \text{when } -1 \leq x \leq 1 \qquad (n = 1, 2, \ldots).$$

Differentiating both sides of equation (2) and then noting that $|P_{n-1}'(x)| < n^2$, $|P_{n-3}'(x)| < n^2$, and so on, when $-1 \leq x \leq 1$, we see by the method used above that

$$(4) \qquad |P_n''(x)| \leq n^4 \qquad \text{when } -1 \leq x \leq 1 \qquad (n = 1, 2, \ldots).$$

Similarly, $|P_n^{(k)}(x)| \leq n^{2k}$ when $k = 3, 4, \ldots$ and $-1 \leq x \leq 1$.

We collect our order properties of P_n as follows.

Theorem 1. *For each positive integer n and at all points of the interval $-1 \leq x \leq 1$ the values of each of the functions*

$$|P_n(x)|, \qquad \frac{1}{n^2}|P_n'(x)|, \qquad \frac{1}{n^4}|P_n''(x)|, \ldots$$

never exceed unity. Also (Sec. 87), for each fixed number x such that $-1 < x < 1$ it is true that

$$|P_n(x)| < \frac{M}{\sqrt{n}} \qquad (n = 1, 2, \ldots),$$

where the value of M depends only on the choice of x.

89. Legendre Series

In Sec. 86 we saw that the set of polynomials

$$(1) \qquad \phi_n(x) = \sqrt{\frac{2n + 1}{2}} \, P_n(x) \qquad (n = 0, 1, 2, \ldots)$$

is orthonormal on the interval $(-1,1)$. The Fourier constants with respect to that set, for a function f defined on the interval $(-1,1)$, are

$$c_n = (f, \phi_n) = \sqrt{\frac{2n + 1}{2}} \int_{-1}^{1} f(x)P_n(x) \, dx;$$

and the generalized Fourier series corresponding to f is

$$\sum_{n=0}^{\infty} c_n \phi_n(x) = \sum_{n=0}^{\infty} \frac{2n+1}{2} P_n(x) \int_{-1}^{1} f(s) P_n(s) \, ds.$$

That is,

(2) $$f(x) \sim \sum_{n=0}^{\infty} A_n P_n(x) \qquad (-1 < x < 1)$$

where

(3) $$A_n = \frac{2n+1}{2} \int_{-1}^{1} f(x) P_n(x) \, dx \qquad (n = 0, 1, 2, \ldots).$$

Series (2) with coefficients (3) is a *Legendre series*. In the following section we shall prove that it converges to $f(x)$ under conditions stated in this theorem.

Theorem 2. *Let f denote a piecewise continuous function on the interval $-1 < x < 1$. At each point x in that interval where f is continuous and has derivatives from the right and left the Legendre series (2) converges to $f(x)$; that is,*

(4) $$f(x) = \sum_{n=0}^{\infty} A_n P_n(x) \qquad (-1 < x < 1)$$

where the coefficients A_n are given by equation (3).

Note that if f and its derivative f' are both piecewise continuous on the interval $-1 < x < 1$, representation (4) must then be valid at any point x in that interval where f is continuous.

The proof of the theorem can be extended to show that the series converges to the mean of the values $f(x+)$ and $f(x-)$ when f has a jump at the point x, provided both one-sided derivatives exist there.† That result will not be needed here.

If f is an *even* function, the product $f(x) P_n(x)$ is even when n is even, and odd when n is odd. Hence $A_{2n+1} = 0$ $(n = 0, 1, 2, \ldots)$ and

(5) $$A_{2n} = (4n + 1) \int_{0}^{1} f(x) P_{2n}(x) \, dx \qquad (n = 0, 1, 2, \ldots).$$

Thus, *if we apply Theorem 2 to the even extension of a function f which is piecewise continuous on the interval $0 < x < 1$ and which has one-sided derivatives at points of continuity x, we find that*

(6) $$f(x) = \sum_{n=0}^{\infty} A_{2n} P_{2n}(x) \qquad (0 < x < 1)$$

where the coefficients A_{2n} have the values (5).

---

† See pp. 65 ff of the book by Jackson (1941) that is listed in the Bibliography.

Similarly, *when the odd extension of f is taken, we may write*

(7) $$f(x) = \sum_{n=0}^{\infty} A_{2n+1} P_{2n+1}(x) \qquad (0 < x < 1)$$

where

(8) $$A_{2n+1} = (4n + 3) \int_0^1 f(x) P_{2n+1} \, dx \qquad (n = 0, 1, 2, \ldots).$$

Both of the sets $\{P_{2n}\}$ and $\{P_{2n+1}\}$, when normalized on the interval $(0,1)$, are therefore closed in the sense of pointwise convergence (Sec. 28).

90. Convergence of the Series

To prove Theorem 2, let $x \ (-1 < x < 1)$ denote a point at which the function f is continuous and has one-sided derivatives. At x the value of the sum of the first $m + 1$ terms of the Legendre series (2), Sec. 89, can be written

(1) $$S_m(x) = \sum_{n=0}^{m} A_n P_n(x) = \int_{-1}^{1} f(s) K_m(s,x) \, ds$$

where

(2) $$K_m(s,x) = \frac{1}{2} \left[1 + \sum_{n=1}^{m} (2n + 1) P_n(s) P_n(x) \right].$$

The recurrence relation

(3) $$(2n + 1) x P_n(x) = (n + 1) P_{n+1}(x) + n P_{n-1}(x) \quad (n = 1, 2, \ldots),$$

which is equation (8), Sec. 86, leads to a more compact expression for K_m in the following way. Multiply each side of relation (3) by $-P_n(s)$ and each side of the same relation in s,

$$(2n + 1) s P_n(s) = (n + 1) P_{n+1}(s) + n P_{n-1}(s),$$

by $P_n(x)$ and then add corresponding sides. Using the notation

(4) $$R_n(s,x) = P_n(s) P_{n-1}(x) - P_{n-1}(s) P_n(x),$$

we find that

$$(2n + 1) P_n(s) P_n(x) = \frac{(n + 1) R_{n+1}(s,x) - n R_n(s,x)}{s - x} \qquad (s \neq x).$$

When this expression for $(2n + 1) P_n(s) P_n(x)$ is used in the sum (2), the terms of that sum cancel in pairs, giving

(5) $$K_m(s,x) = \frac{m + 1}{2} \frac{P_{m+1}(s) P_m(x) - P_m(s) P_{m+1}(x)}{s - x} \qquad (s \neq x).$$

The numerator of the last fraction, which is $R_{m+1}(s,x)$, is a polynomial in s that vanishes when $s = x$; and so it contains $s - x$ as a factor. Therefore the fraction has a finite limit as $s \to x$. We have assumed that $m \geq 1$; but note that equation (5) yields $K_0(s,x) = \frac{1}{2}$, which agrees with expression (1) when $m = 0$.

According to equations (1), (4), and (5),

$$(6) \qquad S_m(x) = \frac{m+1}{2} \int_{-1}^{1} f(s) \frac{R_{m+1}(s,x)}{s-x} \, ds,$$

the integrand here being piecewise continuous. Consider for the moment the special case $f(s) = 1 = P_0(s)$. Then $A_0 = 1$ and $A_n = 0$ $(n = 1, 2, \ldots)$; and, in view of equation (1), $S_m(x) = 1$. Thus, from equation (6),

$$(7) \qquad 1 = \frac{m+1}{2} \int_{-1}^{1} \frac{R_{m+1}(s,x)}{s-x} \, ds.$$

We now multiply each side of equation (7) by $f(x)$ and subtract from the corresponding sides of equation (6) to write

$$S_m(x) - f(x) = \frac{m+1}{2} \int_{-1}^{1} \frac{f(s) - f(x)}{s-x} R_{m+1}(s,x) \, ds,$$

or

$$(8) \qquad S_m(x) - f(x) = \frac{m+1}{2} \int_{-1}^{1} F(s)[P_{m+1}(s)P_m(x) - P_m(s)P_{m+1}(x)] \, ds$$

where

$$F(s) = \frac{f(s) - f(x)}{s - x}.$$

Since $F(x+)$ and $F(x-)$ are the one-sided derivatives of the given function f at the point x, $F(s)$ is piecewise continuous on the interval $-1 < s < 1$. Let C_n denote the Fourier constants of F corresponding to the orthonormal set of functions $\phi_n(x) = \sqrt{(2n+1)/2}\, P_n(x)$. Then equation (8) can be written

$$(9) \qquad S_m(x) - f(x) = \frac{m+1}{2\sqrt{m+\frac{3}{2}}} P_m(x)C_{m+1} - \frac{m+1}{2\sqrt{m+\frac{1}{2}}} P_{m+1}(x)C_m.$$

But $|P_m(x)| < M/\sqrt{m}$ $(m = 1, 2, \ldots)$, where M is independent of m (Theorem 1). Thus it follows from equation (9) that

$$|S_m(x) - f(x)| < M \frac{m+1}{2m}(|C_{m+1}| + |C_m|) \qquad (m = 1, 2, \ldots);$$

and, since $C_m \to 0$ as $m \to \infty$ (Sec. 27), it is true that

$$(10) \qquad \lim_{m \to \infty} S_m(x) = f(x) \qquad (-1 < x < 1).$$

This completes the proof of Theorem 2.

PROBLEMS

1. Use Theorem 2 and results found in Problems 2 and 5, Sec. 86, to show that if $f(x) = 0$ when $-1 < x < 0$ and $f(x) = 1$ when $0 < x < 1$, then

$$f(x) = \frac{1}{2}P_0(x) + \frac{1}{2}\sum_{n=0}^{\infty}[P_{2n}(0) - P_{2n+2}(0)]P_{2n+1}(x)$$

$$= \frac{1}{2} + \frac{3}{4}x + \sum_{n=1}^{\infty}(-1)^n\left(\frac{4n+3}{4n+4}\right)\frac{(2n)!}{2^{2n}(n!)^2}P_{2n+1}(x)$$

when $-1 < x < 1$ and $x \neq 0$. Show that if $f(0)$ is defined as $1/2$, the expansion is also valid when $x = 0$.

2. Suppose that $f(x) = 0$ when $-1 < x \le 0$ and $f(x) = x$ when $0 < x < 1$.
(a) State why f is represented by its Legendre series (4), Sec. 89, at each point of the interval $-1 < x < 1$.
(b) Show that $A_{2n+1} = 0$ ($n = 1, 2, \ldots$) in the series in part (a).
(c) Find the first four nonvanishing terms of the series in part (a).

$$\text{Ans.} \quad (c) \ f(x) = \frac{1}{4} + \frac{1}{2}P_1(x) + \frac{5}{16}P_2(x) - \frac{3}{32}P_4(x) + \cdots \quad (-1 < x < 1).$$

3. Prove that for all x

$$(a) \ x^2 = \frac{1}{3}P_0(x) + \frac{2}{3}P_2(x); \quad (b) \ x^3 = \frac{3}{5}P_1(x) + \frac{2}{5}P_3(x).$$

4. Expand the function $f(x) = 1 \ (0 < x < 1)$ into a series of Legendre polynomials of odd degree on the interval $(0,1)$. What function does the same series represent on the interval $-1 < x < 0$?

$$\text{Ans.} \quad 1 = \sum_{n=0}^{\infty}(-1)^n\left(\frac{4n+3}{2n+2}\right)\frac{(2n)!}{2^{2n}(n!)^2}P_{2n+1}(x) \quad (0 < x < 1).$$

5. Obtain the first three nonzero terms in the series of Legendre polynomials of even degree representing the function $f(x) = x \ (0 < x < 1)$ to show that

$$x = \frac{1}{2}P_0(x) + \frac{5}{8}P_2(x) - \frac{3}{16}P_4(x) + \cdots \quad (0 < x < 1).$$

Point out why this expansion remains valid when $x = 0$, and state what function the series represents on the interval $-1 < x < 1$.

6. State why it is true that when x is a fixed number such that $-1 < x < 1$, then

$$\lim_{n \to \infty} P_n(x) = 0.$$

7. State why

$$\lim_{n \to \infty} \sqrt{4n+1} \int_0^1 f(x)P_{2n}(x) \, dx = 0$$

when f is piecewise continuous on the interval $(0,1)$.

8. Give details in the discussion starting at the beginning of Sec. 87 and leading up to identity (1), Sec. 87, which states that

$$q(x) = \frac{1}{2\pi} \int_{-\pi}^{\pi} q(x + ye^{i\phi}) \, d\phi$$

for any polynomial $q(x)$.

9. Use integration by parts to obtain the reduction formula (4) in Sec. 87.

Suggestion: Write the first integral appearing in that formula as

$$\int_{-\pi}^{\pi} e^{-i(k+1)\phi} \frac{d}{d\phi} \left[\frac{1}{iy} p^{(n-k-1)}(x + ye^{i\phi}) \right] d\phi.$$

10. Using the fact that x^k ($k = 0, 1, 2, \ldots$) can be written as a finite linear combination of the Legendre polynomials $P_k(x)$, $P_{k-2}(x)$, $P_{k-4}(x)$, $\ldots$ (Sec. 88), point out why it is true that

$$\int_{-1}^{1} P_n(x)p(x) \, dx = 0$$

where $P_n(x)$ is a Legendre polynomial of degree n ($n = 1, 2, \ldots$) and $p(x)$ is any polynomial whose degree is less than n.

11. Give details in the derivation of expression (5), Sec. 90, for $K_m(s,x)$.

12. Let n have any one of the values $n = 1, 2, \ldots$.

(*a*) By recalling the result in Problem 6(*a*), Sec. 86, point out why it is true that $P_n(x)$ must change sign at least once in the open interval $-1 < x < 1$. Then let $x_1, x_2, \ldots, x_k$ denote the totality of distinct points in that interval where $P_n(x)$ changes sign. Since any polynomial of degree n has at most n zeros, we know that $1 \le k \le n$.

(*b*) *Assume* that the number of points $x_1, x_2, \ldots, x_k$ in part (*a*) is such that $k < n$, and consider the polynomial

$$p(x) = (x - x_1)(x - x_2) \cdots (x - x_k).$$

With the aid of the result in Problem 10, point out why the integral

$$\int_{-1}^{1} P_n(x)p(x) \, dx$$

has value zero; and, after noting that $P_n(x)$ and $p(x)$ change sign at precisely the same points in the interval $-1 < x < 1$, state why the value of the integral cannot be zero. Having reached this contradiction, conclude that $k = n$ and hence that *the zeros of a Legendre polynomial $P_n(x)$ are all real and distinct and lie in the open interval $-1 < x < 1$.*

13. Show in the following way that for each value of n ($n = 0, 1, 2, \ldots$) the Legendre function of the second kind $Q_n(x)$ and its derivative $Q'_n(x)$ fail to be a pair of continuous functions on the closed interval $-1 \le x \le 1$. Suppose that there is an integer N such that $Q_N(x)$ and $Q'_N(x)$ are continuous on the stated interval. The functions $Q_N(x)$ and $P_n(x)$ ($n \neq N$) are then eigenfunctions corresponding to different eigenvalues of the Sturm-Liouville problem (1), Sec. 85. Point out how it follows that

$$\int_{-1}^{1} Q_N(x)P_n(x) \, dx = 0 \qquad (n \neq N)$$

and then use Theorem 2 to show that $Q_N(x) = A_N P_N(x)$ where A_N is some constant. This is, however, impossible since $P_N(x)$ and $Q_N(x)$ are linearly independent.

91. Dirichlet Problems in Spherical Regions

For our first application of Legendre series, we shall determine the harmonic function V in the region $r < c$ such that V assumes prescribed values $F(\theta)$ on the spherical surface $r = c$ (Fig. 28). Here r, ϕ, and θ are spherical coordinates, and V is independent of ϕ. Thus V satisfies Laplace's equation

$$\text{(1)} \qquad r \frac{\partial^2}{\partial r^2}(rV) + \frac{1}{\sin\theta}\frac{\partial}{\partial\theta}\left(\sin\theta\frac{\partial V}{\partial\theta}\right) = 0 \quad (r < c, 0 < \theta < \pi)$$

and the condition

$$\text{(2)} \qquad \lim_{\substack{r \to c \\ r < c}} V(r,\theta) = F(\theta) \qquad (0 < \theta < \pi);$$

also, V and its partial derivatives of the first and second order are to be continuous throughout the interior $(0 \le r < c, 0 \le \theta \le \pi)$ of the sphere.

  Physically, the function V may denote steady temperatures in a solid sphere $r \le c$ whose surface temperatures depend only on θ; that is, the surface temperatures are uniform over each circle $r = c$, $\theta = \theta_0$. Also, V represents electrostatic potential in the space $r < c$, which is free of charges, when $V = F(\theta)$ on the boundary $r = c$.

  Consider now a solution of equation (1) of the form $V = R(r)\Theta(\theta)$ that satisfies the continuity requirements. Separation of variables shows that, for some constant λ, the function $R(r)$ must satisfy the differential equation

$$\text{(3)} \qquad r(rR)'' - \lambda R = 0 \qquad (r < c)$$

and be continuous when $0 \le r < c$. Also, for the same constant λ,

$$\text{(4)} \qquad \frac{1}{\sin\theta}\frac{d}{d\theta}\left(\sin\theta\frac{d\Theta}{d\theta}\right) + \lambda\Theta = 0 \qquad (0 < \theta < \pi),$$

where Θ, Θ', and Θ'' are to be continuous on the closed interval $0 \le \theta \le \pi$.

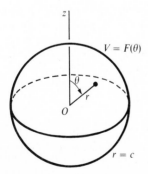

$V = F(\theta)$

$r = c$ **Figure 28**

If in equation (4) we make the substitution $x = \cos\theta$, so that

$$\sin\theta\,\frac{d\Theta}{d\theta} = (1 - \cos^2\theta)\frac{1}{\sin\theta}\frac{d\Theta}{d\theta} = -(1 - x^2)\frac{d\Theta}{dx},$$

it follows readily that

(5)
$$\frac{d}{dx}\left[(1 - x^2)\frac{d\Theta}{dx}\right] + \lambda\Theta = 0 \qquad (-1 < x < 1),$$

where Θ and its first two derivatives with respect to x are continuous on the entire closed interval $-1 \leq x \leq 1$. Equation (5) is Legendre's equation in self-adjoint form; and from Sec. 85 we know that, except when Θ is identically zero, λ must be one of the eigenvalues

$$\lambda = n(n + 1) \qquad (n = 0, 1, 2, \ldots)$$

and that $\Theta = P_n(x)$. Then $\Theta(\theta) = P_n(\cos\theta)$.

Writing equation (3) in the form

$$r^2 R'' + 2rR' - \lambda R = 0,$$

we see that it is a Cauchy-Euler equation, which reduces to a differential equation with constant coefficients after the substitution $r = e^s$ is made (see Problem 13, Sec. 34). When $\lambda = n(n + 1)$, its general solution is

(6)
$$R = C_1 r^n + C_2 r^{-n-1},$$

as is easily verified. The continuity of R at the origin $r = 0$ requires that $C_2 = 0$.

The functions $r^n P_n(\cos\theta)$ $(n = 0, 1, 2, \ldots)$ therefore satisfy Laplace's equation (1) and the continuity conditions accompanying that equation. Formally, their generalized linear combination

(7)
$$V(r,\theta) = \sum_{n=0}^{\infty} B_n r^n P_n(\cos\theta) \qquad (r \leq c)$$

is a solution of our boundary value problem if the constants B_n are such that $V(c,\theta) = F(\theta)$.

We find those constants by introducing a new function f defined on the interval $-1 < x < 1$. In terms of the principal values of the inverse cosine, we write $f(x) = F(\cos^{-1} x)$, so that $F(\theta) = f(\cos\theta)$. It is then evident that the constants B_n are related to the coefficients (Sec. 89)

(8)
$$A_n = \frac{2n + 1}{2}\int_{-1}^{1} f(x)P_n(x)\,dx \qquad (n = 0, 1, 2, \ldots)$$

in the Legendre series representation

(9)
$$f(x) = \sum_{n=0}^{\infty} A_n P_n(x) \qquad (-1 < x < 1)$$

by means of the equation $A_n = B_n c^n$. We assume that f and f' are piecewise continuous on the interval $(-1,1)$, so that representation (9) is valid at each point in the interval at which f is continuous (Theorem 2).

The harmonic function $V(r,\theta)$ can then be written in terms of the coefficients (8) as

$$
(10) \qquad V(r,\theta) = \sum_{n=0}^{\infty} A_n \left(\frac{r}{c}\right)^n P_n(\cos\,\theta) \qquad\qquad (r \le c).
$$

A full verification is given below. A compact form of expression (10) is, of course,

$$
V(r,\theta) = \sum_{n=0}^{\infty} \frac{2n+1}{2} \left(\frac{r}{c}\right)^n P_n(\cos\,\theta) \int_{-1}^{1} f(s)P_n(s)\,ds \qquad (r \le c).
$$

Before proceeding to the verification, we note that *the harmonic function W in the unbounded region $r > c$*, exterior to the spherical surface $r = c$, which assumes the values $f(\cos\,\theta)$ on that surface and which is bounded as $r \to \infty$ can be found in like manner. Here $C_1 = 0$ in equation (6) if R is to remain bounded as $r \to \infty$, and our solutions of equation (1) are $r^{-n-1}P_n(\cos\,\theta)$. Thus

$$
(11) \qquad W(r,\theta) = \sum_{n=0}^{\infty} \frac{B_n}{r^{n+1}} P_n(\cos\,\theta) \qquad\qquad (r \ge c)
$$

where the B_n are this time related to the coefficients (8) by means of the equation $A_n = B_n/c^{n+1}$. That is,

$$
(12) \qquad W(r,\theta) = \sum_{n=0}^{\infty} A_n \left(\frac{c}{r}\right)^{n+1} P_n(\cos\,\theta) \qquad\qquad (r \ge c).
$$

Verification. For convenience, let us define the function $v(r,x)$ which is related to $V(r,\theta)$ by the equation $x = \cos\,\theta$ $(0 \le \theta \le \pi)$. In equation (1) the second term, which is a differential form in θ, can be written

$$
\frac{1}{\sin\,\theta}\frac{\partial}{\partial\theta}\left[(1 - \cos^2\,\theta)\frac{1}{\sin\,\theta}\frac{\partial V}{\partial\theta}\right] = \frac{\partial}{\partial x}\left[(1 - x^2)\frac{\partial v}{\partial x}\right].
$$

If we write $F(\theta) = f(\cos\,\theta)$, as we did earlier, then $v(r,x)$ is required to satisfy the conditions

$$
(13) \qquad r(rv)_{rr} + [(1 - x^2)v_x]_x = 0 \qquad (r < c,\; -1 < x < 1),
$$

$$
(14) \qquad v(c-,x) = f(x) \qquad\qquad (-1 < x < 1);
$$

and v and its derivatives of the first and second order must be continuous functions of r and x when $0 \le r < c$, $-1 \le x \le 1$. Solution (10) becomes

$$
(15) \qquad v(r,x) = \sum_{n=0}^{\infty} A_n \left(\frac{r}{c}\right)^n P_n(x) \qquad\qquad (r \le c);
$$

and, by the method used in Sec. 56, we now verify it as a solution of this alternative form of our boundary value problem.

To prove that $v(r,x)$ satisfies boundary condition (14) for each fixed x where $f(x)$ is continuous, we first note that, in view of the Legendre series representation (9), series (15) converges to $f(x)$ when $r = c$. But the sequence of functions $(r/c)^n$ $(n = 0, 1, 2, \ldots)$ is bounded and monotonic with respect to n. Hence Abel's test, which is developed later on in Chap. 10, shows that series (15) is uniformly convergent with respect to r $(0 \le r \le c)$. Therefore v is a continuous function of r, and $v(c-,x) = v(c,x) = f(x)$.

When $n \ge 1$, each term in series (15) can be written as the product of the three factors A_n/n, $P_n(x)$, and $n(r/c)^n$. The first two factors are bounded for all n and x involved here. The third is nonnegative and not greater than $n(r_0/c)^n$ if $0 \le r \le r_0$. When $r_0 < c$, the series with terms $n(r_0/c)^n$ converges. Therefore series (15) converges uniformly with respect to r and x when $0 \le r \le r_0$ and $-1 \le x \le 1$. Thus v is continuous when $r < c$; and it is bounded when $r \le r_0 < c$ and $-1 \le x \le 1$.

But the series with terms $n^k(r_0/c)^n$ also converges whenever $0 < r_0 < c$, for each fixed positive value of k. Since $P'_n(x)/n^2$ and $P''_n(x)/n^4$ $(-1 \le x \le 1,$ $n = 1, 2, \ldots)$ are bounded, according to Theorem 1, it follows readily that series (15) is twice differentiable with respect to r and x when $r < c$ and that the derivatives of v are continuous when $r < c$. Since each term of the series satisfies equation (13), the sum of the series satisfies that equation.

This completes the verification of solution (10) in its alternative form (15). The solution (12) for the harmonic function W in the external region can be established in the same manner. By writing s/c for c/r in the alternative form

$$w(r,x) = \sum_{n=0}^{\infty} A_n \left(\frac{c}{r}\right)^{n+1} P_n(x) \qquad (r \ge c)$$

of solution (12), we find from the above discussion of series (15) that $rw(r,x)$ is bounded for large values of r $(s \le s_0 < c)$ and for all x $(-1 \le x \le 1)$.

92. Steady Temperatures in a Hemisphere

The base $r < 1$, $\theta = \pi/2$ of a solid hemisphere $r \le 1$, $0 \le \theta \le \pi/2$, part of which is shown in Fig. 29, is insulated. The flux of heat inward through the hemispherical surface is kept at prescribed values $f(\cos \theta)$. In order that temperatures can be steady, those values are such that the resultant rate of flow through the hemispherical surface is zero. That is, f satisfies the condition

$$\int_0^{\pi/2} f(\cos \theta) 2\pi \sin \theta \, d\theta = 0,$$

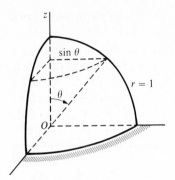

Figure 29

which can also be written

(1) $$\int_0^1 f(x)\,dx = 0.$$

If u denotes temperatures as a function of r and θ, the condition that the base be insulated is

$$\frac{1}{r}\frac{\partial u}{\partial \theta} = 0 \qquad\qquad \text{when } \theta = \frac{\pi}{2};$$

that is,

(2) $$u_\theta(r,\pi/2) = 0 \qquad\qquad (0 < r < 1).$$

The boundary value problem in $u(r,\theta)$ consists of Laplace's equation

(3) $$r\frac{\partial^2}{\partial r^2}(ru) + \frac{1}{\sin\theta}\frac{\partial}{\partial\theta}\left(\sin\theta\frac{\partial u}{\partial\theta}\right) = 0 \qquad (r < 1, 0 < \theta < \pi/2),$$

condition (2), and the flux condition

(4) $$Ku_r(1,\theta) = f(\cos\theta) \qquad\qquad (0 < \theta < \pi/2),$$

where K is thermal conductivity. Writing $x = \cos\theta$, we assume that $f(x)$ and $f'(x)$ are piecewise continuous on the interval $(0,1)$ and that $f(x)$ satisfies condition (1). Also, u is to satisfy the usual continuity conditions when $r < 1$ and $0 \le \theta \le \pi/2$.

Separating variables by replacing u in equations (2) and (3) by $R(r)\Theta(\theta)$, we obtain the equations

(5) $$r(rR)'' - \lambda R = 0 \qquad\qquad (r < 1),$$

where R must be continuous when $0 \le r < 1$, and

(6) $$\frac{1}{\sin\theta}\frac{d}{d\theta}\left(\sin\theta\frac{d\Theta}{d\theta}\right) + \lambda\Theta = 0 \qquad (0 < \theta < \pi/2),$$

$$\Theta'(\pi/2) = 0,$$

where Θ and Θ' are to be continuous when $0 \le \theta \le \pi/2$.

The substitution $x = \cos \theta$ transforms equations (6) into the singular Sturm-Liouville problem consisting of Legendre's equation (see Sec. 91)

$$\frac{d}{dx}\left[(1 - x^2)\frac{d\Theta}{dx}\right] + \lambda\Theta = 0 \qquad (0 < x < 1)$$

and the condition that

$$\frac{d\Theta}{dx} = 0 \qquad \text{when } x = 0,$$

where Θ and $d\Theta/dx$ are to be continuous when $0 \le x \le 1$. According to Sec. 85, this problem has eigenvalues $\lambda = 2n(2n + 1)$, where $n = 0, 1, 2, \ldots$, and eigenfunctions $\Theta = P_{2n}(x)$; hence $\Theta(\theta) = P_{2n}(\cos \theta)$. The corresponding solution of the Cauchy-Euler equation (5) is $R = r^{2n}$.

Formally, then,

$$u(r,\theta) = \sum_{n=0}^{\infty} B_n r^{2n} P_{2n}(\cos \theta)$$

if the constants B_n are such that condition (4) is satisfied. That condition requires that

(7) $$2K \sum_{n=1}^{\infty} nB_n P_{2n}(x) = f(x) \qquad (0 < x < 1),$$

where $x = \cos \theta$. This is the representation for $f(x)$ on the interval $(0,1)$ in a series of Legendre polynomials of even degree (Sec. 89) if $2KnB_n = A_{2n}$ where

(8) $$A_{2n} = (4n + 1)\int_0^1 f(x)P_{2n}(x)\, dx \qquad (n = 1, 2, \ldots)$$

and if f is such that $A_0 = 0$, which is precisely condition (1). Thus B_0 is left arbitrary, and

(9) $$u(r,\theta) = B_0 + \frac{1}{2K}\sum_{n=1}^{\infty}\frac{1}{n}A_{2n}r^{2n}P_{2n}(\cos \theta) \qquad (r \le 1, 0 \le \theta \le \pi/2)$$

where the coefficients A_{2n} have the values (8).

The constant B_0 is the temperature at the origin $r = 0$. Solutions of such problems with just Neumann conditions (Sec. 10) are determined only up to such an arbitrary additive constant because all the boundary conditions prescribe values of derivatives of the unknown harmonic functions.

93. Other Orthogonal Sets

Laplace's equation for a function V of all three spherical coordinates r, ϕ, θ is

(1)
$$r \frac{\partial^2}{\partial r^2}(rV) + \frac{1}{\sin^2 \theta}\frac{\partial^2 V}{\partial \phi^2} + \frac{1}{\sin \theta}\frac{\partial}{\partial \theta}\left(\sin \theta \frac{\partial V}{\partial \theta}\right) = 0.$$

If V and its derivatives are to be continuous throughout the interior $r < c$ of a sphere, those functions must be periodic in ϕ with period 2π. Then separation of variables leads to the solutions

(2)
$$r^n(a_m \cos m\phi + b_m \sin m\phi)P_n^m(x) \quad (m,n = 0, 1, 2, \ldots)$$

where $x = \cos \theta$ and where the functions $y = P_n^m(x)$, called *associated Legendre functions*, satisfy the differential equation

(3)
$$\frac{d}{dx}\left[(1 - x^2)\frac{dy}{dx}\right] + \left[n(n + 1) - \frac{m^2}{1 - x^2}\right]y = 0.$$

The functions $P_n^m(x)$ are generalizations of Legendre polynomials and are related to them as follows:

(4)
$$P_n^m(x) = (1 - x^2)^{m/2}\frac{d^m}{dx^m}P_n(x) \quad (m,n = 0, 1, 2, \ldots).$$

For each fixed m the set $P_n^m(x)$ $(n = 0, 1, 2, \ldots)$ is orthogonal on the interval $(-1,1)$ with weight function unity.[†] Note that $P_n^0(x) = P_n(x)$ and that $P_n^m(x) = 0$ if $m > n$.

Other generalizations of Legendre polynomials include the *Jacobi polynomials* $P_n^{(\alpha,\beta)}(x)$ $(n = 0, 1, 2, \ldots)$, where $\alpha > -1$ and $\beta > -1$; they are defined as

$$P_n^{(\alpha,\beta)}(x) = \frac{(-1)^n}{2^n n!}(1 - x)^{-\alpha}(1 + x)^{-\beta}\frac{d^n}{dx^n}[(1 - x)^{\alpha+n}(1 + x)^{\beta+n}].$$

When α and β are fixed, those polynomials are orthogonal on the interval $(-1,1)$ with weight function $(1 - x)^\alpha(1 + x)^\beta$. When $\alpha = \beta = 0$, they reduce to $P_n(x)$ (see Sec. 86).[‡]

[†] For treatments of associated Legendre functions, see Chap. 15 of the book by Whittaker and Watson (1963) or Chap. 3 of the one by Hobson (1955); applications are given in Secs. 9.11 and 9.12 of the book by Carslaw and Jaeger (1959) on heat conduction. All of these references are listed in the Bibliography.

[‡] An introduction to Jacobi polynomials and also to Hermite and Laguerre polynomials, which are orthogonal on unbounded intervals, is given in the book by Jackson (1941) that is listed in the Bibliography. Also see the book on special functions by Rainville (1972) and the one by Lebedev (1972), listed there.

PROBLEMS

1. Suppose that V is harmonic throughout the regions $r < c$ and $r > c$, that $V \to 0$ as $r \to \infty$, and that $V = 1$ on the spherical surface $r = c$. Show from results found in Sec. 91 that $V = 1$ when $r \le c$ and $V = c/r$ when $r \ge c$.

2. Suppose that, for all ϕ, the steady temperatures $u(r,\theta)$ in a solid sphere $r \le 1$ are such that $u(1,\theta) = 1$ when $0 < \theta < \pi/2$ and $u(1,\theta) = 0$ when $\pi/2 < \theta < \pi$. Obtain the expression

$$u(r,\theta) = \frac{1}{2} + \frac{1}{2} \sum_{n=0}^{\infty} [P_{2n}(0) - P_{2n+2}(0)] r^{2n+1} P_{2n+1}(\cos \theta)$$

for those temperatures.

3. The base $r < 1$, $\theta = \pi/2$ of a solid hemisphere $r \le 1$, $0 \le \theta \le \pi/2$ is kept at temperature $u = 0$, while $u = 1$ on the hemispherical surface $r = 1$, $0 < \theta < \pi/2$. Derive the expression

$$u(r,\theta) = \sum_{n=1}^{\infty} (-1)^n \left(\frac{4n + 3}{2n + 2} \right) \frac{(2n)!}{2^{2n}(n!)^2} r^{2n+1} P_{2n+1}(\cos \theta)$$

for the steady temperatures in that solid.

4. The base $r < c$, $\theta = \pi/2$ of a solid hemisphere $r \le c$, $0 \le \theta \le \pi/2$ is insulated. The temperature distribution on the hemispherical surface is $u = f(\cos \theta)$. Derive the expression

$$u(r,\theta) = \sum_{n=0}^{\infty} (4n + 1) \left(\frac{r}{c} \right)^{2n} P_{2n}(\cos \theta) \int_0^1 f(s) P_{2n}(s) \, ds$$

for the steady temperatures in the solid. Also, show that $u(r,\theta) = 1$ when $f(\cos \theta) = 1$.

5. A function V is harmonic and bounded in the unbounded region $r > c$, $0 \le \phi \le 2\pi$, $0 \le \theta < \pi/2$. Also, $V = 0$ everywhere on the flat boundary $r > c$, $\theta = \pi/2$ and $V = f(\cos \theta)$ on the hemispherical boundary $r = c$, $0 < \theta < \pi/2$. Derive the expression

$$V(r,\theta) = \sum_{n=0}^{\infty} (4n + 3) \left(\frac{c}{r} \right)^{2n+2} P_{2n+1}(\cos \theta) \int_0^1 f(s) P_{2n+1}(s) \, ds.$$

6. The flux of heat $K u_r(1,\theta)$ into a solid sphere at its surface $r = 1$ is a prescribed function $f(\cos \theta)$, where f is such that the net time rate of flow of heat into the solid is zero. Thus (see Sec. 92)

$$\int_{-1}^1 f(x) \, dx = 0.$$

Assuming that $u = 0$ at the center $r = 0$, derive the expression

$$u(r,\theta) = \frac{1}{2K} \sum_{n=1}^{\infty} \frac{2n + 1}{n} r^n P_n(\cos \theta) \int_{-1}^1 f(s) P_n(s) \, ds$$

for the steady temperatures throughout the entire sphere $0 \le r \le 1$.

7. Let $u(r,\theta)$ denote steady temperatures in a hollow sphere $a \le r \le b$ when $u(a,\theta) = f(\cos \theta)$ and $u(b,\theta) = 0$ $(0 < \theta < \pi)$. Derive the expression

$$u(r,\theta) = \sum_{n=0}^{\infty} A_n \frac{b^{2n+1} - r^{2n+1}}{b^{2n+1} - a^{2n+1}} \left(\frac{a}{r} \right)^{n+1} P_n(\cos \theta)$$

where

$$A_n = \frac{2n + 1}{2} \int_{-1}^1 f(x) P_n(x) \, dx.$$

8. Let $u(x,t)$ represent temperatures in a nonhomogeneous insulated bar $-1 \leq x \leq 1$ along the x axis, and suppose that the thermal conductivity is proportional to $1 - x^2$. The heat equation takes the form

$$\frac{\partial u}{\partial t} = b \frac{\partial}{\partial x}\left[(1 - x^2)\frac{\partial u}{\partial x}\right] \qquad (b > 0).$$

Here b is constant since we assume that the thermal coefficient c_0 used in Sec. 6 and in Problem 8, Sec. 7, is constant. Note that the ends $x = \pm 1$ are insulated because the conductivity vanishes there. Assuming that $u(x,0) = f(x)$ $(-1 < x < 1)$, derive the expression

$$u(x,t) = \sum_{n=0}^{\infty} \frac{2n + 1}{2} \exp\left[-n(n + 1)bt\right]P_n(x) \int_{-1}^{1} f(s)P_n(s)\,ds.$$

9. Show that if $f(x) = x^2$ $(-1 < x < 1)$ in Problem 8, then

$$u(x,t) = \frac{1}{3} + \left(x^2 - \frac{1}{3}\right)\exp\left(-6bt\right).$$

10. Suppose that the initial temperature distribution in a solid sphere $r \leq 1$ is a prescribed function $f(\cos\theta)$ which is independent of the spherical coordinates r and ϕ. If $k = 1$, the temperature function $u(r,\theta,t)$ satisfies the heat equation

$$r^2\frac{\partial u}{\partial t} = r\frac{\partial^2}{\partial r^2}(ru) + \frac{1}{\sin\theta}\frac{\partial}{\partial\theta}\left(\sin\theta\frac{\partial u}{\partial\theta}\right)$$

and the usual conditions of continuity. Given that $u = 0$ when $r = 1$, separate variables in the problem to obtain solutions $R(r)\Theta(\theta)T(t)$ of the homogeneous equations, showing that Θ, taken as a function of the variable $x = \cos\theta$, satisfies Legendre's equation where $\lambda = n(n + 1)$ and that

$$r^2R'' + 2rR' + [\mu^2r^2 - n(n + 1)]R = 0, \qquad R(1) = 0.$$

Verify that $R = r^{-1/2}J_{n+1/2}(\mu_j r)$ where $\mu = \mu_j$ are roots of the equation $J_{n+1/2}(\mu) = 0$. Thus, without completing the solution for u, show that

$$R(r)\Theta(\theta)T(t) = r^{-1/2}J_{n+1/2}(\mu_j r)P_n(\cos\theta)\exp\left(-\mu_j^2 t\right).$$

The functions $J_{n+1/2}(x)$ are called *spherical Bessel functions*.

TEN

UNIQUENESS OF SOLUTIONS

94. Cauchy Criterion for Uniform Convergence

In this chapter we examine in greater detail the question of verifying solutions of boundary value problems of certain types and, more particularly, the question of establishing that a solution of a given problem is the only possible solution. A multiplicity of solutions may actually arise when the statement of the problem does not demand adequate continuity or boundedness of the unknown function and its derivatives. This was illustrated in Problem 18, Sec. 69.

Abel's test for uniform convergence (Theorem 1) will enable us to determine further continuity properties of solutions obtained in the form of series, properties that are useful both in verifying solutions and in proving a solution unique. The remaining theorems give conditions under which a solution is unique. They apply only to specific types of problems, and their applications are further limited because they require a rather high degree of regularity of the functions involved.

We begin with a brief discussion of the Cauchy criterion for uniform convergence. Let $s_n(x)$ denote the sum of the first n terms of a series of functions $X_i(x)$ which converges to the sum $s(x)$:

$$(1) \qquad s_n(x) = \sum_{i=1}^{n} X_i(x), \qquad s(x) = \lim_{n \to \infty} s_n(x).$$

Suppose that the series converges uniformly with respect to x for all x in some interval. Then, as noted in Sec. 14, for each positive number ε there exists a positive integer n_ε, independent of x, such that

$$|s(x) - s_n(x)| < \frac{\varepsilon}{2} \qquad\qquad \text{whenever } n > n_\varepsilon$$

for every x in the interval. Let j denote any positive integer. Then

$$|s_{n+j} - s_n| = |s_{n+j} - s + s - s_n| \leq |s - s_{n+j}| + |s - s_n| < \varepsilon,$$

provided $n > n_\varepsilon$.

Thus *a necessary condition for uniform convergence of the series is that, for all positive integers j,*

(2) $|s_{n+j}(x) - s_n(x)| < \varepsilon$ whenever $n > n_\varepsilon$.

Condition (2) is a *sufficient* condition, known as the *Cauchy criterion,* for convergence of the series for each fixed x even if n_ε is not independent of x. Hence it implies that the sum $s(x)$ exists. Then, for any fixed n and x and for the given number ε, an integer $j_\varepsilon(x)$ exists such that

(3) $|s(x) - s_{n+j}(x)| < \varepsilon$ whenever $j > j_\varepsilon(x)$.

To show that *condition (2) implies uniform convergence,* let n denote a particular integer greater than n_ε, where n_ε corresponds to the given number ε in the sense that condition (2) is satisfied for all x. Then, for each particular x, condition (3) is satisfied when $j > j_\varepsilon(x)$; and, since

$$|s - s_n| = |s - s_{n+j} + s_{n+j} - s_n| \le |s - s_{n+j}| + |s_{n+j} - s_n|,$$

it follows from conditions (3) and (2) that

(4) $|s(x) - s_n(x)| < 2\varepsilon$

when $j > j_\varepsilon(x)$ and $n > n_\varepsilon$. Thus $|s(x) - s_n(x)|$, which is independent of j, is arbitrarily small for each x when $n > n_\varepsilon$. Since n_ε is independent of x, the uniform convergence is established.

Note that x here may equally well denote elements $(x_1, x_2, \ldots, x_N)$ of some set in N-dimensional space. The uniform convergence is then with respect to all N variables $x_1, x_2, \ldots, x_N$.

95. Abel's Test for Uniform Convergence

We now establish a test for the uniform convergence of infinite series whose terms are products of functions of certain classes. Applications of the test were made earlier (Secs. 56, 81, and 91) for the purpose of verifying formal solutions of boundary value problems.

The functions in a sequence of functions $T_i(t)$ $(i = 1, 2, \ldots)$ are *uniformly bounded* for all points t in an interval if a constant K, independent of i, exists such that

(1) $|T_i(t)| < K$ $(i = 1, 2, \ldots)$

for all t in the interval. The sequence is *monotonic with respect to i* if for every t in the interval either

(2) $T_{i+1}(t) \le T_i(t)$ $(i = 1, 2, \ldots)$

or else, for every t,

(3) $T_{i+1}(t) \ge T_i(t)$ $(i = 1, 2, \ldots)$.

The following generalized form of a test known as *Abel's test* shows that when the terms of a uniformly convergent series are multiplied by functions $T_i(t)$ of the type just described, the new series is uniformly convergent.

Theorem 1. *The series*

$$(4) \qquad \sum_{i=1}^{\infty} X_i(x) T_i(t)$$

converges uniformly with respect to the two variables x and t together in a region R of the xt plane provided that (a) the series

$$\sum_{i=1}^{\infty} X_i(x)$$

converges uniformly with respect to x for all x such that (x,t) is in R and (b) the functions $T_i(t)$ are uniformly bounded and monotonic with respect to i ($i = 1, 2, \ldots$) for all t such that (x,t) is in R.

To start the proof, we let S_n denote partial sums of series (4):

$$S_n(x,t) = \sum_{i=1}^{n} X_i(x) T_i(t).$$

As indicated in the preceding section, the uniform convergence of that series will be established if we prove that to each positive number ε there corresponds an integer n_ε, independent of x and t, such that

$$|S_m(x,t) - S_n(x,t)| < \varepsilon \qquad \text{whenever } n > n_\varepsilon,$$

for all integers $m = n + 1, n + 2, \ldots$ and for all points (x,t) in R.

We write s_n for the partial sum

$$s_n(x) = X_1(x) + X_2(x) + \cdots + X_n(x).$$

Then for each pair of integers m, n ($m > n$), $S_m - S_n$ can be written

$$X_{n+1} T_{n+1} + X_{n+2} T_{n+2} + \cdots + X_m T_m$$

$$= (s_{n+1} - s_n) T_{n+1} + (s_{n+2} - s_{n+1}) T_{n+2} + \cdots + (s_m - s_{m-1}) T_m$$

$$= (s_{n+1} - s_n) T_{n+1} + (s_{n+2} - s_n) T_{n+2} - (s_{n+1} - s_n) T_{n+2}$$

$$+ \cdots + (s_m - s_n) T_m - (s_{m-1} - s_n) T_m.$$

By pairing alternate terms here, we find that

$$(5) \qquad S_m - S_n = (s_{n+1} - s_n)(T_{n+1} - T_{n+2}) + (s_{n+2} - s_n)(T_{n+2} - T_{n+3})$$

$$+ \cdots + (s_{m-1} - s_n)(T_{m-1} - T_m) + (s_m - s_n) T_m.$$

Suppose now that the functions T_i are nonincreasing with respect to i, so that they satisfy condition (2), and that they also satisfy the uniform bound-edness condition (1). Then the factors $T_{n+1} - T_{n+2}$, $T_{n+2} - T_{n+3}$, and so on, in equation (5) are nonnegative and $|T_i(t)| < K$. Since the series with terms $X_i(x)$ converges uniformly, an integer n_ε exists such that

$$|s_{n+j}(x) - s_n(x)| < \frac{\varepsilon}{3K} \qquad \text{whenever } n > n_\varepsilon,$$

for all positive integers j, where ε is any given positive number and n_ε is independent of x (Sec. 94). Then if $n > n_\varepsilon$ and $m > n$, it follows from equation (5) that

$$|S_m - S_n| < \frac{\varepsilon}{3K}\left[(T_{n+1} - T_{n+2}) + (T_{n+2} - T_{n+3}) + \cdots + |T_m|\right]$$

$$= \frac{\varepsilon}{3K}(T_{n+1} - T_m + |T_m|)$$

$$\leq \frac{\varepsilon}{3K}(|T_{n+1}| + 2|T_m|).$$

Therefore

$$|S_m(x,t) - S_n(x,t)| < \varepsilon \qquad \text{whenever } m > n > n_\varepsilon,$$

and the uniform convergence of series (4) is established.

The proof is similar when the functions T_i are nondecreasing with respect to i.

When x is kept fixed, the series with terms X_i is a series of constants; and the only requirement placed on that series is that it be convergent. Then the theorem shows that when T_i are bounded and monotonic, the series of terms $X_i T_i(t)$ is uniformly convergent with respect to t.

Extensions of the theorem to cases in which X_i are functions of x and t, or where X_i and T_i are functions of several variables, become evident when it is observed that our proof rests on the uniform convergence of the series of terms X_i and the bounded monotonic nature of the functions T_i.

96. Uniqueness of Solutions of the Heat Equation

Let D denote the domain consisting of all points interior to a closed surface S; and let $\bar{D}$ be the closure of that domain, consisting of all points in D and all points on S. We assume always that the closed surface S is *piecewise smooth*. That is, it is a continuous surface consisting of a finite number of parts over each of which the outward-drawn unit normal vector varies con-tinuously from point to point. Then if W is a function of x, y, and z which is continuous in $\bar{D}$, together with its partial derivatives of the first and second order, an extended form of Gauss' divergence theorem states that

$$(1) \qquad \iint_S W \frac{dW}{dn} \, dA = \iiint_D (W \, \nabla^2 W + W_x^2 + W_y^2 + W_z^2) \, dV.$$

Here dA is the area element on S, dV represents $dx \, dy \, dz$, and dW/dn is the derivative in the direction of the outward-drawn normal to S.†

Consider a homogeneous solid whose interior is the domain D and whose temperatures at time t are denoted by $u(x,y,z,t)$. A fairly general problem in heat conduction is the following:

$$(2) \qquad u = k \, \nabla^2 u + \phi(x,y,z,t) \qquad [(x,y,z,t) \text{ in } D, t > 0],$$

$$(3) \qquad u(x,y,z,0) = f(x,y,z) \qquad [(x,y,z) \text{ in } \bar{D}],$$

$$(4) \qquad u = g(x,y,z,t) \qquad [(x,y,z) \text{ on } S, t \geq 0].$$

This is the problem of determining temperatures in a body, with prescribed initial temperatures $f(x,y,z)$ and surface temperatures $g(x,y,z,t)$, interior to which heat may be generated continuously at a rate per unit volume proportional to $\phi(x,y,z,t)$.

Suppose that the problem has two solutions

$$u = u_1(x,y,z,t), \qquad u = u_2(x,y,z,t)$$

where both u_1 and u_2 are continuous functions in the closed region $\bar{D}$ when $t \geq 0$, while their derivatives of the first order with respect to t and of the first and second order with respect to x, y, and z are continuous in $\bar{D}$ when $t > 0$. Since u_1 and u_2 satisfy the linear equations (2), (3), and (4), their difference

$$w(x,y,z,t) = u_1(x,y,z,t) - u_2(x,y,z,t)$$

satisfies the homogeneous problem

$$(5) \qquad w_t = k \, \nabla^2 w \qquad [(x,y,z) \text{ in } D, t > 0],$$

$$(6) \qquad w(x,y,z,0) = 0 \qquad [(x,y,z) \text{ in } \bar{D}],$$

$$(7) \qquad w = 0 \qquad [(x,y,z) \text{ on } S, t \geq 0].$$

Moreover, w and its derivatives have the continuity properties of u_1 and u_2 assumed above.

We shall show now that $w = 0$ in D when $t > 0$, so that the two solutions u_1 and u_2 are identical. It will follow that not more than one solution of the boundary value problem in u can exist if the solution is required to satisfy the continuity conditions stated.

---

† Equation (1) is found by applying the basic divergence theorem used in Sec. 6 to the vector field W grad W. See Taylor and Mann (1972, pp. 520–521), listed in the Bibliography.

The continuity of w with respect to x, y, z, and t together in the closed region $\bar{D}$ when $t \geq 0$ implies that the integral

$$(8) \qquad\qquad I(t) = \frac{1}{2} \iiint_D [w(x,y,z,t)]^2 \, dV$$

is a continuous function of t when $t \geq 0$. According to equation (6), $I(0) = 0$. In view of the continuity of w_t when $t > 0$ and equation (5), we can write

$$I'(t) = \iiint_D w w_t \, dV = k \iiint_D w \, \nabla^2 w \, dV \qquad (t > 0).$$

The divergence formula (1) applies to the last integral because of the continuity of the derivatives of w when $t > 0$; thus

$$(9) \qquad \iiint_D w \, \nabla^2 w \, dV = \iint_S w \frac{dw}{dn} \, dA - \iiint_D (w_x^2 + w_y^2 + w_z^2) \, dV$$

when $t > 0$. But $w = 0$ on S; therefore

$$I'(t) = -k \iiint_D (w_x^2 + w_y^2 + w_z^2) \, dV \leq 0.$$

The mean value theorem applies to $I(t)$. Thus for each positive t a number t_1 $(0 < t_1 < t)$ exists such that

$$I(t) - I(0) = t I'(t_1);$$

and, since $I(0) = 0$ and $I'(t_1) \leq 0$, it follows that $I(t) \leq 0$. However, definition (8) of the integral shows that $I(t) \geq 0$. Therefore

$$I(t) = 0 \qquad\qquad (t \geq 0),$$

and so the nonnegative integrand w^2 cannot have a positive value at any point in D; for if so, the continuity of w^2 requires that w^2 be positive throughout a neighborhood of the point and then $I(t) > 0$. Consequently

$$w(x,y,z,t) = 0 \qquad\qquad [(x,y,z) \text{ in } \bar{D}, \, t \geq 0],$$

and the following theorem on uniqueness is established.

Theorem 2. *Let u satisfy these conditions of regularity: (a) it is a continuous function of the variables x, y, z, and t, taken together, when the point (x,y,z) is in the closed region $\bar{D}$ and $t \geq 0$; (b) those derivatives of u present in the heat equation (2) are continuous in the same sense when $t > 0$. Then if u is a solution of the boundary value problem (2) through (4), it is the only possible solution satisfying conditions (a) and (b).*

The condition that u be continuous in $\bar{D}$ when $t = 0$ restricts the usefulness of our theorem. It is clearly not satisfied if the initial temperature function f in condition (3) fails to be continuous throughout $\bar{D}$, or if at some point on S the initial value $g(x,y,z,0)$ of the prescribed surface temperature differs from the value $f(x,y,z)$. The continuity requirement at $t = 0$ can be relaxed in some cases.†

When conditions (a) and (b) in Theorem 2 are added to the requirement that u is to satisfy the heat equation and the boundary conditions, our boundary value problem is completely stated, provided it has a solution; for that will be the only possible solution.

The proof of Theorem 2 required that the integral

$$\iint_S w \, \frac{dw}{dn} \, dA$$

in equation (9) should either vanish or have a negative value. It vanished because $w = 0$ on S. But it is never positive if condition (4) is replaced by the boundary condition

(10) $$\frac{du}{dn} + hu = g(x,y,z,t) \qquad [(x,y,z) \text{ on } S, t > 0],$$

where $h \geq 0$. For in that case $dw/dn = -hw$ on S, and $w \, dw/dn \leq 0$. Thus our theorem can be modified as follows.

Theorem 3. *The conclusion in Theorem 2 is true if boundary condition (4) is replaced by condition (10), or if condition (4) is satisfied on part of the surface S and condition (10) is satisfied on the rest.*

97. Example

In the problem of temperature distribution in a slab with insulated faces $x = 0$ and $x = \pi$ and initial temperatures $f(x)$ [Sec. 57(b)], assume that f is continuous and f' is piecewise continuous on the interval $0 \leq x \leq \pi$. Then the Fourier cosine series for f converges uniformly to $f(x)$ on that interval.

Let $u(x,t)$ denote the sum of the series

(1) $$\frac{1}{2} a_0 + \sum_{n=1}^{\infty} a_n \exp\left(-n^2 kt\right) \cos nx \qquad (0 \leq x \leq \pi, t \geq 0),$$

obtained as a formal solution of the problem, a_0 and a_n ($n = 1, 2, \ldots$) being the coefficients in the Fourier cosine series for f.

---

† Integral transforms can sometimes be used to prove uniqueness of solutions of certain types of boundary value problems. This is illustrated in Sec. 79 of Churchill (1972), listed in the Bibliography.

We can see from Abel's test (Theorem 1) that series (1) converges uniformly with respect to x and t together in the region $0 \leq x \leq \pi, t \geq 0$; thus u is continuous there. When $t \geq t_0$, where t_0 is any positive number, the series obtained by differentiating series (1) term by term, any number of times with respect to x or t, is uniformly convergent according to the Weierstrass M-test. It follows readily that u not only satisfies all equations in the boundary value problem (compare Sec. 56), but also that u_t, u_x, and u_{xx} are continuous functions in the region $0 \leq x \leq \pi, t > 0$. Thus u satisfies the regularity conditions (a) and (b) in Theorem 2.

The temperature problem for the slab is the same as the problem for a prismatic bar with its bases in the planes $x = 0$ and $x = \pi$ and with its lateral surface, parallel to the x axis, insulated $(du/dn = 0)$. Let the domain D consist of all interior points of the prism. Then Theorem 3 applies to show that the sum $u(x,t)$ of series (1) is the only solution that satisfies the conditions (a) and (b).

PROBLEMS

1. In the problem of temperature distribution in the hemisphere $r \leq 1, 0 \leq \theta \leq \pi/2$, solved formally in Sec. 92, assume that f and f' are piecewise continuous. With the aid of Abel's test, prove that, for each fixed x $(0 < x < 1, x = \cos \theta)$ where f is continuous, the series in the solution (9), Sec. 92, namely

$$B_0 + \frac{1}{2K} \sum_{n=1}^{\infty} \frac{1}{n} A_{2n} r^{2n} P_{2n}(x) \qquad (x = \cos \theta),$$

converges uniformly with respect to r $(0 \leq r \leq 1)$. Also prove that the series is differentiable with respect to r and that

$$Ku_r(1-, \theta) = f(\cos \theta)$$

where $u(r,\theta)$ is the sum of the series.

2. In Problem 10, Sec. 57, on temperatures $u(x,t)$ in a slab initially at temperatures $f(x)$, throughout which heat is generated at a constant rate, assume that f is continuous and f' is piecewise continuous $(0 \leq x \leq \pi)$ and that $f(0) = f(\pi) = 0$. Prove that the function $u(x,t)$ given there is the only solution of the problem which satisfies the regularity conditions (a) and (b) stated in Theorem 2.

3. Verify the solution of Problem 14, Sec. 57, and prove that it is the only possible solution satisfying the regularity conditions (a) and (b) stated in Theorem 2. Note that in this case the Weierstrass M-test suffices for all proofs of uniform convergence.

4. In Sec. 81, given that the initial temperature function $f(\rho)$ for the cylinder is continuous on the interval $0 \leq \rho \leq c$ and is such that the Fourier-Bessel series used there converges uniformly to $f(\rho)$ on that interval, prove that the solution established there is the only one that satisfies our regularity conditions.

98. Solutions of Laplace's or Poisson's Equation

Let W be a harmonic function in a domain D of three-dimensional space bounded by a continuous closed surface S that is piecewise smooth. Assume also that W and its partial derivatives of the first order are continuous in the closure $\bar{D}$ of the domain. Then, since

$$(1) \qquad\qquad \nabla^2 W(x,y,z) = 0 \qquad\qquad [(x,y,z) \text{ in } D],$$

form (1), Sec. 96, of Gauss' divergence formula becomes

$$(2) \qquad \iint_S W \frac{dW}{dn} dA = \iiint_D (W_x^2 + W_y^2 + W_z^2)\, dV.$$

This equation is valid for our function W even though we have required the derivatives of the second order to be continuous only in D, and not in the closed region $\bar{D}$. It is not difficult to prove that because $\nabla^2 W = 0$, modification of the usual conditions in the divergence theorem is possible.†

Suppose that $W = 0$ at all points on the surface S. Then the first integral in equation (2), and therefore the second, vanishes. But the integrand of the second is nonnegative and continuous in $\bar{D}$. It must therefore vanish throughout $\bar{D}$; that is,

$$(3) \qquad\qquad W_x = W_y = W_z = 0 \qquad\qquad [(x,y,z) \text{ in } \bar{D}].$$

Consequently, $W(x,y,z)$ is constant; but it is zero on S and continuous in $\bar{D}$, and therefore $W = 0$ throughout $\bar{D}$.

Suppose that dW/dn, instead of W, vanishes on S; or, to make the condition more general, suppose that

$$(4) \qquad\qquad \frac{dW}{dn} + hW = 0 \qquad\qquad [(x,y,x) \text{ on } S]$$

where $h \geq 0$ and h is either a constant or a function of x, y, and z. Then on S

$$W \frac{dW}{dn} = -hW^2 \leq 0,$$

so that the first integral in equation (2) is less than or equal to zero. But the second integral is greater than or equal to zero; hence it must vanish, and again conditions (3) follow. So W is constant in $\bar{D}$.

If W vanishes over part of S and satisfies condition (4) on the rest of that surface, our argument still shows that W is constant in $\bar{D}$. In this case the constant must be zero.

---

† See the book by Kellogg (1953, p. 119) that is listed in the Bibliography.

Now let U denote a function which is continuous in $\bar{D}$, together with its partial derivatives of the first order. Suppose that it also has continuous derivatives of the second order in D and satisfies these conditions:

(5) $$\nabla^2 U(x,y,z) = f(x,y,z) \qquad [(x,y,z) \text{ in } D],$$

(6) $$p \frac{dU}{dn} + hU = g \qquad [(x,y,z) \text{ on } S].$$

Here f, p, h, and g denote prescribed constants or functions of (x,y,z). We assume that $p \geq 0$ and $h \geq 0$.

Equation (5) is known as *Poisson's equation* and is a generalization of Laplace's equation, which occurs when $f(x,y,z)$ is identically zero throughout D. Boundary condition (6) includes important special cases. When $p = 0$ on S, or on part of S, the value of U is assigned there. When $h = 0$, the value of dU/dn is assigned. Of course p and h must not vanish simultaneously.

If $U = U_1(x,y,z)$ and $U = U_2(x,y,z)$ are two solutions of this problem, their difference

$$W = U_1 - U_2$$

satisfies Laplace's equation in D and the condition

$$p \frac{dW}{dn} + hW = 0$$

on S. Moreover, W satisfies the conditions of regularity required of U_1 and U_2. Thus it is harmonic in D, and W and its derivatives of the first order are continuous in $\bar{D}$. It follows from the results established above for harmonic functions that W must be constant throughout $\bar{D}$. Thus $dW/dn = 0$ on S. If $h \neq 0$ at some point on S, then W vanishes there and $W = 0$ throughout $\bar{D}$. For the harmonic function W vanishes over the closed surface S and hence cannot have values other than zero interior to S.

We have now established the following uniqueness theorem for problems in electrostatic or gravitational potential, steady temperatures, or other boundary value problems for Laplace's or Poisson's equation.

Theorem 4. *Let $U(x,y,z)$ satisfy these conditions of regularity in a domain D bounded by a closed surface S: (a) it is continuous, together with its derivatives of the first order, in $\bar{D}$, and (b) its derivatives of the second order are continuous in D. Then if U is a solution of the boundary value problem consisting of equations (5) and (6), it is the only solution satisfying conditions (a) and (b), except possibly for $U + C$ where C is an arbitrary constant. Unless $h = 0$ at every point on S, $C = 0$ and the solution is unique.*

It is possible to show that this theorem also applies when D is the unbounded domain exterior to the closed surface S, provided U satisfies the additional requirement that the absolute values of rU, r^2U_{xx}, r^2U_{yy}, and r^2U_{zz} be bounded for all r greater than some fixed number, where r is the distance from (x,y,z) to the origin.† Then, since U vanishes as $r \to \infty$, the constant C is zero and the solution is unique. But note that S is a closed surface, so that this extension of the theorem does not apply, for instance, to unbounded domains between two planes or inside a cylinder.

Condition (a) in Theorem 4 is severe because it requires U and its derivatives of the first order to be continuous on the surface S. For problems in which $p = 0$ on S, so that U is prescribed on the entire boundary, the condition can be relaxed so as to require only the continuity of U itself in $\bar{D}$ if derivatives are continuous in D. This follows directly from a fundamental result in potential theory: *if a function other than a constant is harmonic in D and continuous in $\bar{D}$, its maximum and minimum values are assumed at points on S, and never in D.*‡

99. An Application

To illustrate the use of Theorem 4, consider the problem in Sec. 58 of determining steady temperatures $u(x,y)$ in a rectangular plate with three edges at temperature zero and an assigned temperature distribution on the fourth edge. The faces of the plate are insulated. For convenience, we shall consider the plate to be square with edge length π. As long as $du/dn = 0$ on the faces, the thickness of the plate does not affect the problem.

The domain D is the interior of the region bounded by the planes $x = 0$, $x = \pi$, $y = 0$, $y = \pi$, $z = z_1$, and $z = z_2$, where z_1 and z_2 are constants. Then S is the boundary of that domain. The required function u is harmonic in D. It vanishes on the three parts $x = 0$, $x = \pi$, and $y = \pi$ of S; and $u = f(x)$ on the part $y = 0$. Also, $u_z = 0$ on the parts $z = z_1$ and $z = z_2$. Thus Theorem 4 applies if u and its derivatives of the first order are continuous in $\bar{D}$.

First, suppose u is independent of z. Then

$$(1) \qquad u_{xx}(x,y) + u_{yy}(x,y) = 0 \quad (0 < x < \pi, 0 < y < \pi),$$

$$(2) \qquad u(0,y) = u(\pi,y) = 0 \qquad (0 \le y \le \pi),$$

$$(3) \qquad u(x,0) = f(x), \qquad u(x,\pi) = 0 \qquad (0 \le x \le \pi).$$

---

† See the book by Kellogg (1953) referred to earlier in this section.

‡ Physically, the theorem seems evident since it states that steady temperatures cannot have maximum or minimum values interior to a solid in which no heat is generated. For a proof in three dimensions, see the book by Kellogg (1953) cited earlier in this section and, for two dimensions, see Churchill, Brown, and Verhey (1974, Sec. 54).

The formal solution found in Sec. 58 becomes

(4)
$$u(x,y) = \sum_{n=1}^{\infty} b_n \frac{\sinh n(\pi - y)}{\sinh n\pi} \sin nx$$

where b_n are the coefficients in the Fourier sine series for f on the interval $(0,\pi)$.

To show that the function (4) satisfies the regularity conditions, let us require that f and f' be continuous and f'' piecewise continuous $(0 \le x \le \pi)$ and that

$$f(0) = f(\pi) = 0.$$

Results found in Chap. 5 then show that

(5)
$$f(x) = \sum_{n=1}^{\infty} b_n \sin nx, \qquad f'(x) = \sum_{n=1}^{\infty} nb_n \cos nx,$$

and both series converge uniformly on the interval $0 \le x \le \pi$. The second series, obtained by differentiating the first, is the Fourier cosine series for f'; and, since f' is continuous and f'' is piecewise continuous, not only is that series uniformly convergent but the series of absolute values $|nb_n|$ of its coefficients converges. It follows from the Weierstrass M-test that the series

(6)
$$\sum_{n=1}^{\infty} nb_n \sin nx \qquad\qquad (0 \le x \le \pi)$$

also converges uniformly with respect to x.

Let us show that for each fixed y the sequence of functions

(7)
$$\frac{\sinh n(\pi - y)}{\sinh n\pi} \qquad (n = 1, 2, \ldots; 0 \le y \le \pi)$$

appearing in series (4) is monotonic and nonincreasing as n increases. This is evident when $y = 0$ and when $y = \pi$. It is true when $0 < y < \pi$ if the function

$$T(t) = \frac{\sinh \beta t}{\sinh \alpha t} \qquad (t > 0, \alpha > \beta > 0)$$

always decreases as t grows. To see that this is so, write

$$T'(t) \sinh^2 \alpha t = \beta \sinh \alpha t \cosh \beta t - \alpha \sinh \beta t \cosh \alpha t$$

$$= -\frac{1}{2}(\alpha - \beta) \sinh (\alpha + \beta)t + \frac{1}{2}(\alpha + \beta) \sinh (\alpha - \beta)t$$

$$= -\frac{\alpha^2 - \beta^2}{2} \left[\frac{\sinh (\alpha + \beta)t}{\alpha + \beta} - \frac{\sinh (\alpha - \beta)t}{\alpha - \beta} \right]$$

$$= -\frac{\alpha^2 - \beta^2}{2} \sum_{n=0}^{\infty} \frac{(\alpha + \beta)^{2n} - (\alpha - \beta)^{2n}}{(2n + 1)!} t^{2n+1}.$$

The terms of this series are positive, so that $T'(t) < 0$; therefore $T(t)$ decreases as t grows.

Likewise, the positive-valued functions

$$(8) \qquad \frac{\cosh n(\pi - y)}{\sinh n\pi} \qquad (0 \leq y \leq \pi),$$

arising in the series for u_y, never increase in value as n grows because their squares can be written

$$(9) \qquad \frac{1}{\sinh^2 n\pi} + \left[\frac{\sinh n(\pi - y)}{\sinh n\pi}\right]^2$$

and each term is nonincreasing.

The values of functions (7) clearly vary only from zero to unity for all n and y involved. The functions (8) are also uniformly bounded. Therefore those functions can be used in Abel's test for uniform convergence. From the uniform convergence of the series in equations (5) and of series (6), on the interval $0 \leq x \leq \pi$, we conclude not only that series (4) converges uniformly with respect to x and y together in the region $0 \leq x \leq \pi, 0 \leq y \leq \pi$ but also that the uniform convergence holds true for the series obtained by differentiating series (4) once termwise with respect to either x or y.

Consequently, series (4) is differentiable with respect to x and y; also, its sum $u(x,y)$ and u_x and u_y are continuous in the closed region $0 \leq x \leq \pi$, $0 \leq y \leq \pi$. Clearly $u(x,y)$ satisfies boundary conditions (2) and (3).

The derivatives of the second order, with respect to either x or y, of the terms in series (4) have absolute values not greater than

$$(10) \qquad n^2 |b_n| \frac{\sinh n(\pi - y_0)}{\sinh n\pi}$$

when $0 \leq x \leq \pi$ and $y_0 \leq y \leq \pi$, where $y_0 > 0$. Let B be chosen such that $|b_n| < B$ for all n. Then, from the inequalities

$$2 \sinh n(\pi - y_0) < \exp n(\pi - y_0), \qquad 2 \sinh n\pi \geq e^{n\pi}(1 - e^{-2\pi})$$

it follows that the numbers (10) are less than

$$\frac{B}{1 - \exp(-2\pi)} n^2 \exp(-ny_0).$$

The series with these terms converges, according to the ratio test, since $y_0 > 0$; and so the Weierstrass M-test ensures the uniform convergence of the series of second-order derivatives of terms in series (4) when $y_0 \leq y \leq \pi$. Thus series (4) is twice differentiable; also, u_{xx} and u_{yy} are continuous in the region $0 \leq x \leq \pi, 0 < y \leq \pi$.

Since the terms in series (4) satisfy Laplace's equation, the sum $u(x,y)$ of that series satisfies condition (1). Thus u is established as a solution of our boundary value problem. Moreover, u satisfies our regularity conditions, even with respect to z since it is independent of z and $u_z = 0$ everywhere, on the parts $z = z_1$ and $z = z_2$ of S in particular. According to Theorem 4, the function defined by series (4) is the only possible solution that satisfies the regularity conditions.

100. Solutions of a Wave Equation

Consider the following generalization of the problem solved in Sec. 51 for the transverse displacements in a stretched string:

$$ (1) \qquad y_{tt}(x,t) = a^2 y_{xx}(x,t) + \phi(x,t) \qquad\qquad (0 < x < c,\, t > 0), $$

$$ (2) \qquad y(0,t) = p(t), \qquad y(c,t) = q(t) \qquad\qquad (t \geq 0), $$

$$ (3) \qquad y(x,0) = f(x), \qquad y_t(x,0) = g(x) \qquad\qquad (0 \leq x \leq c). $$

But we now require y to be of class C^2 in the region $R: 0 \leq x \leq c, t \geq 0$, by which we shall mean that y and its derivatives of the first and second order, including y_{xt} and y_{tx}, are to be continuous functions in R. As indicated by examples in Chap. 6, the prescribed functions ϕ, p, q, f, and g must be restricted if the problem is to have a solution of class C^2.

Suppose there are two solutions $y_1(x,t)$ and $y_2(x,t)$ of that class. Then the difference $z = y_1 - y_2$ is of class C^2 in R, and it satisfies the homogeneous problem

$$ (4) \qquad z_{tt}(x,t) = a^2 z_{xx}(x,t) \qquad\qquad (0 < x < c,\, t > 0), $$

$$ (5) \qquad z(0,t) = 0, \qquad z(c,t) = 0 \qquad\qquad (t \geq 0), $$

$$ (6) \qquad z(x,0) = 0, \qquad z_t(x,0) = 0 \qquad\qquad (0 \leq x \leq c). $$

We shall prove that $z = 0$ throughout R; thus $y_1 = y_2$.

The integrand of the integral

$$ (7) \qquad I(t) = \frac{1}{2} \int_0^c \left(z_x^2 + \frac{1}{a^2} z_t^2 \right) dx \qquad\qquad (t \geq 0) $$

satisfies conditions such that we can write

$$ (8) \qquad I'(t) = \int_0^c \left(z_x z_{xt} + \frac{1}{a^2} z_t z_{tt} \right) dx. $$

Since $z_{tt} = a^2 z_{xx}$, the integrand here can be written

$$ z_x z_{tx} + z_t z_{xx} = \frac{\partial}{\partial x} (z_x z_t). $$

So in view of equations (5), from which it follows that

$$ z_t(0,t) = 0, \qquad z_t(c,t) = 0, $$

we can write

(9) $$I'(t) = z_x(c,t)z_t(c,t) - z_x(0,t)z_t(0,t) = 0.$$

Hence $I(t)$ is a constant. But definition (7) shows that $I(0) = 0$ because $z(x,0) = 0$, and so $z_x(x,0) = 0$; also $z_t(x,0) = 0$. Thus $I(t) = 0$. The nonnegative continuous integrand of that integral must therefore vanish; that is,

$$z_x(x,t) = z_t(x,t) = 0 \qquad (0 \le x \le c, t \ge 0).$$

So z is constant; in fact, $z(x,t) = 0$ because $z(x,0) = 0$.

Thus *the boundary value problem consisting of equations* (1), (2), *and* (3) *cannot have more than one solution of class* C^2 *in R.*

If y_x, instead of y, is prescribed at the end point in either or both of conditions (2), the proof of uniqueness is still valid because condition (9) is again satisfied.

The requirement of continuity on derivatives of y is severe. Solutions of many simple problems in a wave equation have discontinuities in their derivatives.

PROBLEMS

1. In the Dirichlet problem for a rectangle treated in Sec. 58(a), let f and f' be piecewise continuous on the interval $(0,a)$. Verify that the formal solution found there satisfies the condition $u(x,0+) = f(x)$ when $0 < x < a$ if $f(x)$ is defined as the mean of $f(x+)$ and $f(x-)$ at its points of discontinuity.

2. Formulate a complete statement of the boundary value problem for steady temperatures in a square plate with insulated faces when the edges $x = 0$, $x = \pi$, and $y = 0$ are insulated and the edge $y = \pi$ is kept at temperatures $u = f(x)$. Show that if f, f', and f'' are continuous on the interval $0 \le x \le \pi$ and $f'(0) = f'(\pi) = 0$, the problem has the unique solution

$$u(x,y) = \frac{1}{2} a_0 + \sum_{n=1}^{\infty} a_n \frac{\cosh ny}{\cosh n\pi} \cos nx$$

where

$$a_n = \frac{2}{\pi} \int_0^\pi f(x) \cos nx \, dx \qquad (n = 0, 1, 2, \ldots).$$

3. Replace the infinite cylinder in Sec. 60(a) by a cylinder bounded by the surfaces $\rho = 1$, $z = z_1$, and $z = z_2$, where $u_z = 0$ on the last two parts. Also let $f(\phi)$ be a periodic function of period 2π with a continuous derivative of the second order everywhere. Then show that the function u given by equation (5), Sec. 60, is the unique solution, satisfying our conditions of regularity, of the problem in steady temperatures.

4. Use the uniqueness that was established in Sec. 100 to show that the solution $y = A \sin \pi x \cos \pi a t$ of Problem 1, Sec. 52, is the only solution of class C^2 in the region $0 \le x \le 1, t \ge 0$.

5. Show that the solution of Problem 3, Sec. 55, is unique in the class C^2.

6. In Sec. 51, let $f(x)$ be such that its odd periodic extension $F(x)$ has a continuous derivative $F''(x)$ for all $x(-\infty < x < \infty)$; then show that the solution (9) there is unique in the class C^2.

BIBLIOGRAPHY

The following list of books and articles for supplementary study of the various topics that have been introduced is far from exhaustive. Further references are given in the ones mentioned here.

Abramowitz, M., and I. A. Stegun (Eds.): "Handbook of Mathematical Functions with Formulas, Graphs, and Mathematical Tables," Dover Publications, Inc., New York, 1965.

Apostol, T. M.: "Mathematical Analysis," 2d ed., Addison-Wesley Publishing Company, Reading, Massachusetts, 1974.

Bell, W. W.: "Special Functions for Scientists and Engineers," D. Van Nostrand Company, Ltd., London, 1968.

Berg, P. W., and J. L. McGregor: "Elementary Partial Differential Equations," Holden-Day, Inc., San Francisco, 1966.

Bowman, F.: "Introduction to Bessel Functions," Dover Publications, Inc., New York, 1958.

Boyce, W. E., and R. C. DiPrima: "Elementary Differential Equations," 3d ed., John Wiley & Sons, Inc., New York, 1977.

Buck, R. C.: "Advanced Calculus," 3d ed., McGraw-Hill Book Company, New York, 1978.

Budak, B. M., A. A. Samarskii, and A. N. Tikhonov: "A Collection of Problems on Mathematical Physics," Pergamon Press, Ltd., Oxford, 1964.

Byerly, W. E.: "Fourier's Series and Spherical Harmonics," Dover Publications, Inc., New York, 1959.

Carslaw, H. S.: "Theory of Fourier's Series and Integrals," 3d ed., Dover Publications, Inc., New York, 1930.

————, and J. C. Jaeger: "Conduction of Heat in Solids," 2d ed., Oxford University Press, London, 1959.

Churchill, R. V.: "Operational Mathematics," 3d ed., McGraw-Hill Book Company, New York, 1972.

————, J. W. Brown, and R. F. Verhey: "Complex Variables and Applications," 3d ed., McGraw-Hill Book Company, New York, 1974.

Coddington, E. A., and N. Levinson: "Theory of Ordinary Differential Equations," McGraw-Hill Book Company, Inc., New York, 1955.

Courant, R., and D. Hilbert: "Methods of Mathematical Physics," Interscience Publishers, Inc., New York, vol. 1, 1953; "Partial Differential Equations," vol. 2, 1962.

Davis, H. F.: "Fourier Series and Orthogonal Functions," Allyn and Bacon, Inc., Boston, 1963.

Erdélyi, A. (Ed.): "Higher Transcendental Functions," McGraw-Hill Book Company, Inc., New York, vols. 1, 2, 1953; vol. 3, 1955.

Farrell, O. J., and B. Ross: "Solved Problems in Analysis as applied to Gamma, Beta, Legendre, and Bessel Functions," Dover Publications, Inc., New York, 1971.

Fourier, J.: "The Analytical Theory of Heat," translated by A. Freeman, Dover Publications, Inc., New York, 1955.

Franklin, P.: "A Treatise on Advanced Calculus," Dover Publications, Inc., New York, 1964.

Grattan-Guinness, I.: "Joseph Fourier, 1768–1830," The MIT Press, Cambridge, Massachusetts and London, England, 1972.

Gray, A., and G. B. Mathews: "A Treatise on Bessel Functions and their Applications to Physics," 2d ed. (with T. M. MacRobert), Dover Publications, Inc., New York, 1966.

Greenspan, D.: "Introduction to Partial Differential Equations," McGraw-Hill Book Company, Inc., New York, 1961.

Hobson, E. W.: "The Theory of Spherical and Ellipsoidal Harmonics," Chelsea Publishing Company, Inc., Bronx, New York, 1955.

Hochstadt, H.: "The Functions of Mathematical Physics," John Wiley & Sons, New York, 1971.

Ince, E. L.: "Ordinary Differential Equations," Dover Publications, Inc., New York, 1956.

Jackson, D.: "Fourier Series and Orthogonal Polynomials," Carus Mathematical Monographs, no. 6, Mathematical Association of America, 1941.

Jahnke, E., F. Emde, and F. Lösch: "Tables of Higher Functions," 6th ed., McGraw-Hill Book Company, Inc., New York, 1960.

Kaplan, W.: "Advanced Calculus," 2d ed., Addison-Wesley Publishing Company, Inc., Reading, Massachusetts, 1973.

Kellogg, O. D.: "Foundations of Potential Theory," Dover Publications, Inc., New York, 1953.

Lanczos, C.: "Discourse on Fourier Series," Hafner Publishing Company, New York, 1966.

Langer, R. E.: Fourier's Series: The Genesis and Evolution of a Theory, Slaught Memorial Papers, no. 1, *Amer. Math. Monthly*, vol. 54, no. 7, part 2, pp. 1–86, 1947.

Lebedev, N. N.: "Special Functions and their Applications," Dover Publications, Inc., New York, 1972.

————, I. P. Skalskaya, and Y. S. Uflyand: "Problems of Mathematical Physics," Prentice-Hall, Inc., Englewood Cliffs, New Jersey, 1965.

McLachlan, N. W.: "Bessel Functions for Engineers," 2d ed., Oxford University Press, London, 1955.

MacRobert, T. M.: "Spherical Harmonics," 3d ed. (with I. N. Sneddon), Pergamon Press, Ltd., Oxford, 1967.

Magnus, W., F. Oberhettinger, and R. P. Soni: "Formulas and Theorems for the Special Functions of Mathematical Physics," 3d ed., Springer-Verlag, New York, Inc., 1966.

Oberhettinger, F.: "Fourier Expansions: A Collection of Formulas," Academic Press, New York and London, 1973.

Powers, D. L.: "Boundary Value Problems," Academic Press, Inc., New York, 1972.

Rainville, E. D.: "Special Functions," Chelsea Publishing Company, Inc., Bronx, New York, 1972.

————, and P. E. Bedient: "Elementary Differential Equations," 5th ed., Macmillan Publishing Co., Inc., New York, 1974.

Rogosinski, W.: "Fourier Series," 2d ed., Chelsea Publishing Company, Inc., Bronx, New York, 1959.

Sagan, H.: "Boundary and Eigenvalue Problems in Mathematical Physics," John Wiley & Sons, Inc., New York, 1961.

Sansone, G.: "Orthogonal Functions," rev. ed., Robert E. Krieger Publishing Co., Inc., Huntington, New York, 1975.

Seeley, R. T.: "An Introduction to Fourier Series and Integrals," W. A. Benjamin, Inc., New York, 1966.

Smirnov, V. I.: "Complex Variables—Special Functions," Addison-Wesley Publishing Company, Inc., Reading, Massachusetts, 1964.

Sneddon, I. N.: "Elements of Partial Differential Equations," McGraw-Hill Book Company, Inc., New York, 1957.

————: "Fourier Transforms," McGraw-Hill Book Company, Inc., New York, 1951.

————: "Special Functions of Mathematical Physics and Chemistry," 2d ed., Oliver and Boyd, Ltd., Edinburgh, 1961.

————: "The Use of Integral Transforms," McGraw-Hill Book Company, Inc., New York, 1972.

Taylor, A. E., and W. R. Mann: "Advanced Calculus," 2d ed., Xerox College Publishing, Lexington, Massachusetts, 1972.

Titchmarsh, E. C.: "Eigenfunction Expansions Associated with Second-order Differential Equations," Oxford University Press, London, part I, 2d ed., 1962; part II, 1958.

————: "Theory of Fourier Integrals," 2d ed., Oxford University Press, London, 1967.

Tolstov, G. P.: "Fourier Series," Dover Publications, Inc., New York, 1976.

Tranter, C. J.: "Bessel Functions with Some Physical Applications," Hart Publishing Company, Inc., New York, 1969.

Van Vleck, E. B.: The Influence of Fourier's Series upon the Development of Mathematics, *Science*, vol. 39, pp. 113–124, 1914.

Watson, G. N.: "A Treatise on the Theory of Bessel Functions," 2d ed., Cambridge University Press, London, 1952.

Weinberger, H. F.: "A First Course in Partial Differential Equations," Xerox College Publishing, Lexington, Massachusetts, 1965.

Whittaker, E. T., and G. N. Watson: "A Course of Modern Analysis," 4th ed., Cambridge University Press, London, 1963.

Zygmund, A.: "Trigonometric Series," 2d ed., 2 vols. in one, Cambridge University Press, London, 1968.

INDEX